Purification and Analysis of Recombinant Proteins

Bioprocess Technology

Series Editor

W. Courtney McGregor

Xoma Corporation
Berkeley, California

Volume 1 Membrane Separations in Biotechnology, *edited by W. Courtney McGregor*

Volume 2 Commercial Production of Monoclonal Antibodies: A Guide for Scale-Up, *edited by Sally S. Seaver*

Volume 3 Handbook on Anaerobic Fermentations, *edited by Larry E. Erickson and Daniel Yee-Chak Fung*

Volume 4 Fermentation Process Development of Industrial Organisms, *edited by Justin O. Neway*

Volume 5 Yeast: Biotechnology and Biocatalysis, *edited by Hubert Verachtert and René De Mot*

Volume 6 Sensors in Bioprocess Control, *edited by John V. Twork and Alexander M. Yacynych*

Volume 7 Fundamentals of Protein Biotechnology, *edited by Stanley Stein*

Volume 8   Yeast Strain Selection, *edited by Chandra J. Panchal*

Volume 9   Separation Processes in Biotechnology, *edited by Juan A. Asenjo*

Volume 10 Large-Scale Mammalian Cell Culture Technology, *edited by Anthony S. Lubiniecki*

Volume 11 Extractive Bioconversions, *edited by Bo Mattiasson and Olle Holst*

Volume 12 Purification and Analysis of Recombinant Proteins, *edited by Ramnath Seetharam and Satish K. Sharma*

Volume 13 Drug Biotechnology Regulation: Scientific Basis and Practices, *edited by Yuan-yuan H. Chiu and John L. Gueriguian*

Volume 14 Protein Immobilization: Fundamentals and Applications, *edited by Richard F. Taylor*

Additional Volumes in Preparation

Purification and Analysis of Recombinant Proteins

edited by

Ramnath Seetharam
*E. I. du Pont de Nemours & Company, Inc.
Newark, Delaware*

Satish K. Sharma
*The Upjohn Company
Kalamazoo, Michigan*

MARCEL DEKKER, INC.   New York • Basel • Hong Kong

ISBN 0-8247-8277-1

This book is printed on acid-free paper.

Copyright © 1991 by Marcel Dekker, Inc. All Rights Reserved.

Neither this book nor any part may be reproduced or transmitted in any form or by any means, electronic or mechanical, including photocopying, microfilming, and recording, or by any information storage and retrieval system, without permission in writing from the publisher.

Marcel Dekker, Inc.
270 Madison Avenue, New York, New York 10016

Current printing (last digit):
10 9 8 7 6 5 4 3 2

PRINTED IN THE UNITED STATES OF AMERICA

Series Introduction

Bioprocess technology encompasses all the basic and applied sciences as well as the engineering required to fully exploit living systems and bring their products to the marketplace. The technology that develops is eventually expressed in various methodologies and types of equipment and instruments built up along a bioprocess stream. Typically in commercial production, the stream begins at the bioreactor, which can be a classical fermentor, a cell culture perfusion system, or an enzyme bioreactor. Then comes separation of the product from the living systems and/or their components followed by an appropriate number of purification steps. The stream ends with bioproduct finishing, formulation, and packaging. A given bioprocess stream may have some tributaries or outlets and may be overlaid with a variety of monitoring devices and control systems. As with any stream, it will both shape and be shaped with time. Documenting the evolutionary shaping of bioprocess technology is the purpose of this series.

Now that several products from recombinant DNA and cell fusion techniques are on the market, the new era of bioprocess technology is well established and validated. Books of this series represent developments in various segments of bioprocessing that have paralleled progress in the life sciences. For obvious proprietary reasons, some developments in industry, although validated, may be published only later, if at all. Therefore, our continuing series will follow the growth of this field as it is available from both academia and industry.

W. Courtney McGregor

Preface

The phenomenal advances made in recombinant DNA technology during recent years have enabled us to produce large quantities of proteins and peptides for a variety of uses. These advances have opened up new and exciting areas of research related to the study of proteins. As a result, there has been a flurry of activity aimed at devising new ways to produce and recover proteins and peptides using genetic engineering techniques. This, in turn, has given rise to a number of new challenges and opportunities related to the purification and analyses of proteins. This book addresses some of these issues in chapters written by experienced industrial scientists, intimately involved in developing novel strategies for the production, purification, and analysis of proteins. These chapters deal with many of the practical aspects of protein production and recovery. However, they also describe the theoretical basis of the approaches used, wherever appropriate.

The book is organized into four sections. The first section, consisting of two chapters, along with the Introduction, introduces the book and surveys some of the challenges and problems faced by scientists in the field. The four chapters in the second section deal with some important issues related to purification and recovery of recombinant proteins in the active form. The third section is comprised of four chapters that deal briefly with specific examples of proteins obtained from bacterial and nonbacterial hosts. The fourth section deals with some recent trends in the field. The book illustrates that these are, indeed, very exciting times for the scientist involved in protein research!

This book is intended to serve as a reference volume for scientists in academia and industry. It will be of value as a comprehensive guide to managers directing research in the area of proteins. Graduate students interested in protein research may use it as advanced reading material.

We express our sincere appreciation to all the contributors for their excellent chapters. We are grateful to the series editor, Dr. Courtney McGregor (Vice President of Process Development and Engineering, Xoma Corporation), and to Marcel Dekker for giving us the opportunity to compile this volume. We thank Dr. Richard C. Burgess (Director of the Biotechnology Center, University of Wisconsin) and James K. Gierse (Senior Scientist, Monsanto Company) for their encouragement and support during the planning stages of the book. We also thank all our friends, colleagues, and the management at the Monsanto Company, The Upjohn Company, and E. I. du Pont Company for their input and support. We appreciate the efforts of Drs. Robert L. Heinrikson and F.J. Kezdy of The Upjohn Company for their thought-provoking Introduction. Finally, Ramnath Seetharam would like to thank his wife and colleague, Shobha Seetharam, who has been a constant source of support, suggestions, and encouragement throughout the course of the book.

Ramnath Seetharam
Satish K. Sharma

Introduction

Until only a few years ago, the structural analysis of proteins was confined to preparations derived from natural sources. Not surprisingly, the first proteins to yield to sequence analysis were small, stable, and abundant. Insulin, ribonuclease, lysozyme, myoglobin, and chymotrypsin, to name but a few, served as paradigms for the development of methods such as ion-exchange chromatography for amino acid analysis and the Edman degradation for sequential removal of amino acids from the NH_2-terminus of polypeptide chains. As the field of protein chemistry developed in the 1960s and 1970s, procedures for the compositional and sequence analysis of proteins were streamlined, with the result that they became more and more sensitive and efficient. The application of high-performance liquid chromatography (HPLC) to the purification of proteins and peptides greatly extended the range of sensitivity to include many of the rare and biologically important molecules of cellular regulation. HPLC was soon adapted, as well, to the separation of phenylthiohydantoin amino acids produced by Edman chemistry and to resolution of other classes of derivatized amino acids for the purpose of compositional analysis. Refinements in automation and scale reduction of the Edman degradation led to sequencing capabilities at the picomole level, and structural information began to be derived for low-abundance proteins such as growth factors, receptors, and cytokines. By 1975, everything was in place, from the point of view of protein analysis, for the advent of molecular biology.

INTRODUCTION

To say that the first descriptions in the mid-1970s of gene sequencing and chemical and enzymatic manipulations of the genetic material were received with astonishment would be putting it mildly. Structural genes and complementary DNA could now be isolated with relative ease, and the newly established methods for determining nucleotide sequences created a cascade of derived protein sequences, often for proteins not even known to exist. Oligonucleotide sequencing was not only more rapid, but generally speaking, more accurate than classic methods of protein structural analysis, and many published protein sequences underwent some level of revision when the sequences of the corresponding cDNA were determined. Site-directed mutagenesis held the powerful potential of being able to alter a protein structure at will, without the deleterious interpretative drawbacks that plague chemical modification procedures. Proteins rare in natural sources could be expressed in heterologous host cell lines and, therefore, could be produced in mass quantities. These methods could create new proteins for new functions, as well as provide limitless amounts of the rarest of proteins that might serve as therapeutic agents, or as targets for drug development.

Like any other method, however, molecular biology has inherent limitations, the most important being its inability to predict or identify posttranslational modifications, such as proteolytic processing, disulfide bond formation, covalent or noncovalent attachment of prosthetic groups, glycosylation, carboxylation, acylation, phosphorylation, and the formation of quaternary structures. In brief, unless a newly derived protein sequence shows homology to a structurally and functionally well-characterized molecule, it does not provide much information about the chemistry and biochemistry of the protein. Furthermore, the recombinant proteins derived from this technology are rarely, if ever, expressed in the same cells as the ones from which the original protein has been isolated. It is then necessary to establish whether the posttranslational processing of the naturally occurring protein is truly duplicated in the artificial system instead of being replaced by a new system of reactions. At the beginning, the various advantages of the new molecular biology, as compared to the more traditional methods of protein analysis, were seen by many as the end of the line for protein chemistry. In fact, just the opposite has proven to be true. Protein chemists have been given a whole new set of challenges and opportunities as a result of genetic engineering and the host of new recombinant proteins that are now available for detailed study. In a real sense, that is what this book is all about.

The first step in solving most problems by methods of molecular biology is itself dependent on some knowledge of the protein of interest. This means that some, albeit limited, amount of sequence information is required for synthesis of oligonucleotide probes and isolation of the appropriate genetic message. This phase of the study places heavy demands on the micro methods men-

INTRODUCTION

tioned above for protein isolation and sequencing. After this first step, isolation, sequence analysis, and cloning of the cDNA is usually a predictable operation; the result is a recombinant protein, and the ball is back in the court of the protein chemist. A number of distinctive problems surround the purification and characterization of these recombinant proteins. All relate in one way or another to the new medium of expression and to the quality of the molecules produced. First, the new medium introduces a large number of new components to cope with during the purification, either as new impurities, i.e., molecules with properties similar to those of the target protein which were not present during the purification of the native protein, or as chemical modifiers, such as tenacious bacterial proteases. In some cases the new medium may lack components essential for the synthesis of a functional protein, such as chaperone molecules, modifying enzymes, or special prosthetic groups. Once the purification of the protein, as such, is accomplished, a host of questions inherent to recombinant proteins still have to be answered: Are they folded correctly? Have they been subjected to adventitious processing? How pure are they chemically? What proportion of the total population is in the native conformation? These questions are the subject of the present text.

Many of the chapters deal with *Escherichia coli* as a host for production of recombinant proteins. The advantages relate to the capability of producing huge quantities cheaply. Moreover, since *E. coli* do not glycosylate proteins, the resulting molecules may be more amenable to crystallographic analysis. Proteins made in *E. coli* are often deposited as insoluble inclusion bodies. This may facilitate their purification from bacterial contaminants, but it also often poses a problem for dissolution of the protein and for refolding it from denaturing solvents. Protein refolding is still in a highly empirical stage; some proteins are cooperative, others not. Alternative hosts such as yeast have a high capacity for production and produce folded proteins, but these organisms often hyperglycosylate the recombinant molecules with carbohydrate moieties distinct from those found in the native counterparts. This can cause problems for purification, and if the proteins are to be used therapeutically, they may be highly immunogenic. Mammalian cell line expression systems are often the only viable alternative insofar as they secrete folded, biologically active proteins; they suffer from a lower capacity than can be obtained with the microorganisms.

Another problem that must be considered is proteolysis during the workup; this subject is also treated in the present text. As the field advances, we are learning more and more about proteinases of the popular expression systems, and how to take measures to block their activity during purification.

Finally, it is important to establish that the purified protein of interest is in its native conformation. Titration of the protein with ligands that bind tightly to catalytic or other sites is a useful way to evaluate the functional integrity of the recombinant preparations as they relate to their native protein counterparts.

INTRODUCTION

Ever since the beginning of protein chemistry as a science, the answer to its fundamental quest remained elusive: even today, we still try to define what a protein really is, and in the light of every newly developed method we have to revise our way of looking at proteins. Recombinant proteins not only generated a new understanding of proteins, but they also posed a host of new questions concerning the identity, structure, chemical function, and biological activity of novel polypeptides. It should be clear from the foregoing that the job of the protein chemist has been intensified, rather than made obsolete, by molecular biology. It is not the game, but the rules, that have changed—and for the better. Protein chemists no longer muse over strategies for total sequence analysis of a protein, but concern themselves with the prelude and postlude sections of a collaborative venture in molecular biology. This new alliance has already been highly productive. In fact, the catalog of protein sequences now available from the prodigious output of nucleic acid sequencing far exceeds our capacity to deal with it effectively. One of the major secrets left to be unlocked concerns how the information in the primary structure of proteins is translated to folding and function at the level of tertiary structure. This is a fundamental problem that will continue to occupy a central position in research. In the meantime, the host of new and rare proteins of importance in health and disease that are now being made available for mechanistic and structural analysis provides a wealth of opportunity for research and challenges aplenty for the future benefit of mankind.

Robert L. Heinrikson
Ference J. Kezdy
The Upjohn Company
Kalamazoo, Michigan

Contents

Series Introduction *W. Courtney McGregor* iii
Preface v
Introduction *Robert L. Heinrikson and Ference J. Kezdy* vii
Contributors xiii

I. INTRODUCTION

1. **Purification and Characterization of Recombinant Proteins: Opportunities and Challenges** 3
   Robert L. Heinrikson and Alfredo G. Tomasselli

2. **Purification and Production of Therapeutic Grade Proteins** 29
   Vipin K. Garg, Maureen A. C. Costello, and Barbara A. Czuba

II. TECHNICAL ISSUES RELATED TO RECOVERY OF RECOMBINANT PROTEINS

3. **Physical and Chemical Cell Disruption for the Recovery of Intracellular Proteins** 57
   T. R. Hopkins

4. **Proteases During Purification** 85
   Georg-B. Kresze

5. Properties of Recombinant Protein-Containing
   Inclusion Bodies in *Escherichia coli* — 121
   James F. Kane and Donna L. Hartley

6. Methods for Removing N-Terminal Methionine
   from Recombinant Proteins — 147
   Arie Ben-Bassat

III. PURIFICATION OF RECOMBINANT PROTEINS FROM *Escherichia coli*, YEAST, AND MAMMALIAN CELLS

7. Purification of Secreted Recombinant Proteins
   from *Escherichia coli* — 163
   Hung V. Le and Paul P. Trotta

8. Purification of Recombinant Proteins from Yeast — 183
   Roger G. Harrison, Jr.

9. Production of Recombinant Proteins in the
   Methylotrophic Yeast *Pichia pastoris* — 193
   M. J. Skogen Hagenson

10. Purification of Monoclonal Antibodies — 213
    Tom C. Ransohoff and Howard L. Levine

IV. RECENT TRENDS IN THE AREA OF RECOMBINANT PROTEIN PURIFICATION AND ANALYSES

11. Engineering Proteins to Enable Their Isolation
    in a Biologically Active Form — 239
    Stephen J. Brewer, Barry L. Haymore, Thomas P. Hopp,
    and Helmut M. Sassenfeld

12. Practical Aspects of Receptor Affinity Chromatography — 267
    Pascal S. Bailon, David V. Weber, and John E. Smart

13. Recombinant DNA Technology and Crystallography:
    A New Alliance in Unraveling Protein Structure–
    Function Relationships — 285
    Alfredo G. Tomasselli, Robert L. Heinrikson,
    and Keith D. Watenpaugh

Index — 317

Contributors

Pascal S. Bailon Roche Research Center, Hoffmann-La Roche, Inc., Nutley, New Jersey

Arie Ben-Bassat Cetus Corporation, Emeryville, California

Stephen J. Brewer Monsanto Company, St. Louis, Missouri

Maureen A. C. Costello Bio-Response, Inc., Hayward, California

Barbara A. Czuba* Bio-Response, Inc., Hayward, California

Vipin K. Garg[†] Bio-Response, Inc., Hayward, California

M. J. Skogen Hagenson Phillips Petroleum Company, Bartlesville, Oklahoma

Roger G. Harrison, Jr. School of Chemical Engineering and Materials Science, University of Oklahoma, Norman, Oklahoma

---

Present affiliation:
* Baxter, Hyland Division, Hayward, California.
[†] Sepracor Inc., Marlborough, Massachusetts.

Donna L. Hartley Centre International de Recherche Daniel Carasso, Le Plessis-Robinson, France

Barry L. Haymore Monsanto Company, St. Louis, Missouri

Robert L. Heinrikson The Upjohn Company, Kalamazoo, Michigan

T. R. Hopkins* Phillips Petroleum Company, Bartlesville, Oklahoma

Thomas P. Hopp Immunex Corporation, Seattle, Washington

James F. Kane Monsanto Company, St. Louis, Missouri

Georg-B. Kresze Biochemical Research Center, Boehringer Mannheim GmbH, Penzberg, Federal Republic of Germany

Hung V. Le Schering-Plough Research, Bloomfield, New Jersey

Howard L. Levine Xoma Corporation, Berkeley, California

Tom C. Ransohoff[†] Xoma Corporation, Berkeley, California

Helmut M. Sassenfeld Immunex Corporation, Seattle, Washington

John E. Smart Roche Research Center, Hoffmann-La Roche, Inc., Nutley, New Jersey

Alfredo G. Tomasselli The Upjohn Company, Kalamazoo, Michigan

Paul P. Trotta Schering-Plough Research, Bloomfield, New Jersey

Keith D. Watenpaugh The Upjohn Company, Kalamazoo, Michigan

David V. Weber Roche Research Center, Hoffmann-La Roche, Inc., Nutley, New Jersey

---

Present affiliation:
* BioSpec Products, Inc., Bartlesville, Oklahoma.
† Dorr-Oliver, Inc., Milford, Connecticut.

Purification and Analysis of Recombinant Proteins

I
INTRODUCTION

1

Purification and Characterization of Recombinant Proteins
Opportunities and Challenges

Robert L. Heinrikson and Alfredo G. Tomasselli

The Upjohn Company
Kalamazoo, Michigan

I. INTRODUCTION

The explosion of protein sequence information generated since the introduction of recombinant DNA technology in the mid-1970s has presented us with an unexpected treasure of knowledge that would never have been obtainable by classical methods of protein chemistry. In a real sense, this treasury has been an embarrassment of riches; the primary structural information remains locked in an unsolved and apparently insoluble code. We can discern similarities of the new sequences with known proteins and, hence, infer structural relationships that provide important new insights as to the function of the "new" protein. A timely example may be found in the protease of retroviruses the structure of which, deciphered by genetic analysis, has been shown to be homologous to members of the aspartyl proteinases. With this information and detailed knowledge of the structures of several aspartyl proteinases, a rather detailed description of the as yet uncharacterized retroviral protease can be formulated (1). But without such detailed knowledge as to the folding pattern of homologous proteins, we are quite at a loss to know how a new protein molecule will fold. The mechanisms by which primary structural information is translated into a biologically relevant molecule with a precisely defined three-dimensional structure remains a mystery. For the time being, therefore, we must be content with rather laborious, computer intensive X-ray crystallographic methods of determining tertiary structure, and then to use this ever-expanding body of knowledge to develop and refine folding algorithms that may be applied generally.

The onus placed on experimental means to establish three-dimensional structures of proteins and to characterize them in functional detail means that we must produce the proteins in large quantities and, again, recombinant methods come to the rescue for expression of rare messages in heterologous cell lines that can be grown easily in culture. The new methods of molecular biology therefore have not been the death knell for protein analysis (2) but, to the contrary, the avenue to greatly enhanced opportunities and challenges for the protein chemist. A whole new world of possibilities exists now that the means are available to produce any protein—in principle, in quantities sufficient for the most exacting structural and functional analysis.

The title of the present chapter reflects this current wave of enthusiasm among proteinologists. Indeed, the opportunities and challenges come to bear at every point along the way from identification of the new protein, to choice of the expression vector and host, to assay, purification, and characterization of the recombinant product, and, finally, to design of mutated derivatives that will help shed light on the structure and function of the protein in question. Future decades will see protein chemists and molecular biologists working in closer harmony to solve the formidable problem of how a protein's structure is related to its function. This coalition has already translated into new therapeutic agents in the pharmaceutical marketplace, and the presentation which follows is slanted somewhat toward an industrial perspective.

II. PURIFICATION AND CHARACTERIZATION OF THE RECOMBINANT PROTEIN; OBJECTIVE, STRATEGIES, AND TECHNIQUES

At the outset it should be noted that purification of a recombinant protein offers select advantages in ease of isolation relative to the native counterpart. Most of this advantage follows from the relative abundance of the recombinant product with respect to contaminating proteins, especially when bacterial or yeast expression systems are employed. Moreover, secreting cell lines can be exploited which greatly reduce contaminants. By the same token, it must be remembered that isolation of a protein from its natural tissue source may differ substantially from the protocol needed for its recombinant counterpart expressed in different cell lines. A yeast cell will, for example, present a whole different set of proteins compared to a kidney cell, and the nature of glycosylation may also contribute to an altered behavior with respect to a well-established protocol. Nevertheless, the balance is in favor of the recombinant product for ease of purification; we begin with an advantage.

Most of the new and interesting proteins are scarce in natural biological sources and have not been defined in terms of kinetics (if an enzyme), physical-chemical properties, or three-dimensional structure. Therefore, only the recombinant counterpart will be available for characterization. Information from these studies is highly desirable from an economic point of view if we want the framework to design and produce second-generation products with improved pharmaceutical, industrial, or agricultural performance. The recombinant protein may serve as the drug itself, and thus be subject to stringent quality control relative to purity and safety. On the other hand, these proteins may provide important tools in delineating metabolic regulatory pathways, or in developing strategies for rational drug design. In the following we will deal with the issues of purification and characterization of recombinant proteins within the framework of ideas presented above.

III. OBJECTIVE

A. Purity and Safety of the Product

The objective, of course, is to purify the protein away from contaminating protein, nucleic acid, lipid, carbohydrate, and a host of metabolites and cell growth media components. The degree of purity of the product has to be related to its intended use. The scientist purifying a protein at the laboratory bench of an academic institution does not have the same concerns relative to purity and safety as one who is preparing a product for use in human subjects. It should be stressed that there is no such thing as a 100% pure protein. We all live with a degree of impurity that is acceptable for our work. Concern for purity will vary considerably depending on whether the protein is to be used for kinetic, sequence, or crystallographic studies, or is to be injected into children. Clearly, safety is of primary importance when the products are destined for the clinic. There is always the risk that the final product contains fragments of viral or cellular DNA, with potential oncogenic capacity, contaminating proteins that may give immunogenic responses, or pyrogenic contaminants of viral or microbial origin. Purification techniques aimed at assuring the elimination of these contaminants must be devised and maintained with maximal accuracy and reproducibility. These questions were probed at a recent Cold Spring Harbor Symposium, and the reader may be referred to the recent publication of papers on the subject (3). The dictum "purity consistent with intended use" may well be followed in the case of pharmaceutical agents as with research reagents. Indeed, purity should not be pursued for its own sake, but with common sense with respect to the product, its dosage, and the condition of the persons receiving treatment.

B. Cost

Another objective of great importance, both to the manufacturer and to the client, is the cost of the final product. A lengthy purification procedure is going to be expensive. Therefore, the purification steps must be thoughtfully integrated in such a way that the sample from one step is suitable for the next without further handling. Cutting steps means cutting time and cost and increasing the final yield of biologically active product.

IV. STRATEGIES FOR PURIFICATION

Protein purification encompasses a vast repertoire of methods; it is the strategy as to how to put these techniques together that is of paramount interest to the general reader. The process can be totally empirical and, in fact, this was the only avenue open for purification strategies prior to development and introduction of affinity methods. The latter, tailored to correspond uniquely to the protein of interest through use of specific antibodies or ligands, have found increased application over the years and have greatly facilitated protein purification in general.

Before expounding on the general and specific approaches that have been followed in purification protocols, it is appropriate to say a few words about the assay and quantitation of the overall procedure. It is no exaggeration to say that the assay, or means of detection and quantitation of the protein at each phase of the purification, is of paramount importance to success in this endeavor. Every student of biochemistry knows this, but it is surprising how often, in practice, assay development is relegated to a secondary status in the rush to purify the protein of interest. Here again, recombinant technology has had an enormous impact. Rare proteins, for which complex bio- or radioimmunoassays were the only routes available for detection, can now be produced in quantities sufficient for direct biochemical assay. An example here is the enzyme renin, the prime mover in a cascade of reactions leading to the production of angiotensin II, a potent vasoconstrictor and stimulus of aldosterone secretion. Renin exists only in picogram quantities in plasma and assays depend on measurement of the release of angiotensin I (the product of renin activity) by a radioimmunoassay. This indirect method is clearly unnecessary for assay of recombinant renin produced in milligram quantities, and we (4) based our purification and characterization of renin on the direct measurement of proteolysis of model peptide substrates by high-performance liquid chromatography (HPLC). In the case of proteins which have no catalytic activity, behavior on gel or column separation systems may serve to identify and quantify the protein of interest. In any case, access to simplified assays greatly expedites the purification process of any protein, and this is one of the great advantages of recombinant products.

Hand in hand with the assay is a step-by-step assessment of the purification protocol, summarized in a table which presents specific activities, total units, total protein, and yields at each step of the procedure. Specific examples of assays and purification tables will be presented later as they relate to discussions of a variety of recombinant proteins, but for now we will assume that methods for detection and quantitation of the protein under study are well in hand.

A. Cell Extraction

Ideally, the recombinant protein is recovered from the cell culture medium after secretion and proper posttranslational processing. This situation most often obtains in animal cells and yeast, and the culture medium is conveniently separated from the producing host by centrifugation or filtration, the latter being more suitable for large volumes.

However, a number of proteins, especially those expressed in bacterial cells, are produced in high yields as dense, insoluble masses called inclusion bodies. Extraction and purification procedure schemes are presented in Figs. 1 and 2 for secreted and inclusion body recombinant proteins, respectively. The remainder of this section deals with bacterial expression as cytoplasmic or inclusion body proteins, although secretion vectors have been employed to promote release of the recombinant proteins into the cellular environment. In many cases, the inclusion bodies contain the recombinant protein in a denatured but almost pure form. In less fortunate cases, the inclusion bodies are variably contaminated by the host biological material. If the protein is not secreted, one needs to open up the cells by a variety of lytic procedures, including shearing force, variation in ionic strength, pH, or solvent dielectric constant, or by enzymatic digestion. After disruption of the cell wall, inclusion bodies may be pelleted by centrifugation, or the cell supernatant can be subjected to further purification. Careful disruption of the cell wall is preferable to allow cellular proteins to stay in the supernatant following low-speed centrifugation to sediment the inclusion bodies. Repetition of the disruption process may be necessary if broken cell fragments, detectable under the phase contrast microscope, appear in the pellet. The amount of recombinant protein in the pellet may constitute from about 40% to more than 60% of the inclusion body proteins. If the protein is found partially associated with inclusion bodies and partially as a soluble product, treatment with acid or small amounts of organic solvents may be used to precipitate the soluble fraction of the recombinant product.

B. Solubilization of Inclusion Bodies and Refolding of the Recombinant Protein

Before treating the general case of secreted products, it is useful to expand on the manipulations that have proven successful with the "intractable" inclusion bodies. Dissolution of these insoluble masses of protein requires denaturing agents such as 6–8 M guanidinium chloride or detergents like Triton or sodium dodecyl sulfate (SDS) in the presence of thiol compounds to reduce disulfide bridges and convert the polypeptide to a totally random structure form. The trick then is to be able to remove the perturbing influence and restore the protein to its native and fully biologically active conformation. Because of the relatively high level of purity of the recombinant protein in solubilized inclusion bodies, molecular sieve chromatography is often desirable at this point. If the denaturing agent is ionic (e.g., guanidinium chloride), then it has to be exchanged by dialysis against some nonionic agent (e.g., urea) in order to allow application of ion-exchange or neutral adsorption chromatographic purification techniques. Of course, it is always preferable to use the minimal amount of denaturant required to keep the recombinant protein in solution. It is also advisable to add a mixture of oxidized and reduced forms of

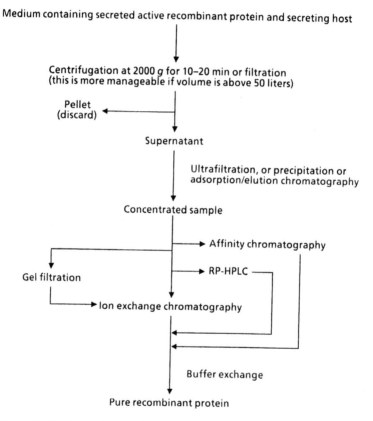

Figure 1 A schematic procedure to purify recombinant proteins secreted into the culture medium.

OPPORTUNITIES AND CHALLENGES

the thiol compound during refolding of the polypeptide. Refolding is a subject unto itself and has been treated extensively in the literature (cf. 5–7 for further information). The process is still largely empirical and usually requires extended evaluation of a variety of denaturants, reducing agents, and conditions of removal of the same. Generally speaking, the smaller the protein and the fewer the disulfide bridges, the greater is the likelihood of success in

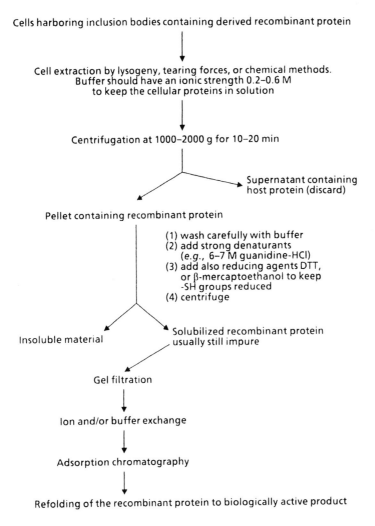

Figure 2 A schematic procedure to purify recombinant proteins produced in *E. coli* as inclusion bodies.

recovering biologically active protein from inclusion bodies. Further details with regard to this subject may be found in a recent review on the recovery of genetically engineered proteins from *Escherichia coli* (8).

C. Concentration of Secreted Proteins

Bacterial expression systems like *E. coli* offer the advantage of being able to produce large amounts of recombinant protein, but from the foregoing discussion it is clear that the problems associated with recovery of active product may dictate use of a host that will secrete folded, processed, and active material. With proteins that are secreted into a large extracellular volume of growth medium we must deal with the problem of concentrating this solution before proceeding to purification.

1. Ultrafiltration

With this method, the solution is passed under pressure through a porous membrane. Molecules whose molecular weights are above a certain threshold, referred to as the membrane cutoff, are retained, while smaller molecules pass through with the aqueous solvent. Membrane binding due to precipitation is limited by gentle stirring or by use of tangential flow concentration systems.

2. Precipitation

Salts, most notably ammonium sulfate, or organic solvents such as ethanol and acetone have been used extensively to precipitate proteins from solution. These agents may require extra purification steps for their removal or, worse, have negative effects on product stability.

3. Adsorption/Elution Chromatography

Methods described thus far are effective for concentration but provide very limited purification. Therefore, it is desirable, when possible, to use a technique that combines the concentration step with a partial purification step; in the best cases the purification can be more than 90%. Adsorption/elution chromatography is one such technique. It exploits the property of proteins to interact, through electrostatic and/or hydrophobic forces, with certain ligands covalently bound to solid supports. Agarose, dextran, and polyacrylamides have been widely employed as supports. Some of the ligands used are lectins, for binding of glycoproteins; lysines, for particular serine proteases; protein A, for binding IgG and IgA molecules; the dye Cibacron Blue which binds enzymes carrying the dinucleotide-fold moiety; and metal chelate agarose which has been useful in the purification of several proteins including tissue plasminogen activator (tPA) and human interferon-β. There are, of course, a host of other columns that have been described. The advantage of a weak binding specificity is that elution may be performed under mild conditions with high recoveries of biologically active products. Most of the adsorption columns mentioned above are available commercially, but in any case they can be made

OPPORTUNITIES AND CHALLENGES 11

easily in the laboratory, so that the general approach is quite flexible and convenient for development of new applications.

Once the volume of the cell extract has been reduced to a manageable size, the real purification starts. As mentioned above, there are no fixed protocols when purifying a completely new protein, and the procedure required to purify the recombinant protein may be similar to or quite different from that devised for the natural counterpart. Knowledge of the methodology for the latter, however, is always useful when embarking on isolation of the recombinant product.

D. Column Chromatographic Separation Techniques

1. Gel Filtration

Separation by gel filtration or molecular sieve chromatography is achieved due to differences in molecular size. For good resolution, the gel must not interact with the molecules to be separated; such interactions can be minimized by operation of the columns at ionic strength values above 0.1 M. The ideal separation takes place on the basis of the distribution of the proteins between the mobile liquid phase and the liquid located in the pores of the gel. A variety of products from Pharmacia, most notably a series of gels prepared by crosslinking dextran with epichlorohydrin bearing the trademark name Sephadex, and crosslinked acrylamide gels called Bio-Gels from Bio-Rad, have been employed for this purpose. These products are available in a series of matrices encompassing a wide range of separation potential based on molecular weight.

A more recent development in gel filtration chromatography is the introduction of a hydrophilic, vinyl polymer-based resin marketed under the tradename Toyo Pearl. This material is totally porous, semirigid, and particularly well suited for medium- and low-pressure liquid chromatography. It provides high resolution even in fast flow liquid chromatography and is therefore especially useful for large-scale purification of proteins. Gel filtration can be used at any step of the purification as long as the volume delivered to the column is reasonably small (preferably <2% of the total column volume). It offers the advantage of being able to transfer the protein quickly and efficiently to a new solvent, but this process always leads to sample dilution, which requires an additional concentration step. Nevertheless, because they provide a means of purification and solvent transfer, information as to molecular size, and near-quantitative recovery, the gel filtration methods are practically indispensable in protein purification and analysis.

2. Adsorption/Elution Chromatography

This approach, already mentioned above, can be considered under the more specific titles ion-exchange, hydrophobic interaction, and affinity chromatography. In these applications, the interaction between the protein and the insoluble support is strong enough to enable a high degree of purification.

Ion Exchange Chromatography This procedure exploits the amphoteric character of a protein, namely, the property of a protein to exist in either cationic or anionic form depending on the pH of the media. An important parameter to know when performing protein purification by this method is the isoelectric point of the protein, i.e., the pH at which the net electric charge is zero. The strength of the binding is determined by the number of interactions between ionic groups on the protein and on the ion exchanger, and the ionic strength of the mobile phase. Thus, elution of proteins is achieved by a gradient of increasing salt concentration, a pH gradient, or a combination of the two. Resolution is usually improved by slowing down the elution flow rate and making shallower gradients. The end of such optimization often leads to isocratic elution conditions, which provides maximal resolving power. With the proper choice of exchanger, recoveries can be nearly quantitative. The charged groups themselves have undergone few changes in the past 25 years. Both strong and weak versions of anion and cation exchangers have been employed. Sulfonic acid- and quaternary ammonium-containing exchangers have the advantage of being fully ionized over the whole pH range. Weaker cationic exchangers include those containing carboxyl groups with pK = 4.8, or phosphates which ionize over a wider range of pH from 4 to 8. DEAE tertiary ammonium exchangers continue to be most widely employed for protein separations in the pH range 7–9.

Most of the development in this field has been with respect to the nature of the backbone. Polystyrene matrices are useful for separation of small molecules like amino acids, but proteins denature easily on such columns and are best recovered from matrices having a hydrophilic character such as cellulose, polydextrans, polyacrylamides, and vinyl polymer-based resins employed in the Toyo Pearl series. The advent of HPLC technology has led to the development of HPLC ion exchange columns in which the aformentioned functionalities are attached to silica and other matrices (see below).

Hydrophobic-Interaction Chromatography (HIC) Based on the interaction between hydrophobic chains on the surface of the protein with a nonpolar stationary phase, HIC makes use of phenyl- or octyl-Sepharose adsorbants where proteins are bound at high ionic strength and eluted with a gradient to low salt concentrations. The approach, therefore, is the reverse of ion exchange methodology. A variant of the method called reversed phase (RP) chromatography has become very popular in conjunction with high-performance liquid chromatographic separations (HPLC), and it is this procedure we want to describe in some detail. Before the advent of HPLC, almost all of the chromatographic applications were carried out at low pressure and with beaded particles of a size range between 50 and 400 μm. The chromatographic method can be improved dramatically in terms of separation, performance, and reproducibility if we use the same ligands attached to high-pressure-resistant materials with beaded particle size ranging between 5 and 50 μm tightly packed in small stainless steel columns. Such a configuration lends itself nicely to automation by introducing high-pressure fittings (up to 3000 psi), precision systems for pumping small volumes of eluant that will provide constant flow at high pressure, and electronic devices to automate the system. This technique is generally referred to as HPLC, and it is dominating more and more the field of protein chemistry at both the preparative and the analytical levels.

RP-HPLC is hydrophobic chromatography in which separation is based on differential binding to heavily alkylated stationary phases (usually silica but also polystyrene). This method allows proteins to adhere even in aqueous media of low ionic strength. Elution, carried out with gradients of increasing concentrations of an organic solvent, exploits a competitive interaction between the protein, the matrix, and solvent. Silica gel supports modified by alkylation are commonly used in RP-HPLC. The chain length of the alkyl substituent usually varies in length between octadecyl (C_{18}) and methyl depending on the size of the protein or peptide to be chromatographed; C_4, C_8, and C_{18} have been the most popular. Dilute trifluoroacetic acid has been used very often as an ion pair-forming compound, and offers the advantages of being a good solvent for proteins, transparent in the far ultraviolet where peptide bonds absorb (205–220 nm), and volatile. A variety of other chemicals have been employed with the constraint that with silica-based supports the pH must be kept below neutrality to avoid dissolution to the matrix. The organic component employed for elution of the proteins is more limited in the range of choices because it must possess sufficient polarity to keep proteins in solution and sufficient apolarity to elute the protein from the matrix. Acetonitrile and alcohols have been the predominant choices for this purpose. RP-HPLC separations are rapid to perform, very reproducible, and, in view of recent scale-up applications, becoming economically attractive. However, the acidic conditions often used may not always be compatible with the stability of the proteins to be purified. Moreover, serious losses of proteins may be encountered as the molecular weight increases due to irreversible adsorption to the matrix. Generally, the method is best applied to small, stable proteins.

Affinity Chromatography With the development of affinity chromatography, an element of rational design was introduced into an otherwise empirical and serendipitous area of research. This method takes advantage of some recognitional property that is unique to the protein of interest. The power of the approach lies in the fact that in the ideal circumstance, purification of the desired protein can be achieved in a single step. In principle, the single-step purification can be reached using as a ligand a monoclonal antibody to the protein of interest, a receptor specific for the protein, or a potent inhibitor or substrate of the enzyme of interest. The purity of the products obtained in this way may be adequate for basic research, but additional purification steps are often required: (a) multimeric forms and/or fragments of the recombinant protein produced by incomplete translation are also adsorbed specifically; (b) small amounts of contaminants are adsorbed nonspecifically; and (c) losses of the ligand from the column can occur if harsh conditions of pH are required for elution. We must also keep in mind that monoclonal antibodies are products of transformed cell lines and their use as immunoaffinity adsorbents could be potentially harmful if the protein purified thereby is to be used therapeutically. Furthermore, since the antibody ligands are themselves proteins, they are susceptible to proteolytic fragmentation. Despite these reservations, affinity chromatographic methods are at the forefront of purification strategies, and their importance will be evident as we consider specific applications. As an example, we have purified recombinant human renin directly from cell growth media by a single passage through a column containing an enzyme inhibitor attached to Sepharose (4). The resulting enzyme was greater than 95% pure in the protein component, although carbohydrate heterogeneity was extensive.

E. Scaling Up the Purification Process

Pilot preparations of recombinant proteins are usually performed in the medium-scale range producing quantities that can satisfy only basic research needs. Production for commercial purposes requires a scaling up of all operations and the leap to a large-scale operation brings with it a number of new and complex problems:

1. Reduction of the volume of the initial extract to a manageable size may be a long process, and cold temperatures may be difficult to maintain. Moreover, particulate materials must be removed to assure proper function of columns. Large-scale equipment for ultrafiltration and centrifugation have become available to circumvent these problems in scale-up.
2. It is difficult to pack huge separation columns uniformly; nonuniform packing causes nonuniform flow distribution and consequent losses of material and poor separation.
3. High losses of product are intolerable because the overall purpose is to maximize yields and still keep a high-quality product. Maximization is not a major problem in medium-scale operations where, for basic research purposes, we can sacrifice quantity for purity.

Cooperation of both basic research and production scientists is crucial for efficient scale-up of the production process. Ranges of conditions rather than specific details of purification must be explored by the R&D scientist and this information passed on to the production line in order to implement the overall process efficiently. This includes ranges of pH, temperature, buffer concentration, and flow rate which do not affect the final yields of pure biologically active material. Information about cleansability, disinfection, and sterilization in situ of the materials used must be included. Finally, input of the quality control scientist is essential to assure that the product meets the specifications required for commercial use. Integration of a wide variety of expertise is fundamental to the successful industrial scale production of proteins as therapeutic agents.

F. Assessment of Purity

We have already stated that there is no such thing as a "pure" protein; it is only a question of how impure it might be. There are a battery of tests that can be applied to assess contaminants in a purified protein. Techniques such as amino acid analysis, RP-HPLC, SDS-PAGE, isoelectric focusing and other native gel electrophoretic procedures, and N-terminal sequence analysis are effective, easy to perform, and by these means one can demonstrate that a protein is greater than 99.9% pure. Particular concerns arise when nucleic acids with potential oncogenicity are present. Hybridization with properly radiolabeled probes is recommended in this case; such methods allow detection on the order of 10 ng of nucleic acid. Putatively pure proteins may be contaminated by various compounds and, with respect to the drug industry, every country has its own set of rules by which purity must be documented. Without going into further detail, we cite the paper of Duncan et al. (9) and the Banbury Conference text (3) for more information regarding regulatory requirements for licensing medicinal products of biotechnology.

Since our discussion has centered on recombinant proteins, the question must always be asked as to whether or not the engineered protein is identical to its native counterpart. To answer this question in depth, we need to have characterized the native protein in detail, a task not possible in most cases. If enough information about the native protein is available, direct comparisons can be made. Otherwise, the only way to evaluate the recombinant product is to test its biological activity directly in animal models. As experience with recombinant proteins grows, it is becoming more and more accepted that the engineered product can suffice to set the standards relative to units of activity and structure that are not determinable for the natural protein.

V. PRODUCTION OF PROTEINS OF MEDICAL INTEREST BY RECOMBINANT DNA TECHNOLOGY

A. Human Interferons

These are proteins with antiviral, antiproliferative, and immunoregulatory activities. As antiviral agents, they protect their target cells rather than attack viruses directly. As immunoregulators, they activate macrophages and natural killer (NK) cells. Antiproliferative activity is shown by inhibition of the growth of cancerous cells. The purification procedure described here concerns human interferon-β (hIFN-β), and is taken from the work of Utsumi et al. [(10), Fig. 3].

E. coli cells were lysed in the presence of polyethyleneimine and NaCl, and after centrifugation a supernatant containing the recombinant hINF-β was obtained. The protein was concentrated by precipitation with ammonium sulfate. Extraction and concentration steps were followed by a purification procedure consisting of column chromatography on immunoaffinity and zinc chelate matrices. The availability of native and recombinant hIFN-β allowed characterization of both products, thereby permitting a comparison of several properties of the two proteins. Purified hIFN-β from E. coli and fibroblasts gave M_r values of 19,000 and 23,000, respectively, on SDS-PAGE. As expected, the carbohydrate moiety (about 4000 D) is present only in the natural hIFN-β; otherwise, the molecules have the same amino acid sequence.

Comparison of the CD spectra of the two proteins showed very similar secondary structures containing about 80% α helix and a low content of β sheet. The 1H-NMR spectra of both molecules indicated a similar folding pattern. By measuring the retention times on RP-HPLC, it was found that the hydrophobicity of the r-hIFN-β was slightly higher than that of the naturally occurring protein due in all likelihood to the presence of hydrophilic carbohydrate in the latter. The specific activity of purified r-hIFN-β was lower than that of the native polypeptide. Cause-and-effect relationships are not easily ascertained in comparisons of natural and recombinant proteins; the difference in activity could be due to carbohydrate, folding anomalies, or a variety of factors.

B. Monoclonal Antibodies

Monoclonal antibodies are immunoglobulins made by a single hybridoma cell line (11). The industrial interest in monoclonal antibodies in increasing rapidly due to their various applications in cancer therapy, diagnosis, tumor imaging, and immunopurification.

Two techniques have been developed to produce monoclonal antibodies (12). The first is an in vivo method based on the injection of hybridoma cells intraperitoneally into mice. Cells proliferate and secrete monoclonal antibodies into the ascites fluid (3–15 g of antibody is present per liter of fluid). The monoclonal antibody thus obtained is usually less than 30% pure; major contaminants include nonhybridoma IgG, albumin, transferrin, and a high content of lipids. The second is an in vitro method consisting in the culture of hybridoma cells in nutrient media into which the monoclonal is eventually secreted in a rather diluted form [0.01–0.05 g/liter if airlift fermentors are used; 0.1–10 g/liter if cells are immobilized in hollow fibers]. Contaminants are the same as in the in vivo method and, in addition, the proteins from the culture media, notably from fetal calf serum.

The purification scheme described in Fig. 3 refers to murine-derived monoclonal antibody (13). When the material is produced in vitro, it usually needs to be concentrated. Ultrafiltration by tangential flow or hollow fiber is used to avoid membrane blocking by the copious amount of lipids present.

1. Affinity chromatography on immobilized protein A often provides a single-step procedure to obtain a nearly pure product in high yield that is suitable for diagnostic and/or therapeutic applications (14). According to Ostlund (13), the procedure can be scaled up to process 100 g of mono-

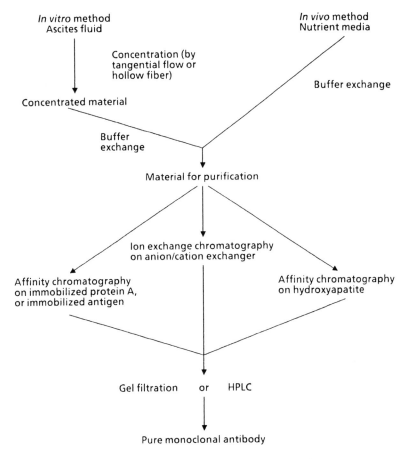

Figure 3 A scheme for monoclonal antibody purification (13).

clonal antibody in 3 hr. A potential alternative is affinity chromatography where the affinity ligand is the antigen. This, however, presents obvious problems of scaling up if the antigen is limited in quantity. Furthermore, the antigen–antibody complex can be so tight that harsh elution conditions may be required to remove the protein from the column.

2. Ion exchange chromatography, especially on cation exchangers, is an excellent method applicable to purification of gram quantities of monoclonal antibodies. The exchanger is less expensive than protein A and it can be regenerated in situ with NaOH. It is not as selective as protein A, however, and the product may elute with proteinase contaminants.

3. Stanker et al. (15) achieved a one-step purification of mouse monoclonal antibodies from ascites fluid by hydroxylapatite chromatography. However, Manil et al. (16) reported that adsorption chromatography on hydroxylapatite gave results inferior to methods described above. Both the ion exchange and adsorption chromatographic strategies can be coupled to HPLC for medium-scale purification of monoclonal antibodies with reasonably good results (17, 18).

C. Insulin

Insulin is a hormone composed of two polypeptide chains (A with 21, and B with 30 residues) linked by two disulfide bridges. It is synthesized in the islet cell of the pancreas as preproinsulin precursor having a leader N-terminal peptide and a connecting peptide (C peptide) that joins the carboxyl end of chain B to the amino terminus of chain A. Processed insulin is secreted into the blood in response to increasing concentrations of glucose. Insulin is used extensively in the treatment of diabetes and is an important recombinant product in the pharmaceutical industry. Two methods are described here for the engineering of recombinant human insulin. The first procedure for the production of insulin by gene cloning was that of Goeddel et al. (19), who chemically synthesized two genes coding for human insulin A and B chains. Each gene was fused to an *E. coli* β-galactosidase gene and expressed. These authors estimated that approximately 20% of the total cellular protein was the expressed β-galactosidase–insulin A- or B-chain hybrid. The hybrid proteins composed about 50% of the total protein precipitate of the inclusion bodies. After extraction of the precipitate with guanidinium hydrochloride, centrifugation, and dialysis of the supernatant, the two solubilized chains from the fusion proteins were liberated by cleavage with cyanogen bromide. This was made possible by a construction wherein a methionine residue preceded the N terminus of both chains, thus taking advantage of the fact that there is no methionine in insulin. Purification of the two chains so generated, and reconstitution of the two into mature insulin is described elsewhere (19).

The more recent system for production of recombinant insulin is that of Cockle et al. [(20), Fig. 4] who expressed multiple joined proinsulin genes in *E. coli*. Synthesis was under the control of an inducible lac promoter to direct the production of fused peptides comprising several proinsulin moieties. Such precursors containing about 90% proinsulin were produced in high yield and accumulated as insoluble aggregates. The precursor consisted of single-chain

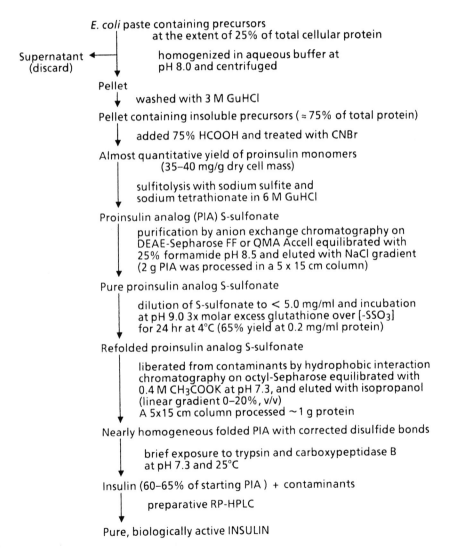

Figure 4 Purification of insulin expressed by a polyproinsulin construct (20).

proinsulin containing 86 amino acids and three disulfide bridges, and had the following sequence:

$$\text{MITDSLAM(PROINSULIN RRNSM)}_4\text{GS}$$

The nine-residue leader corresponds to the first eight residues of *E. coli* β-galactosidase, plus a methionine linker; individual proinsulins are separated by a five-amino-acid linker, also ending in methionine. Therefore, as in the previous example, the stage was set for the use of cyanogen bromide as a means of liberating the proinsulin analogs. Digestion of the analogs with trypsin at the double-arginine site, followed by treatment with carboxypeptidase B gave rise to properly folded insulin, but side products were also generated from internal cleavages which required further purification (Fig. 4).

D. Tissue Plasminogen Activator (tPA)

Human tissue-type plasminogen activator is a serine protease with a fibronectin-like finger region, a growth factor domain, and two kringle structures which converts plasminogen into its active form, plasmin. Plasmin, in turn, can activate more tPA as well as perform its major function in the dissolution of blood clots. tPA is found in very small amounts in kidney, uterus, and blood vessels. The mature form of tPA has 527 amino acids and is glycosylated at three N-linked sites. Recombinant tPA is currently available on the market as a therapeutic in the treatment of myocardial infarct.

The procedure described here for purification of tPA secreted by a cultured human melanoma cell line is that of Rijken and Collen (21). This method, outlined in Fig. 5, employs three column chromatographic steps on zinc chelate agaraose, concanavalin A–agarose, and Sephadex G-150. Analysis of the tPA by SDS-PAGE under nonreducing conditions showed a band migrating with a M_r of 72,000; in the presence of thiols, two bands were observed migrating with M_r of 33,000 and 39,000. A protease present in the culture medium cleaves the monomer between Arg-275 and Ile-276 to produce what is commonly called two-chain tPA in which the halves of the molecule are held together by a single disulfide bond. Two-chain tPA is the activated form of tPA. Inclusion of the proteinase inhibitor aprotinin in the culture medium blocked this activation and a single-chain tPA is recovered under these conditions. A battery of tests, including SDS-PAGE, compositional and sequence analysis, and biological activity measurements, suggested that the recombinant product is indistinguishable from the tPA found in normal tissue.

This general purification protocol has been adopted by others (22) to purify recombinant human tPA produced in *E. coli*, yeast, and mammalian cells into which the vector containing the tPA gene has been introduced by the techniques of molecular cloning.

E. Copper-Zinc Superoxide Dismutase (h-SOD)

Cu, Zn superoxide dismutase is a eukaryotic cytoplasmic enzyme, also present in serum, which scavenges the superoxide radicals produced in a variety of normal and pathological events. SOD has anti-inflammatory properties and is known to reduce reperfusion damage to the heart, kidneys and other organs after transplantation or after dissolution of blood clots. Therefore, while tPA dissolves blood clots, SOD helps to reduce the damage caused by subsequent reperfusion. To be effective, SOD, like tPA, must be administered in large doses.

Hallewell and colleagues (23) obtained efficient transcription of h-SOD in yeast using the promoter for the yeast glycolytic enzyme glyceraldehyde 3-phosphate dehydrogenase (GAPDH). They inserted the linked GAPDH promoter and h-SOD coding sequence into the pC1/1 yeast shuttle vector, and after transformation of yeast strain 2150-2-3 leu- with this construct, they selected recombinant colonies that lacked leucine. With these methods, the

10 liters culture supernatant containing 60 µg/ml protein, 20 IV/ml, 4°C

↓

Affinity chromatography column (5 x 10 cm) on Zn^{2+}-chelating agarose equilibrated with 20 mM Tris-HCl pH 7.5, 1 M NaCl, 0.01% Tween-80. Elution with 0-50 mM imidazole gradient in the above buffer

↓

Affinity chromatography column (0.9 x 25 cm) on concanavalin A-agarose equilibrated with 10 mM phosphate buffer pH 7.5, 1 M NaCl, 0.01% Tween-80. Elution with 0-0.2 M α-d-methylmannoside and 0-1 M KSCN

↓

Gel filtration on Sephadex G-150 column (2.5 x 90 cm) equilibrated with 10 mM sodium phosphate buffer, pH 7.5, 1.6 M KSCN, 0.01% Tween 80

↓

Dialysis against 0.15 M NaCl + 0.01% Tween 80 (v/v)
tPA yield was 46%, and specific activity 90,000 IV/mg

Figure 5 Purification of tPA from human melanoma cells (21).

authors were able to express h-SOD to an extent of 30–70% of total cell protein in stationary phase yeast. The recombinant h-SOD was compared with native enzyme from human erythrocytes and the two proteins were shown to be identical with respect to activity, physical characteristics, and tryptic maps. Remarkably, the two enzymes are both acetylated at the N-terminus, underscoring the capability of yeast to modify recombinant proteins posttranslationally. Production of h-SOD in yeast is an alternative to expression in *E. coli*, an organism that is incapable of N-acetylation (24). Problems with yeast that can arise as a result of hyperglycosylation were not encountered since there is no carbohydrate in SOD. Another interesting feature of this expression system is that yeast SOD is acetylated; therefore, this organism has the machinery to carry out this modification and applies it in response to the proper signal.

F. Reverse Transcriptase from the AIDS Virus

Reverse transcriptase (RT) is an essential enzyme for viral replication and thus represents an important target for antiretroviral drugs. The enzyme is encoded by a portion of the viral *pol* gene (25, 26) and is produced in its mature form by proteolytic processing by viral protease of a large, multiprotein precursor (26). Only small amounts of RT can be produced in this way (27); generation of large amounts of RT by cloning techniques allows for reasonably fast in vitro screening of drugs currently available, and for structure–function analysis to provide a rational basis for drug design.

Larder and colleagues (28) expressed a 66-kDa RT in *E. coli* and devised a simple purification procedure consisting of high-salt fractionation followed by chromatography on DNA-cellulose and fast-performance liquid chromatography (FPLC) on a Mono Q column. With this approach, they were able to obtain large amounts of purified RT (about 10 mg/liter of culture) with authentic RT activity.

Barr et al. (29) expressed the RT in genetically engineered *S. cerevisiae*. For direct expression of the RT domain in yeast, these workers constructed a gene on the basis of the N-terminal sequence of native RT (29) which starts with Pro-156 of the pol gene. Their choice for the C-terminus was Val-691, selected on the basis of *pol* gene products of other retroviruses. They constructed a vector in which the RT gene was flanked by promoter and terminator sequences of the yeast GAPDH gene. The vector was expressed in yeast and the purification procedure was as outlined in Fig. 6. Purified RT showed two biologically active peptides of 66 and 51 kDa that were similar, if not identical, to the 66- and 51-kDa proteins expressed in virus-infected cells (29). This was documented by migration in SDS-PAGE, N-terminal analysis, inhibition by several antiviral agents, and immunological cross-reactivity. Sequence analysis also showed that the N-terminal methionine constructed in the synthetic initiation codon was posttranslationally removed. This, therefore,

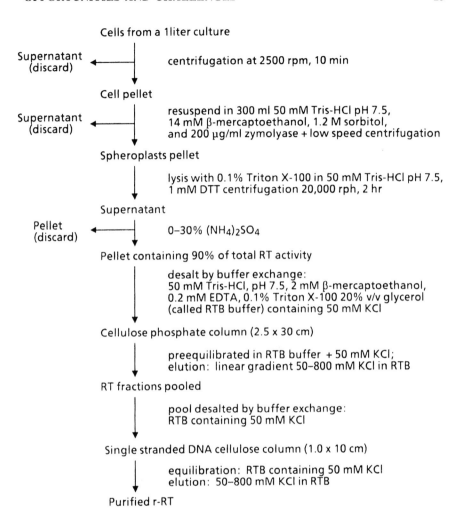

Figure 6 Purification of AIDS virus r-RT (29).

provides another example of the versatility of the yeast expression system in being able to carry out some, if not all, of the posttranslational modifications characteristic of mammalian cell hosts.

VI. PERSPECTIVES: THE ROAD TO RATIONAL DESIGN

The theme of this chapter has been focused on the advantages and challenges in protein purification and characterization offered by the recombinant DNA technology. The advantages are obvious; in principle any protein, however rare its message in living forms, can be generated in limitless quantities, purified and characterized in a detail often not hitherto possible. In so doing, we have cataloged a vast number of nucleic acid-derived amino acid sequences, and new relationships among diverse sets of proteins have been established. The flood of new information is reaching a point where it is almost impossible to keep up with the task of detailed comparative analysis.

All of this is good; where is the challenge? In a sense, protein purification will always present challenges since each protein offers its own unique set of physicochemical peculiarities; there is no set procedure for isolating proteins. Yet, as we have seen in the few examples given in the preceding section, these obstacles can be overcome and proteins will yield one way or another to our arsenal of purification strategies.

We would like to make the point that the real challenge which lies ahead has to do with the time-worn term "structure–function" or "structure–activity" relationships. Indeed, a major goal of contemporary research in the protein field has to do with the creation, by rational design, of new proteins with new biological potentials. This emphasis notwithstanding, our knowledge of structure far surpasses that of function, or mechanism. Admittedly, we have a long way to go in understanding how structure is related to function; how can the recombinant DNA technology help?

For starters, we need to know something about the chemistry of biological function, and we need detailed information about three-dimensional structure before we can put the pieces together to make new proteins that are intelligently designed for new applications. Site-directed mutagenesis provides the tool for construction of any protein we might desire, but a blueprint for such work must have a rational basis. The permutations available for a protein of a size even so small as 100 amino acids are astronomical. We must be able to key in on structural features that are related, in however remote or blurred a fashion, to function. Then those products must be evaluated rigorously for conformation and activity.

As always, we can learn something from nature. Nature has been engineering proteins for 2 billion years, and the results of this "nature-directed mutagenesis" are imprinted in the sequences of the same or closely related proteins as they exist today in the whole biological spectrum of organisms. As they share a commonality in primary structure, they also are similar, if not identical, in function. Comparative sequence analysis of such proteins will reveal portions of the molecules that have "survived" the evolutionary process and which may therefore be considered to be essential for function. Ever-increasing productivity of research in translating genetic messages into protein sequences holds continued promise of gaining new insights into function; the efficiency of the genetic route far surpasses our capabilities in protein sequence analysis.

It is clear, therefore, that our data base of information about protein primary structure will continue to expand exponentially and to shed light on our understanding of the essential aspects of each molecule. Sometimes we forget that the protein sequence is itself a message, providing information not only for folding, but for a conformation optimized for biological activity. When we don't even understand the elementary rules for bringing segments of structure together in a general sense, it is almost beyond the imagination to conceive how the evolutionary-driven process led to generation of functional proteins. Efforts in decoding the secrets locked in the protein sequence have provided algorithms for predicting aspects of secondary structure that are often reasonably accurate and which can be used quite successfully as the basis for understanding function. The leap to tertiary structural prediction, however, continues to pose a formidable challenge. We still rely almost exclusively on X-ray crystallographic analysis for such detailed information on protein conformation.

The X-ray crystallographic approach to the resolution of protein structure has until very recently been highly labor-intensive and costly. The last decade has witnessed major advances in the technology of protein structure determination in areas involving speed of collection and analysis of diffraction data, elegant computer graphic systems to facilitate modeling, and new programs for the calculation of structures at high resolution. Nature is conservative in the use of folding elements in protein architecture; α helices, β sheets, turns, etc., provide all of the structural diversity needed to assemble a vast spectrum of molecules differing in shape, size, physicochemical properties, and function. Thus, every crystal structure solved adds to another expanding data base that will find application as related proteins come to light by gene sequencing. One of the central tenets of molecular biology is that proteins linked by similarity in sequence and function will share common tertiary structures. For example, a new trypsin inhibitor that appears imprinted in the gene of some rare message and that is homologous to pancreatic trypsin inhibitor can, by inference, be assumed to have a three-dimensional architecture and mechanism similar to that of the well-studied prototype.

Crystallographic analysis of protein–ligand systems such as enzyme–substrate complexes provides important new information about function. Such studies supply a set of snapshots of the enzyme in situations that, in the best cases, are close to intermediate stages of the catalytic reaction. However, X-ray crystallography alone does not address directly the question of mechanism. Dynamic measurements embodied in classical enzymology and kinetic analysis provide the fundamental knowledge that must be accommodated in any mechanistic models derived from X-ray studies. To make the enzyme come alive, these and other spectroscopic methods must be performed in conjunction with crystallographic analysis because it is in this way that valuable information about catalytic intermediates can be obtained.

Thus, the road to rational design is paved with information from several sources; those that provide the structural framework for intelligent change, and those that tell us about how the molecule works. Even with the most sophisticated understanding of a particular protein molecule, efforts to produce by recombinant methods a new protein with a carefully conceived new function have been by and large disappointing. This only serves to underscore our rather primitive level of understanding of function relative to structure; the rules for predictive change are not yet in place. Nevertheless, despite this dim perception of cause and effect in protein mutagenesis, the power and utility of protein engineering remain and will be of fundamental importance in establishing those same rules. We can continue to design and purify novel recombinant proteins and then apply a battery of mechanistic and structural studies which will allow us to understand the roles of individual amino acids and to learn in great detail, and step by step, the intricate laws which govern the structural, functional, physical, and chemical properties of proteins (Chapter 13, this volume). Some of the corresponding experimental disciplines and their interrelationships are summarized in Fig. 7. Some day we will be able to use these newly emerging rules to make more accurate predictive changes, to predict tertiary structure from amino acid sequence, to create proteins with novel properties, to design drugs for precise targeting to particular proteins, and to understand the evolutionary process of divergence.

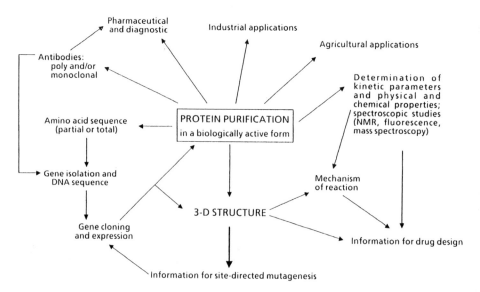

Figure 7 Interrelationships among disciplines involved in elucidation of protein structure and function.

REFERENCES

1. Pearl LH, Taylor WR. A structural model for the retroviral proteases. Nature 1987; 329:351–354.
2. Malcom ADB. Decline and fall of protein chemistry? Nature 1978; 275:90–91.
3. Banbury Report 29. Therapeutic peptides and proteins: Assessing the new technologies. New York: Cold Spring Harbor Laboratory, 1988.
4. Poorman RA, Palermo DP, Post LE, Murakami K, Kinner JH, Smith CW, Reardon I, Heinrikson RL. Isolation and characterization of native human renin derived from chinese hamster ovary cells. Proteins: Struct Funct Genet 1986; 1:139–145.
5. Pain R. Protein folding for pleasure and for profit. TIBS 1987; 12:309–312.
6. Fetrow JS, Zehfus MH, Rose GD. Protein folding: New twists. Biotechnology 1988; 6:167–171.
7. Creighton TE. Toward a better understanding of protein folding pathways. Proc Natl Acad Sci USA 1988; 85:5082–5086.
8. Sharma SK. On the recovery of genetically engineered proteins from *Escherichia coli*. Sep Sci Technol. 1986; 21(8):701–726.
9. Duncan ME, Charlesworth FA, Griffin JP. Regulatory requirements for licensing medicinal products of biotechnology. Tib. Tech. 1987; 5:325–328.
10. Utsumi J, Yamazaki S, Hosoi K, Kimura S, Hanada K, Shimaru T, Shimuzo H. Characterization of *E. coli*-derived recombinant human interferon-β as compared with fibroblast human interferon-β. J. Biochem. 1987; 101:1199–1208.
11. Köhler G. Derivation and diversification of monoclonal antibodies. EMBO J 1985; 4:1359–1365.
12. Commercial production of monoclonal antibodies. A guide for scale-up. Seaver SS. ed. Hygeia Sciences, Newton, MA, 1987.
13. Östlund C, Large-scale purification of monoclonal antibodies. Tibtech Nov. 1986; 4:288–293.
14. Lee SM. Affinity purification of monoclonal antibody from tissue culture supernatant using protein-A-Sepharose CL-4B. pp. 199–216 of Ref 12.
15. Stanker LH, Vanderlaan M, Juarez-Salinas H. One-step purification of mouse monoclonal antibodies from ascites fluid by hydroxylapaitite chromatography. J Immunol Meth 1985; 76:157–169.
16. Manil L, Motte P, Pernas P, Troslen F, Bohuon C, Bellet D. Evaluation of protocols for purification of mouse monoclonal antibodies. Yield and purity in two-dimensional gel electrophoresis. J Immunol Meth 1986; 90:25–37.
17. Strickler MP, Gemski MJ. Single-step purification of monoclonal antibodies by anion exchange high-performance liquid chromatography. pp 217–245 of Ref 12.
18. Juarez-Salinas H, Brooks TL, Ott GS, Peters RC, Stanker L. High-performance liquid chromatography characterization and purification of monoclonal antibody. pp 277–297 of Ref 12.
19. Goeddel DV, Kleid DG, Bolivar F, Heyneker HL, Yamsura DG, Crea R, Hirose T, Kraszewski A, Itakura K, Riggs AD. Expression in *Escherichia coli* of chemically synthesized genes for human insulin. Proc Natl Acad Sci USA 1979; 76:106–110.

20. Cockle S, Lennick M, Shen S-H. Production of peptide hormones in E. coli via multiple joined genes. Protein purification: Micro to macro. New York: Alan R. Liss, 1987:375–381.
21. Rijken DC, Collen D. Purification and characterization of the plasminogen activator secreted by human melanoma cells via culture. J Biol Chem 1981; 256:7035–7041.
22. Rystaro K, Tsuneo U, Shigeki O, Kobayashi S, Tadao S. Process for producing tissue plasminogen activator EP 0208486A2.
23. Hallewell RA, Mills R, Tekamp-Olson P, Blacher R, Rosenberg S, Ötting F, Masiarz FR, Scandells CJ. Amino terminal acetylation of authentic human Cu, Zn superoxide dismutase produced in yeast. Biotechnology 1987; 5:363–366.
24. Hallewell RA, Masiarz FR, Najarian RC, Puma JP, Quiroga MR, Randolph A, Sanchez-Pescador R, Scandells CJ, Smith B, Steomer KS, Mullenbach GT. Human Cu/Zn superoxide dismutase cDNA: Isolation of clones synthesizing high levels of active or inactive enzyme from an expression library. Nucleic Acids Res 1985; 13:2017-2034.
25. Ratner L, Haseltline W, Patarca R, Livak KJ, Starcich B, Josephs SF, Doran ER, Rafalski JA, Whitehorn EA, Baumeister K, Ivanoff L, Petteway SR, Pearson ML, Lautenberger JA, Papas TS, Ghrayeb J, Chang NT, Gallo RC, Wong-Staal F. Complete nucleotide sequence of the AIDS virus, HTLV-III. Nature 1985; 316:277–284.
26. Weiss R, Teich N, Vermus H, Coffin J. RNA tumor viruses. New York: Cold Spring Harbor Laboratory, 1982.
27. diMarzo Veronese FM, Copeland TD, DeVico AL, Rahman R, Oroszlan S, Gallo RC, Sarngadharan MG. Characterization of highly immunogenic p 66/p 51 as the reverse transcriptase of HTLV-III/LAV. Science 1986; 231:1289–1291.
28. Larder B, Purifoy D, Powell K, Darby G. AIDS virus reverse transcriptase defined by high level expression in *Escherichia coli*. EMBO J 1987; 6:3137.
29. Barr PJ, Power MD, Lee-Ng CT, Gibson HL, Lucia PA. Expression of active human immunodeficiency virus reverse transcriptase in *Saccharomyces cerevisae*. Biotechnology 1987; 5:486–489.

2

Purification and Production of Therapeutic Grade Proteins

Vipin K. Garg,* Maureen A. C. Costello, and Barbara A. Czuba[†]

Bio-Response, Inc.
Hayward, California

I. INTRODUCTION

Purified proteins are being increasingly used for therapeutic purposes as a result of the current biotechnological revolution. Such therapeutic proteins include monoclonal antibodies (MAbs), interleukins, interferons, growth factors, tissue plasminogen activator (tPA), erythropoietin, tumor necrosis factor, and the blood-clotting protein factors. These proteins may be produced by hybridoma cell lines, as in the case of MAbs, by genetically engineered cells (microbial or animal cells), or by natural producer lines. This complex array of products along with the complexity of their production technologies has raised unique concerns regarding the safety of these therapeutic protein products. These concerns are resulting in a redefinition of the requirements for the quality control of therapeutic proteins by the regulatory agencies (1–4). For instance, the criteria for establishing protein purity are currently undergoing substantial reexamination. Due to the emergence of rapid, high-resolution chromatographic and electrophoretic techniques, the potential now exists to systematically address the quality control of therapeutic proteins at analytical levels hitherto not feasible (5, 6). Furthermore, it is also possible to develop new purification strategies specifically designed for attaining therapeutic grade purity (7, 8).

The purpose of this chapter is to present an overview of purification-related considerations for producing therapeutic grade proteins. The chapter's theme

---

Present affiliation:
*Sepracor Inc., Marlborough, Massachusetts.
[†]Baxter, Highland Division, Hayward, California.

has been deliberately developed to address general purification safety issues, i.e., validation of protocols, cleaning and hygiene of matrices, and removal of nonproteinaceous contaminants. Specific purification protocols for individual proteins have not been discussed in this chapter. These protocols can be found in other specialized chapters of this book.

II. CONSIDERATIONS FOR DEVELOPING THE PURIFICATION PROTOCOLS

The challenge in protein therapeutics is to acquire large amounts of a purified product under strictly controlled conditions. In order to purify a biopolymer to near homogeneity and satisfy the stringent limits of contaminant detection (Table 1), a battery of sophisticated chromatographic techniques is required by the purification chemists. Schemes developed in research laboratories to purify small amounts of a protein are often not useful because of poor recoveries and the elaborate use of chromatographic and electrophoretic procedures; thus, new preparative schemes must often be developed.

From the point of view of therapeutic grade purity, the purification protocol must be able to remove all the contaminants (both proteinaceous and nonpro-

Table 1 Impurities and Contaminants of Concern and Analytical Techniques Frequently Employed for Their Detection in the QC of Biotechnology-Derived Therapeutic Proteins

Impurities or contaminants	Analytical technique
1. Proteinaceous contaminants (e.g., host cell proteins, other protein impurities)	SDS-PAGE electrophoresis, HPLC, immunoassays (ELISA, etc.)
2. Endotoxin	Rabbit pyrogen test, LAL[a]
3. DNA	DNA dot-blot hybridization
4. Proteolytic degradation products	IEF,[b] SDS-PAGE, HPLC, N- and C-terminus analysis
5. Presence of mutants and other residues	Tryptic mapping, amino acid analysis
6. Deamidated forms	IEF
7. Microbial contamination	Sterility testing
8. Virus	Viral susceptibility assay
9. Mycoplasma	21 CFR method (Code of Federal Regulations)
10. General safety	21 CFR 610.11

[a]Limulus amebocyte lysate.
[b]Isoelectric focusing.

teinaceous in nature) from the crude product solution. At the same time, further contamination during the purification operations has to be minimized. Regulatory guidelines recommend that for cell culture protein products destined for use in vivo, the product should be in excess of 95% pure when examined by all available analytical techniques (3–5). It is also expected that no single impurity should exceed 1% of the total protein and that the identification of all contaminants has been attempted. In addition, the product should be nonpyrogenic and contain less than 10 pg of DNA in each dose received by a patient (4, 9). Examples of therapeutic proteins purified from animal cell culture supernatant in our laboratory are shown in Figs. 1 and 2. Since the market for such injectable protein products is growing rapidly, there is a great demand for developing novel and economical purification protocols in the biotechnology industry today.

Development of a preparative scale purification protocol for therapeutic proteins is a multistep process that requires several considerations. These include the following.

A. Chemical Structure and Physical Properties of the Target Protein

The chemical structure and physical properties of the target protein are the two key parameters used to develop most purification protocols. Isoelectric point (pI), pH stability, and charge density are some of the properties of proteins

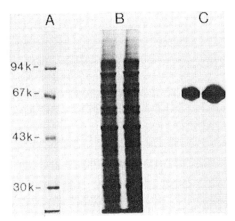

Figure 1 SDS-PAGE analysis of a therapeutic grade protein, tissue plasminogen activator (tPA). A = Molecular weight standards; B = starting material before purification; C = purified tPA (>99% pure). Purified product loads were 50 and 100 μg, respectively.

exploited during purification. A number of separation techniques are capable of resolving proteins on the basis of differences in their net charge. These include gel electrophoresis, free-flow electrophoresis, and, most importantly, ion exchange chromatography (IEC). Ion exchange chromatography is currently the most widely used method of protein purification. Further extensions of the net charge separation approach are found in chromatofocusing and isoelectric focusing.

Physical properties of the proteins, e.g., temperature stability, solubility, and apparent size and shape, are also exploited in several purification approaches. For instance, temperature stability and solubility form the basis of a number of fractionation schemes, whereas molecular size and shape characteristics are used in ultrafiltration, and gel filtration chromatographic separations. Further examples of protein structure and properties that can be utilized in developing purification strategies are accessibility and surface distribution of lipophilic or nonpolar residues, distribution of amino acid residues at the surface of the folded protein, carbohydrate content, content of free sulfhydryl groups, presence of exposed metal ions, functional properties, and antigenicity of the protein molecule.

B. Concentrations and the Nature of Contaminants in the Starting Material

Since the purpose of purification is to separate the target protein from contaminating materials (both proteinaceous as well as nonproteinaceous), it is also important to know the concentrations and the nature of the contaminants in the

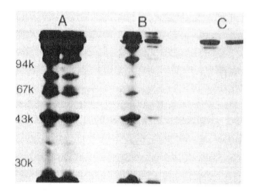

Figure 2 SDS-PAGE analysis of an IgG1 monoclonal antibody purification. A = starting material after concentration; B = antibody after intermediate purification. C = purified therapeutic grade antibody (>98% pure). Purification was performed by a three-step protocol using ion exchange chromatography.

starting material. The composition of the starting material varies significantly from one product to another, and is affected by such factors as the nature of the expression organism (bacterial, primitive eukaryotic, or animal) (10), type of production technology (bioreactor design, etc.), and the characteristics of the individual cell line used in production (11). Genetically engineered bacterial cells have been used for producing nonglycosylated proteins like interferon. However, since bacteria do not generally secrete the protein of interest (engineered target protein), it is necessary to break the cell membrane by mechanical or chemical lysis to obtain the desired product. This may introduce additional contaminants (e.g., bacterial DNA, viruses, lipids, intracellular proteins) into the starting material, thereby complicating the purification process. Another purification-related concern with production in bacterial cells is contamination by endotoxins.

When bacterial expression is not feasible (e.g., for MAbs and tPA), mammalian cells may be selected as the culture system. In mammalian cell culture, the product is secreted into the surrounding medium, thereby eliminating the need for cell disruption. However, this does not necessarily lead to an advantage in the purification process. Mammalian cells require a much richer nutrient medium than bacterial cells (12). Thus, the product stream not only contains complex biological molecules and waste products secreted by the cells but may also include a large number of serum proteins, peptides, carbohydrates, lipids, and organic acids which must be added to the culture medium. For the purpose of downstream processing it is extremely important to minimize the level of serum and thus reduce the major source of contaminant proteins. Many mammalian cell lines can be adapted to grow in serum-free media containing defined levels of albumin, insulin, and transferrin, thereby simplifying purification (12–14). In addition, some cell lines can be cultured in protein-free, chemically defined media (15), which may allow a one- or two-step purification scheme.

In between bacterial and mammalian cells are lower organisms such as yeast and fungi as well as insect cells that have been cultured for specific purposes. These expression organisms have their unique cultivation considerations that should be taken into account for downstream processing. Insect cells, for example, may not always secrete the target protein into the surrounding medium (16). Furthermore, there are regulatory concerns associated with the removal of insect and vector (virus) DNA.

In terms of the production technology, a basic distinction can be made between cell culture systems based on whether they utilize batch or perfusion methods. The concentrations and the nature of the contaminants are greatly affected by the choice of culture conditions (17, 18). Clearly, the choice of production technology should be considered carefully, as it usually has a profound effect on the purification strategy.

C. Selection of the Chromatographic Modes to Be Used in the Protocol

The attainment of very high purities for proteins can only be achieved by the application of multiple chromatographic modes. Clearly, the selection of the exact separation chemistries to be used becomes a crucial consideration. In the past, size exclusion (SEC), ion exchange (IEC) and reversed phase (RPC) chromatography have been the most popular purification modes. Hydrophobic interaction chromatography (HIC) and affinity chromatography (AC) are recent and powerful additions to the available battery of techniques.

As indicated earlier, IEC has become one of the most widely used methods of protein purification. It is a highly interactive adsorption technique, capable of separating molecules with very small differences in charge. The first ion exchange supports made were anion exchangers, primarily because the chemistry for their synthesis tended to be simpler. Many of the ion exchange protein separations have been performed on the weak anion exchanger DEAE. More recently, however, strong anion exchangers, quaternary ammonium (Q), and cation exchangers, carboxymethyl (CM) and sulfopropyl (SP), have also been shown to be suitable for protein purification. Several review articles (19–21) previously discussed methods of protein separation by IEC in detail.

Just as IEC takes advantage of the charge distribution on the surface of the protein molecule, RPC and HIC exploit the accessibility and surface distribution of lipophilic or nonpolar residues. Because of its historical origin the term HIC is frequently attributed to separations affected by a decreasing salt concentration while the term RPC refers to the separations involving an increasing concentration of organic solvent in the eluent. However, the underlying principle for both separation methods is common (22). For most protein purification applications HIC is usually more suited than RPC. A major concern with RPC of proteins is the loss of activity and denaturation during chromatography using the standard acidic and organic mobile phases (23–25). RPC has proven, however, to be an extremely useful technique for the separation of peptides (25, 26) and certain proteins that are soluble in organic solvents (27).

In contrast to RPC, HIC is a mild separation technique which involves nondenaturing physiological buffers and a descending salt gradient for elution. In recent years, HIC has gained increasing popularity as a versatile protein purification tool (28, 29). HIC provides a selectivity very different from ion exchange chromatography thereby providing another dimension of fractionation and simplifying many otherwise difficult separations. Finally, HIC can also be used as a polishing step to remove minor contaminants and to remove nucleic acid (DNA) and pyrogens in some applications.

The next group of separation techniques exploit the asymmetrical distribution of specific amino residues at the surface of folded proteins. For example, access to exposed histidine residues is exploited in metal chelate interaction chromatography (MIC). A number of important mammalian proteins bind selectively to divalent metals held on resin by chelation. Zinc is the most commonly used metal, interacting with interferon (30) and tissue plasminogen activator (31), among others. The adsorbed proteins are easily eluted with a soluble chelating agent such as EDTA. Another form of chelate interaction chromatography can be exploited when a coordination site of a metal ion cofactor is exposed on the surface of the target protein. Since only a limited number of proteins exhibit these types of interactions, MIC may show a good resolving power between the target product and contaminants.

The final group of separation parameters, and the ones which give the highest selectivity, employ the methods based on affinity chromatography. AC separations can be divided into two categories: (a) group-specific AC, and (b) biospecific AC. Examples of some affinity ligands used in the purification of proteins are shown in Table 2. Group-specific AC includes generic ligand interactions such as dye affinity, protein A and protein G affinity (IgG), borate affinity, and lectin affinity. Biospecific AC, on the other hand, exploits functional properties of the target protein such as a specific ligand-binding site or antigenicity (immunoaffinity). With appropriate immobilization chemistries and ligand choice, biospecific AC has the potential to generate separation capacities more than two orders of magnitude greater than observed with adsorption methods based on simple chemical ligands such as the ones employed for ion exchange or reversed phase chromatography. Due to its highly specific interaction, immunoaffinity chromatography can be very effective in isolating the target component from dilute solutions and complex mixtures. This results in a significant advantage when purifying novel therapeutic proteins in that a very high purity product can be purified with a minimum number of processing steps and with much greater product recoveries than those seen using conventional purification methods.

D. Sequence of the Processing Steps in the Purification Protocol

With most biotechnology-derived protein products, the ideal single-step purification is often unobtainable because of low starting concentrations and large initial volumes. Clearly, the specific sequence of processes adopted becomes an important consideration. Generally, the sequence of steps progresses from clarification and filtration to an initial enrichment (concentration) step, followed by intermediate purification steps, and finally purification by a high-resolution method to the standards of purity required for the particular product (8, 11, 32).

Clarification and filtration are essential to remove cell debris and any particulates introduced during the production. Both filtration and centrifugation

Table 2 Examples of Some Affinity Ligands Used in the Purification of Proteins

Affinity ligand	Specificity	Proteins purified
1. Protein A	Fc region of IgG and related molecules	IgG, IgM, and IgA, antibody fragments, insulin-like growth factor (IGF-1)
2. Protein G (recombinant)	Fc region of IgG, binds IgG from most mammalian species	IgG (IgG1, IgG2, and IgG3), used for both polyclonal and monoclonal antibodies
3. Concanavalin A	Terminal D-manno-, D-glucopyranose, or sterically similar residues	Interferon, plasminogen activator, colony-stimulating factor, glycoproteins, glycopeptides
4. Other lectins (e.g., lentil lectin, wheat germ lectin, peanut lectin)	Specific sugars and glycoconjugates	Glycoproteins, membrane glycoproteins, cell surface antigens, viral glycoproteins, nerve growth factor
5. Various dyes (e.g., Cibacron Blue F3GA, Procion Blue, Brown, and Red, blue dextran)	Nucleotide-binding sites of enzymes, and other specificities	α_1-Antitripsin, IgG, interferon, kinases, dehydrogenases, phosphotases, and other enzymatic proteins
6. Phenylboronic acid	*cis*-Diol-containing substances	Glycosylated hemoglobins, phosphodiesterases, hexapeptides
7. Heparin	Enzyme-inhibitor or enzyme-activator interaction; also, ionic interactions	Human α_1-proteinase inhibitor, human antithrombin III, polymerases, nucleases, proteases, and other enzymatic proteins
8. Calmodulin	Calcium-dependent proteins	Phosphodiesterases, ATPase, calcinerin, other calmodulin-binding proteins
9. Benzamidine	Interaction with trypsin and trypsinlike serine proteases	Urokinase, trypsin, tPA, thrombin, plasminogen, kallikrein
10. Arginine and lysine	Biospecific or charge dependent; both stereospecific and electrostatic effects	Plasminogen, plasminogen activator, prothrombin

Table 2 (*Continued*)

Affinity ligand	Specificity	Proteins purified
11. DNA, RNA, nucleosides, and nucleotides	Interaction with complementary sequences, other nucleotide or nucleoside-binding sites	Nucleases, polymerases, kinases, interferon, antibodies to nucleic acids, tubulin assembly protein
12. Gelatin	Interaction with fibronectin	Fibronectin
13. Immunoaffinity	Antigen–antibody interaction	(A) Purification of antibodies using immobilized antigens (e.g., anti-growth factor, antialbumin, anti-tPA antibodies) (B) Purification of antigens using immobilized antibodies (e.g., interferon, interleukins, tPA, urokinase, growth factors, blood-clotting protein factors)

methods can be used. The technical problems and advantages of these methods have been reviewed extensively elsewhere (33). The next step is initial product enrichment, which can be performed by either precipitation or ultrafiltration (34). This product enrichment step is particularly important with animal cell culture, where the concentration of products is generally low, certainly when compared to product concentrations in cultures of bacteria and yeast. In our experience, an ultrafiltration method having a membrane with a pore size smaller than the desired protein is suitable for a 50- to 100-fold product concentration. Precipitation techniques by neutral salts or by organic solvents have been described by several investigators (34, 35). While these precipitation techniques are extensively used in the plasma fractionation industry, they are of limited utility for modern recombinant protein purification due to the low protein (product) concentration found in the spent culture fluid (11, 36).

The intermediate purification steps generally employ chromatographic separations utilizing the chemical and physical properties of the target proteins described above. With many proteins it is possible to combine the initial enrichment (concentration) step with an intermediate purification step in a chromatographic mode. The starting feed material can be loaded on a process scale chromatography column either directly after filtration or with some adjustment of pH or salt concentration. With modern fast-flow resins, packed in large-diameter columns (>20 liters column volume), flow rates of hundreds of liters per hour can be easily achieved (11, 37). This allows convenient processing of large volumes, and combines both the concentration and first purification step in a single operation, thereby reducing losses associated with a multistep enrichment process.

During the intermediate purification steps, several different chromatographic modes are usually required to achieve the desired product purity. As discussed earlier, various chromatographic modes, i.e., IEC, AC, HIC, or MIC, each have their advantages and disadvantages. These should be carefully considered before deciding on a final chromatographic sequence. Generally, ion exchange chromatography can be adopted as a convenient first step. This can be followed by either AC (if available), MIC, HIC, or even additional IEC. As indicated earlier, if bioaffinity chromatography is available as a separation mode, it can be used to yield a significant advantage. In fact, a large number of modern genetically engineered proteins are currently being purified by immunoaffinity chromatography. An important consideration while using bioaffinity chromatography must be the removal of any proteins (ligand) leached from the affinity matrix during purification. Additional purification steps should be included to ensure the removal of these extraneous proteins.

If AC is not available as a separation mode, HIC may be used as an alternative mode. Since proteins are bound to HIC supports in the presence of high (rather than low) ionic strength, HIC is a convenient next step for fractions that are eluted from ion exchange in relatively high ionic strength buffers. In some cases, HIC may even be appropriate following affinity chromatography when the sample is eluted in a high salt concentration. Finally, if the target protein shows interactions with heavy metals (e.g., Zn, Cu), MIC can be employed in place of AC or HIC.

The intermediate purification steps are usually followed by a final purification step, which is specifically designed as a polishing step to meet therapeutic grade purity requirements (Table 1). While most of the chromatographic modes available during this final step are the same as the ones available during the intermediate purification steps, it is possible to use certain high-resolution methodologies during this step. For instance, HPLC can be conveniently employed at this step, since the bulk of the contaminating proteins, which would otherwise reduce the capacity of the HPLC columns, would have been removed by this stage of purification (7, 38, 39). Furthermore, by this time the sample volume has been considerably reduced in comparison to that which existed at the early purification steps. This eliminates long sample loading times or the use of larger, expensive HPLC columns to reduce sample loading time. Similarly, other high-resolution techniques based on composite cartridges and membranes can also be employed at this step.

E. Choice of the Stationary Phase

In addition to the type of chromatographic modes and their sequence in a purification protocol, the choice of the stationary phase (matrix) should also be considered carefully. The nature of the matrix determines several important parameters, such as mechanical strength, flow characteristics, adsorbing capacity, and nonspecific binding. Similarly, the pore size of the matrix determines accessibility and protein penetration, thereby affecting loading capacity and efficiency. Widely used matrices for preparative protein chromatography include either the insoluble soft gels based on polysaccharides and other polymers (e.g., cellulose, dextrans, agarose, and polyacrylamides), or rigid supports based on inorganic substances (e.g., silica and controlled-pore glass).

From a safety standpoint, a major consideration in selecting the stationary phase is cleaning and sterilizing the chromatography matrix. Chemical stability of the matrix to regeneration and sanitization treatment is very important. The matrix should be stripped of any bound proteins and depyrogenated before reuse. With many matrices (e.g., Sepharose, cellulose), regeneration and depyrogenation can be achieved by washing the matrix with 0.5–1.0 M NaOH. However, if the matrix is unstable at high pH (e.g., silica), other methods should be developed for depyrogenation and sanitization. A variety of mobile phases are available for cleaning silica-based matrices (40). Results in our laboratories have also shown that silica-based matrices (36) can be kept pyrogen-free if the chromatography is performed using sterile buffers and the column is periodically cleaned with organic solvents. In addition to cleaning and sterilization, the possibility of leachable compounds from chromatography matrices and their bonded phases should also be considered. The concentration and nature of leachable materials, if any, from bonded phases during their normal usage should be established, and such data should be compiled in Food and Drug Administration master files by the manufacturers of chromatography media.

In terms of the resolution and capacity, silica-based matrices show a considerably superior performance to that of soft gels. This combined with the ability to use higher flow rates allows for substantially higher throughput on silica-based matrices. In the overall scheme of purification, however, both types of matrices (soft gels versus rigid inorganic supports) have their unique applications. In our experience, highly crosslinked Sepharose-type gels are best suited for initial purification steps, in which large volumes and highly impure feed materials are usually processed. As the product becomes concentrated and the purity levels are improved, high-resolution silica-type supports become more appropriate.

As an alternative to gel matrices, several bioprocess companies recently developed membrane and composite structure matrices in cartridge form (41–43). Composite matrix cartridges for large-scale chromatography have been developed by Cuno, Inc. (Meriden, CT) (41), and are available both in ion exchange and affinity modes. These cartridges have several advantages over gel matrices (41) and have been successfully used for protein purification (44). Similarly, membrane-based purification devices have also been developed in both ion exchange and affinity modes (42, 43). A hollow-fiber membrane device, recently developed by Sepracor, Inc. (Marlborough, MA), has shown exceptional promise for affinity purification of proteins (43, 45). Clearly, membrane-based systems will find increasing use in downstream recovery and purification of proteins. Despite its attractiveness, regulatory issues could slow the acceptance of membrane technology in the bioprocess industry.

III. REMOVAL OF NONPROTEINACEOUS CONTAMINANTS

To meet the specifications for therapeutic grade purity, it is important that the purification processes be designed not only to maximize the protein purity, but also to remove any nonproteinaceous contaminants (e.g., nucleic acids, pyrogens, etc.) that may compromise the safety of the purified product. Too often purification chemists concentrate only on the protein purity and ignore the nonproteinaceous contaminants until the final purification steps. In our experience, contaminants such as DNA and pyrogens should be removed gradually as the relative purity of the target protein increases from one purification step to the other. Furthermore, the general principle of avoiding problems downstream by taking appropriate action upstream should always be considered (13).

During the purification process, three potential nonproteinaceous contaminants require special consideration: (a) DNA, (b) pyrogens, and (c) viruses.

A. DNA Removal

Since normal cells under normal controls do not overproduce proteins, new biotechnology-derived products are produced in genetically transformed or "abnormal" cells. These abnormal cells are usually tumorigenic and have potential oncogenes. To eliminate this biological hazard, it is desirable to reduce the amount of DNA to very low levels in a product derived from genetically engineered or transformed cells. Currently, levels of DNA <10 pg of DNA per dose are generally acceptable for therapeutic proteins. At this low level of DNA there is essentially no risk of inducing tumors in recipient patients (46).

DNA levels in the purified products are determined by the DNA dot-blot hybridization method. In this method, the DNA impurities in the product are identified through hybridization with ^{32}P-labeled DNA probes on nitrocellulose paper. The current level of sensitivity of this method is 1–10 pg DNA/mg protein.

Early approaches for the removal of DNA from protein products were exemplified by the production of lymphoblastoid interferon (47) and insulin (48).

In these production schemes, cellular DNA was removed by extraction techniques under strongly denaturing conditions. Since both interferon and insulin are relatively robust proteins, it was possible to show that added DNA did not survive the extraction procedures, while the target protein was recovered without a loss of activity. In contrast to interferon and insulin, however, most other biotechnology-derived proteins (e.g., MAbs and tPA) cannot survive harsh denaturing conditions. In these cases, mild chromatographic procedures are required to remove contaminating cellular DNA. Since DNA is a strongly anionic molecule at a wide pH range (pH 4.0 and above), anion exchange chromatography can be effectively used to remove DNA from proteinaceous solutions (7). Anion exchange chromatography conditions can be developed such that the target protein molecules do not bind to the anion exchanger (and will elute in the void volume), while the majority of the DNA will be bound to the column and thus removed from the target protein (7).

We have systematically evaluated different anion exchange chemistries for the removal of DNA from protein solutions (7, 8). An evaluation of a single step is illustrated in Table 3. The genomic DNA used for spiking in these studies was derived from the mammalian cell line secreting the product to mimic, as closely as possible, the conditions that exist at the start of an actual tual purification. Based on these studies, quaternary amine, strong anion exchangers (QMA and QAE) were most effective in binding DNA at physiological pH (Table 3). Weak anion exchangers, PEI and DEAE, were also able to bind DNA, but to a much lesser extent than QMA and QAE. Anion exchange chromatography is a very effective approach for removing DNA from proteins with isoelectric points (pIs) above pH 6.0. In our experience, best results are obtained when anion exchange steps are repeatedly used, under either binding or nonbinding conditions, throughout the purification process. This allows for both the maximum protein purity and DNA removal during purification.

Table 3 Evaluation of Different Preparative HPLC Matrices for the Removal of DNA and Endotoxin from Protein Solutions[a]

Matrix	Log reduction in DNA[b]	Log reduction in endotoxin[c]	Protein recovery (%)
PEI	1.0	2.0	98
DEAE	1.0	0.5	95
QAE	3.0	2.0	99

[a]Buffer conditions for each column run were adjusted to allow maximum protein recovery in the flow-through.
[b]Protein solutions were spiked with 10 μg of genomic DNA/mg protein.
[c]Endotoxin level was artificially increased to 800 EU/mg protein.

In contrast to anion exchange chromatography, cation exchange chromatography steps are usually not developed for DNA removal. However, this can be a useful approach with some strongly acidic proteins. Cation exchange chromatography is particularly applicable during the early part of a purification process, where the relatively dilute target protein is bound to the cation exchanger, and many impurities, including DNA, are eluted unbound in the flow-through.

If affinity chromatography is used for the purification of the target protein, this step can also produce a significant DNA clearance. In particular, if an immunoaffinity ligand is available, this can be used for the removal of up to 99% of cellular DNA from the feed material. Since there are no ionic interactions involved in an immunoaffinity purification step, there is very little (if any) potential for cellular DNA to bind to the column. The DNA flows right through the column, whereas the target protein stays behind on the column. In our experience, an important consideration for achieving maximum DNA removal during affinity chromatography is to perform extensive column washing (even after a stable baseline is obtained) prior to the elution of the target protein. In some cases, a column can be washed in an intermediate buffer containing high salt (0.2–0.5 M NaCl) to ensure complete removal of any loosely bound DNA to the column. Following this wash, the protein of interest can be eluted from the affinity column with either a low pH elution buffer or a chaotropic agent.

Finally, HIC has also shown a great potential for DNA removal in recent studies (49). In the purification of MAbs, it was shown that mammalian DNA eluted in the unbound fraction from a propyl (C3) silica column in the presence of 700 mM ammonium sulfate, while all of the MAb was bound to the column (49). Once the DNA is separated from the MAb molecules, the MAb can be eluted from the column by lowering the salt concentration. We have shown that HIC can be used in a similar fashion to remove DNA from other therapeutic proteins. One advantage of HIC methodology is that the high-salt conditions used for the loading of columns tend to dissociate DNA–protein binding, thereby facilitating their separation from one another.

B. Pyrogen Removal

The purification of injectable therapeutic proteins requires that the final product by nonpyrogenic. The parenteral administration of medicinal products may result in the induction of fever, chills, histamine release, altered vascular permeability, or other unwanted side reactions in the recipient (50). Such reactions are termed pyrogenic responses, and any component, contaminant or otherwise, of a biological or pharmacological product which induces these reactions if referred to as a pyrogen. Although potential pyrogenic contaminants of pharmaceuticals can be either microbial or nonmicrobial, the one of most concern to pharmaceutical manufacturers is contaminant bacterial endotoxin from the family Enterobacteriacea. Endotoxins are ubiquitous in nature. They are lipopolysaccharide (LPS) components of the cell walls of gram-negative bacteria, organisms that require minimal nutrients for growth. Endotoxins are shed during bacterial growth or cell lysis and are quite stable. They remain biologically active even after autoclaving.

Eliminating endotoxins from a protein solution is a difficult and often expensive undertaking. In the production of biotechnology-derived protein pharmaceuticals the best strategy for generating pyrogen-free parenteral solutions is to maintain aseptic conditions in the manufacturing process and during subsequent downstream processing (8, 11). It is particularly important to control or eliminate pyrogen contamination and leaching during each of the chromatographic purification steps. As described earlier in this chapter, all chromatography resins must be depyrogenated and cleaned prior to each use. The chromatographic steps must be performed at low temperature (2–8°C), and only sterile buffers must be used for column equilibration, washing, and elution. The eluted protein solutions must be quickly sterile-filtered (0.2 μm filtration) and stored at 2–8°C. In some cases elution peaks can be sterile-filtered on-line during the chromatographic procedure. In our experience, the application of the above precautions normally results in a nonpyrogenic, purified protein preparation (8, 11). However, in some cases additional methods may be needed for pyrogen removal from protein solutions.

Traditional methods for depyrogenation have employed two general approaches: (a) inactivation by chemical or heat treatment, or (b) removal by media adsorptive techniques using activated carbon, asbestos, and barium sulfate (51). These methods, however, are not compatible with modern biotechnology-derived protein products. If used with proteinaceous solutions, these methods will either inactivate or randomly adsorb the key components of the product. More recently, other methods have been developed that remove endotoxin on the basis of their molecular size. Commonly used examples of these methods are ultrafiltration (UF) and reverse osmosis (RO). These methods work well for removing pyrogens from pharmaceutical products that have low molecular sizes (52), e.g., peptides and low molecular weight proteins. For most macromolecular proteins, however, even these methods are useless. Clearly, removal of pyrogens from protein solutions represents a difficult problem which may sometimes be impossible to overcome. In most cases, well-optimized and validated chromatographic methods represent the only alternatives for removing pyrogens from protein solutions.

Since LPS are anionic molecules, they can be adsorbed to a positively charged chromatographic resin and therefore be removed by anion exchange chromatography (7, 53). As shown in Table 3, several different anion exchange chemistries can be employed for endotoxin removal from protein solutions. This represents a situation very similar to DNA removal, discussed earlier in this chapter. As with DNA removal, buffer conditions can be optimized such that the target protein molecules do not bind to the anion exchanger, while the LPS molecules will bind to the column and thus be removed from the protein solution (7). For example, in our studies with immunoglobulin purification, we found that at a pH approximately one unit below the immunoglobulin pI, the IgG does not bind to the anion exchanger, while the majority of the endotoxin does bind to the column and is separated from the IgG molecules (7, 11). A comparison of different anion exchange chemistries showed that at pH 6.0–7.0 QAE and PEI chemistries were more effective in binding DNA than DEAE chemistry (Table 3). In addition to removal of endotoxin based on its negative charge, hydrophobic interaction of the lipid A portion of LPS with a hydrophobic surface can also be exploited for endotoxin removal (49, 54).

Since the LPS molecule has also been shown to have binding affinity to certain biological molecules, affinity chromatography can also be used to remove endotoxin from several pharmaceutical products. Methods used have included affinity chromatography using polymixin B (PMB), LAL (Limulus amebocyte lysate), and antibodies. Several studies have shown that immobilized PMB, a cationic antibiotic, can be used to remove endotoxins from solutions (53, 55, 56). A PMB affinity column step has been described for depyrogenation of interferon (56). However, more recent studies have shown that PMB affinity chromatography is not very effective in removing endotoxin from protein-containing solutions in which the endotoxin may be bound to the protein (57). In such cases enhanced endotoxin removal can be achieved by combining PMB affinity with detergent treatment (57). LAL can also be coupled to a chromatographic matrix and used as an affinity column to remove endotoxin. It has been shown that LAL was capable of removing endotoxin from certain protein products, namely, concanavalin A and erythropoietin (58).

Affinity chromatographic approaches based on polyclonal or monoclonal antibodies can also be used for endotoxin removal. It has been suggested that antibodies coupled to a stable matrix could be used to remove either specific endotoxin from a given bacteria or generic endotoxins from all gram-negative bacteria (53). Several studies have demonstrated the ability to produce broad-spectrum antibodies against LPS (53). These antibodies exhibit protection against heterologous endotoxin and could be coupled to a resin for affinity chromatography.

C. Removal of Viruses

In addition to DNA and pyrogens, the question of viral contamination from cells used to produce protein products should also be systematically considered. The possibility that viable viruses might remain in purified products does exist. Virus testing of protein products is a recommendation set forth by regulatory agencies both in the United States and in Europe (1, 9, 59). A major concern is that patients may be immunocompromised due to their disease states and therefore may be more susceptible to potential viral infection from repeated exposure to therapeutic protein products (e.g., MAb) that may contain active viruses or viral particles.

Since the main sources of viral impurities in the final product are endogenous viruses that might be present in the host cell, adequate viral screening and monitoring of the master cell bank is very important. The master cell bank should be free of any viral infestation before the cell line is taken into production (60). This will virtually eliminate the risk of viral contamination in the final purified product.

The types of viruses that must be tested for vary from cell line to cell line, and also from one product to the next. For instance, MAb products derived from ascites are subject to contamination with viruses from the host mouse as well as murine viruses propagated by the hybridoma cells. Hybridomas entirely or partially of human origin also may have human viruses that likewise contaminate the antibody. Similarly, other therapeutic protein products produced in mammalian cells may be contaminated with murine, human, or bovine viruses (60, 61). Many murine viruses can be tested for by the mouse antibody production (MAP) test. Additional specific tests may need to be carried out for lymphocytic choriomeningitis virus (LCMV), mouse cytomegalovirus, mouse rotavirus (EDIM), thymic virus, and lactic dehydrogenase virus (9, 59–61). Similarly, tests capable of detecting murine retroviruses should also be performed. The human virsues that are often tested for are Epstein–Barr virus (EBV), human cytomegalovirus, hepatitis B virus, and retroviruses (9, 59–61). Specific tests for these viruses are described in the literature. Finally, a tissue culture safety test using a series of established cell lines is also recommended because of potential contamination by other viruses not specifically screened for.

With respect to purification, in most cases normal chromatographic protocols (e.g., ion exchange) designed to obtain the desired level of protein purity should actually be able to eliminate any viral contamination in the final product. However, it is recommended that specifically validated procedures, e.g., filtration, UV inactivation, which remove and/or inactivate viruses, be employed during purification (9). Several chemical and physical viral inactivation steps have been described in the literature (62), and should be incorporated during the purification process in order to achieve adequate safety margins even if such steps do not improve the product yield or purity.

IV. EXTRANEOUS PROTEIN REMOVAL

As discussed earlier in this chapter, all proteinaceous contaminants (e.g., serum proteins, cellular proteins, etc.) present in the raw product solution must be separated from the target protein during the purification process. Additionally, any extraneous protein components introduced during the purification process must also be removed from the final product. Of major concern here are the proteins that are linked to matrices to generate affinity columns and may be leached from the matrix during the purification process. Even if there is virtually no risk of ligand leaching from a given affinity matrix, protocols capable of removing any leached ligand should be developed and incorporated in the purification process (63).

A widely discussed example of extraneous protein contamination can be found in antibody purification schemes where protein A affinity chromatography is used (11). The problem of leakage of protein A into the product is of great concern to those developing antibody products for injection into humans. While new coupling chemistries that significantly reduce ligand leakage are currently available from several manufacturers of the chromatography supports, in our view it is still essential to incorporate additional purification steps that will ensure removal of any leached ligand. In our studies, we have found that leaked protein A can be effectively removed from antibody preparations by anion exchange chromatography. In many cases, a final ion exchange polishing step can be used to remove traces of protein A from purified MAb with minimum loss (<10%) of purified final product (11, 64).

In another purification protocol, we have developed chromatographic steps for the removal of any leached ligand from a mammalian protein preparation following affinity chromatography with an immobilized plant protein (63). As shown in Fig. 3, when spiked with extraneous plant protein, the mammalian protein (target protein) can be separated from the plant protein (contaminant) by cation exchange chromatography. The elution profile of ^{125}I label (Table 4) shows that when spiked with extraneous protein, <0.4% of the label coeluted with the mammalian protein of interest (fraction 4). This represents a >2 log clearance of the extraneous protein across this step.

If immunoaffinity chromatography is used in the purification scheme, similar chromatographic methods should be developed for removing any leached antibody from the product. As with protein A, ion exchange methods can be conveniently used for removing traces of antibodies from other protein products. However, in this case individual protocols will have to be developed depending on the properties of the antigen being purified by immunoaffinity chromatography. In some cases, HIC and SEC chromatography can also be used for the removal of extraneous proteins from the protein of interest. Finally, it must be remembered that the removal of free ligand is usually difficult

THERAPEUTIC GRADE PROTEINS

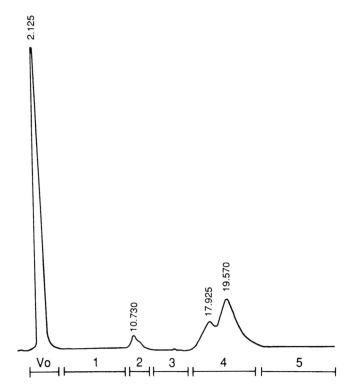

Figure 3 Demonstration of removal of a labeled ^{125}I extraneous protein (a plant protein used in affinity chromatography) from the eluates of a mammalian cell culture product using HPLC. HPLC conditions: Column, Toyo Soda TSK SP-5SW (7.5 × 75 mm); buffer A = 100 mM sodium acetate, pH 5.0; buffer B = 100 mM sodium acetate + 1 M NaCl, pH 7.0; flow rate = 1 ml/min. The column was eluted with a gradient of 0% B to 30% B in 10 min. V_0 = void volume; 1, 3, and 5 = baseline elutions; 2 = extraneous plant protein; 4 = mammalian cell culture protein. The tabulation of the numbers from three different HPLC runs is shown in Table 4.

with affinity ligands which remain tightly bound to the soluble component. Clearly, wherever possible, stable coupling chemistries should be used to minimize ligand leaching.

V. VALIDATION OF PROTOCOLS

In order to consistently meet all specifications, a purification protocol must be validated at different process steps. During this validation, the purification process is examined critically for its ability to allow reduction of risk factors.

Table 4 Elution Profile of ^{125}I-Labeled Extraneous Plant Proteins During HPLC Purification of a Mammalian Cell Culture Protein[a]

Fraction	Column (% of radioactivity in each fraction)		
	1	2	3
V_0 (unretained proteins)	43.80	42.53	52.79
1 (baseline elution)	0.98	0.94	0.96
2 (plant protein)	53.90	55.74	44.71
3 (baseline elution)	0.96	0.75	1.21
4 (mammalian cell culture protein)	0.38	0.36	0.17
5 (baseline elution)	0.00	0.00	0.00

[a]Chromatographic elution profile of different fractions is shown in Fig. 3.

Often this is done by means of "spiking" experiments, in which suitably labeled or recoverable preparations of each risk factor are added to specific process steps (46, 65). These studies should be meticulously planned and executed to represent the actual events during full-scale purification, so that the data and conclusions are above reproach. Also, the validation experiments need to demonstrate the phenomenon of removal or inactivation in repeated trials. Besides cells, viruses, pyrogens, and DNA, other candidates for this exercise are chemicals and reagents used in purification, tissue culture media components, and potential allergens. Clearly, much of the perception of safety of the product rests on these validation studies.

A summary of validation stages for therapeutic proteins produced in mammalian cell culture is shown in Table 5. The levels of protein contaminants, cellular DNA, and endotoxin are determined at stages 1 and 2 in order to establish baseline levels of contaminants of each starting material that must meet specifications. Following this, extensive validation studies are conducted during the chromatography steps that are performed to purify the protein. During chromatography steps, the validation studies are required at two levels. First, the validation studies must be conducted during all in-house protocol development work for a novel protein purification (63). Following this, the validation studies must also be conducted during the scale-up and production phase. As shown in Table 6, analyses must be performed on at least three separate runs of each step. Once appropriate changes are made to the protocol, purification of gram quantities of protein is undertaken with continuous evaluation of purity of product (PAGE, HPLC), DNA levels, and endotoxin content (63).

Table 5 Validation Stages for Therapeutic Grade Proteins Produced in Mammalian Cell Culture (Stages from Production in MCT[a] to Purified Final Product)

Validation	Stage
Analyze starting material for DNA, endotoxin, viruses, and protein composition	1. Pooled MCT[a] culture eluate
Validate filter or column methods used to reduce volume	2. Concentrate
Analyze each step a minimum of three times with and without spiking (DNA, endotoxin, viruses, other proteins)	3. Chromatography (LC, HPLC)
Analyze product in all stages of formulation	4. Formulation (dialysis, filtration, other)
Analyze by DNA hybridization and rabbit pyrogen testing. Determine final purity and conduct final product testing	5. Final product

[a]MCT = mass culturing techniques used at Bio-Response, Inc.

Examples of validation studies during chromatography steps are shown in Tables 4 and 6. In one case (Table 4), successful removal of an extraneous plant protein (used as a purification reagent) has been shown from the target protein in three separate column runs. In another case (Table 6), a single chromatographic step effectively removed a 2 log concentration of cellular DNA. It is important to note that the protein of interest was recovered with minimal loss (>93% recovery) during this DNA removal step. Similar experiments

Table 6 Validation of QAE Chromatography for Removal of Cellular DNA from an Antibody Product

Run		Protein	DNA[a]
Pre-QAE	1	1 g	25 µg cellular DNA/mg protein
	2	1 g	25 µg cellular DNA/mg protein
	3	1 g	25 µg cellular DNA/mg protein
Post-QAE	1	0.95 g	0.22 µg cellular DNA/mg protein
	2	0.93 g	0.20 µg cellular DNA/mg protein
	3	0.98 g	0.23 µg cellular DNA/mg protein

[a]Elutions were performed under nonbinding conditions for antibody protein. Samples were spiked with cellular DNA isolated from mammalian cell lines. Proteins coating the DNA were not removed in order to closely mimic behavior of the DNA that would be found in mammalian cell culture eluates.

must be conducted to evaluate different chromatography matrices with respect to their capability to remove various contaminants.

In the course of validation studies at stages delineated in Table 5, data are gained about the function of procedures and matrices. These data can be used to generate pass/fail criteria at the various stages of purification (63). For instance, pooled culture eluate and concentrate must assay below certain DNA or endotoxin levels in order to proceed to further purification. Similarly, an effective log clearance of endotoxin, DNA, and other proteins must be attained at given chromatographic steps in order to be further processed.

Finally, with regard to viruses, it is important to note that the presence of a virus in cells need not preclude the use of such cells in the production of vaccines or other biologicals (46). The safety of the final product is the appropriate end point in deciding on the acceptability of cell substrates. Clearly, it is important that validation studies demonstrate that viruses are removed or inactivated by the purification process.

VI. CONCLUSION

Product safety is an important goal of purification procedures for the biotechnology-derived products which are intended to be used clinically. The main concern of any regulatory agency is to guarantee, to the best of its ability, that any product intended for human use is safe and will not cause additional disease or complications. In this chapter, the authors have attempted to review the principles of protein purification typical to the manufacture of cell-derived products with emphasis on factors critical to product safety. The importance of repeated testing and validation studies is emphasized, and examples and guidelines for chromatographic considerations for therapeutic protein purification are discussed.

ACKNOWLEDGMENTS

The word-processing support of Diane Franklin is gratefully acknowledged.

REFERENCES

1. Points to consider in the production and testing of new drugs and biologicals produced by recombinant DNA technology (1985). Office of Biologics Research and Review, Center for Drugs and Biologics, Food and Drug Administration, Bethesda, MD, April 10, 1985.
2. Bogdansky FM. Considerations for the quality control of biotechnology products. Pharmaceut Technol Sept 1987; 72–73.

3. Miller HI. FDA regulation of products of the new biotechnology. Am Biotechnol Lab Jan 1988; 38–43.
4. Baker DA, Harkonen S. Regulatory agency concerns in the manufacture and testing of monoclonal antibodies for therapeutic use. In: Tyle P, Ram BP. eds. Targeted diagnosis and therapy. Volume 3: Targeted therapeutic systems. New York: Marcel Dekker; 1990; 75–98.
5. Hancock WS. Significance of purity in the manufacture of recombinant-DNA-derived proteins. Chromatogr Forum, Sept–Oct 1986:57–59.
6. Borman S. Analytical biotechnology of recombinant products. Anal Chem 1987; 59:969A–973A.
7. Garg VK. Use of preparative HPLC in large scale purification of therapeutic grade proteins from mammalian cell culture. 1987, Paper 911, Seventh international symposium on HPLC of proteins, peptides, and polynucleotides, November 2–4, 1987, Washington, D.C.
8. Garg VK, Crane IJ, Czuba BA, Stevenson V, Mangilog M, Branson RE. Current techniques for large scale purification of therapeutic grade proteins. Biotech '88 USA, November 14–16, 1988, San Francisco.
9. Points to consider in the manufacture and testing of monoclonal antibody products for human use (1987). Office of Biologics Research and Review, Center for Drugs and Biologics, Food and Drug Administration, Bethesda, MD, June 1, 1987.
10. Baskin L. Mammalian cell culture techniques allow for new opportunities in commercial scale-up of biotech drugs. The Med Business J July 15, 1988; 198–202.
11. Garg VK. Large scale purification of monoclonal antibodies for therapeutic use. In: Tyle P, Ram BP. eds. Targeted diagnosis and therapy. Volume 3: Targeted therapeutic systems. New York: Marcel Dekker, 1990; 45–73.
12. Griffiths B. Can cell culture costs be reduced? Tibtech. 1986; 4:268–272.
13. Cartwright T. Isolation and purification of products from animal cells. Tibtech 1987; 5:25–30.
14. Brown BL. Reducing costs up front: A method for adapting myeloma and hybridoma cells to an inexpensive chemically defined serum-free medium. In: Seaver SS. ed. The commercial production of monoclonal antibodies. New York: Marcel Dekker, 1987; 35–48.
15. Bliem R, Oakley R, Matsuoka K, Taiariol V, Zitzner L, Branson E. Continuous cell culture using protein-free medium. Biotech '88 USA, November 14–16, 1988, San Francisco.
16. Luckow VA, Summers MD. Trends in the development of baculovirus expression vectors. Biotechnology 1988; 6:47–55.
17. Brown PC, Costello MAC, Oakley R, Lewis JL. Applications of the mass culturing technique (MCT) in the large-scale growth of mammalian cells. In: Feder J, Tolbert WR. eds. Large-scale mammalian cell culture. New York: Academic Press, 1985; 59–71.
18. Seaver SS. Culture method affects anitbody secretion of hybridoma cells. In: Seaver SS. ed. Commercial production of monoclonal antibodies. New York: Marcel Dekker, 1987; 49–71.

19. Himmelhock SR. Chromatography of proteins on ion exchange adsorbents. In: Jacoby WB. ed. Methods in enzymology, Volume 22. New York: Academic Press, 1971; 273–286.
20. Pharmacia. Ion exchange chromatography: Principles and methods. Pharmacia Fine Chemicals publication, 1982, Uppsala, Sweden.
21. Scopes R. Protein purification principles and practice. New York: Springer-Verlag, 1982; 75–101.
22. Hearn MTW. General strategies in the separation of proteins by high-performance liquid chromatographic methods. J Chromatogr 1987; 418:3–26.
23. Luiken J, Van Der Zee R, Welling GW. Structure and activity of proteins after reversed-phase high-performance liquid chromatography. J Chromatogr 1984; 284:482–486.
24. Benedek L, Dong S, Karger BL. Kinetics of unfolding of proteins on hydrophobic surfaces in reversed-phase liquid chromatography. J. Chromatogr 1984; 317:227–243.
25. Gooding KM. High-performance liquid chromatography of proteins: A current look at the state of the technique. Biochromatography 1986; 1:34–40.
26. Hermodson M, Mahoney WC. Separation of peptides by reversed-phase high-performance liquid chromatography. Meth Enzymol 1983; 91:352–359.
27. Regnier FE. High-performance liquid chromatography of proteins. Meth Enzymol 1983; 91:137–190.
28. Gooding DL, Schmuch MN, Gooding KM. Analysis of proteins with new, mildly hydrophobic HPLC packing materials. J Chromatogr 1984; 296:107–114.
29. Kato YT, Kitamura T, Hashimoto T. Operational variables in high-performance hydrophobic interaction chromatography of proteins on TSK gel Phenyl-5 PW. J Chromatogr 1984; 298:407–418.
30. Edy VG, Billiau A, de Somer P. Purification of human fibroblast interferon by zinc chelate affinity chromatography. J Biol Chem 1977; 252:5934–5935.
31. Rijken DC, Collen D. Purification and characterization of the plasminogen activator secreted by human melanoma cells in culture. J Biol Chem 1981; 256:7035–7041.
32. Rosevear A, Lambe C. Downstream processing of animal cell culture products: Recent developments. In: Spier RE, Griffiths JB. eds. Animal cell biotechnology, Volume 3. New York: Academic Press, 1988:397–440.
33. Ball GD. Clarification and sterilization. In: Spier RE, Griffiths JB. eds. Animal cell biotechnology, Volume 2. New York: Academic Press, 1985;87–127.
34. Van Der Marel P. Concentration. In: Spier RE, Griffiths JB. eds. Animal cell biotechnology, Volume 2. New York: Academic Press, 1985:185–216.
35. Deutsch HF, Fahey JL. Purification of antibody. In: Williams CA, Chase MW. eds. Methods in immunology and immunochemistry, Volume 1. New York: Academic Press, 1987; 315–332.
36. Garg VK, Branson RE. Applications of mixed and multimodal chromatography in protein purification. Paper S-66, Society for Industrial Microbiology Annual Meeting, August 9–14, 1987, Baltimore.
37. Scott RW, Duffy SA, Moellering BJ, Prior C. Purification of monoclonal antibodies from large-scale mammalian cell culture perfusion systems. Biotechnol Progr 1987; 3:49–56.

38. Berkowitz SA, Henry MP, Nau DR, Crane LJ. A strategic approach to the use of silica-based chromatography media (bonded phases) for protein purification. Am Lab May 1987:33–42.
39. Kagel RA, Kagel GW, Garg VK. High performance affinity chromatography for commercial protein purification. HPLC '88, Twelfth International Symposium on Column Liquid Chromatography, June 19–24, Washington, D.C.
40. Wehr CT. Sample preparation and column regeneration in biopolymer separations. J Chromatogr 1987; 418:27–50.
41. Hou KC, Mandaro RM. Bioseparation by ion exchange cartridge chromatography. Biotechniques 1986; 4:358–367.
42. Kontes. Kontes fast-chrom membrane chromatography. Bioseparations, March 1988, No. 8:2–9.
43. Brandt S, Goffe RA, Kessler SB, O'Conner JL, Zale SE. Membrane-based affinity technology for commercial scale purifications. Biotechnology 1988; 6:779–782.
44. Menozzi FD, Vanderpoorten P, DeJaiffe C, Miller AOA. One-step purification of mouse monoclonal antibodies by mass ion exchange chromatography on zetaprep. J Immunol Meth 1987; 99:229–233.
45. Bratzler RL. Membrane-based affinity purification at the process scale. Biotech '88 USA, November 14–16, 1988, San Francisco.
46. Petricciani JC. Changing attitudes and actions governing the use of continuous cell lines for the production biologicals. In: Spier RE, Griffiths, JB. eds. Animal cell biotechnology, Volume 3. New York: Academic Press, 1988; 13–25.
47. Zoon KC, Buckler CE, Bridgen PJ, Gurari-Rotman D. Production of human lymphoblastoid interferon by Namalwa cells. J Clin Microbiol 1978; 7:44–51.
48. Johnson IS. Human insulin from recombinant DNA technology. Science 1983; 219:632–637.
49. J. T. Baker Product Information Bulletin. Removal of nucleic acid and pyrogen from IgG with Bakerbond HI-Propyl. J. T. Baker Chemical Company, Phillipsburg, NJ, 1988.
50. Wolff SM. Biological effects of bacterial endotoxins in man. J Infect Dis 1973; 128:Suppl., S259–S264.
51. Pearson FC. Endotoxin: Its structure and function with respect to depyrogenation. In: Depyrogenation, Parenteral Drug Association Technical Report No. 7. Parenteral Drug Association, Philadelphia, 1985:1–14.
52. Wolber P, Dosmar M. Depyrogenation of pharmaceutical solutions by ultrafiltration: aspects of validation. Pharmaceut Technol September, 1987; 1(0):38–44.
53. Weary M, Pearson F III. A manufacturer's guide to depyrogenation. Biopharm 1988; 1(4):22–29.
54. Robinson JR, O'Dell MC, Takacs J, Barnes T, Genovesi C. Depyrogenation by microporous membrane filters. In: Depyrogenation, Parenteral Drug Association Technical Report No. 7. Parenteral Drug Association, Philadelphia, 1985:54–69.
55. Issekutz AC. Removal of gram-negative endotoxin from solutions by affinity chromatography. J Immunol Meth 1983; 61:275–281.
56. Duff GW, Waisman DM, Atkins E. Removal of endotoxin by a polymyxin B affinity column. Clin Res 1982; 30:565.

57. Karplus TE, Ulevitch RJ, Wilson CB. A new method for reduction of endotoxin contamination from protein solutions. J Immunol Met 1987; 105:211–220.
58. Rickles FR, Levin J, Atkins E, Quesenberry P. Endotoxin contamination of biological reagents: Detection and removal of endotoxin with Limulus amoebocyte lysate. In: Cohen E. ed. Biomedical applications of the horseshoe crab (Limulidae). New York: Alan R. Liss, 1979.
59. Guidelines on the production and quality control of monoclonal antibodies of murine origin intended for use in man. Commission of the European Communities notes to applicants for marketing authorizations. Tibtech 6:G5–G8.
60. Joner E, Christiansen GD. Hybridoma technology products: required virus testing. Biopharm 1988; 1(8):50–55.
61. Fenno J, Luczak J, Moore A, Poiley J, Raineri R, Whiteman M. The characterization of transformed cell substrates. Biopharm 1988; 1(8):36–42.
62. Doel TR. Inactivation of viruses produced in animal cell cultures. In: Spier RE, Griffiths JB. eds. Animal cell biotechnology, Volume 2. New York: Academic Press, 1985:129–149.
63. Costello MA, Reed E, Dias R, Krishnan R, Czuba B, Garg V. Purification of therapeutic proteins produced from mass culture of mammalian cells in gram quantities: Validation of protocol with respect to DNA, endotoxin, and extraneous protein removal. Miami Biotechnology Winter Symposium, February 8–12, 1988, Miami.
64. Kenny AC. Production and recovery of monoclonal antibodies produced in airlift reactors. Large scale production of monoclonal antibodies, Dec 1986, Society of Chemical Industry, London.
65. Lubiniecki AS. Safety of recombinant biologics: Issues and emerging answers. In: Spier RE, Griffiths JB. eds. Animal cell biotechnology, Volume 3. New York: Academic Press, 1988:3–12.

II

TECHNICAL ISSUES RELATED TO RECOVERY OF RECOMBINANT PROTEINS

3

Physical and Chemical Cell Disruption for the Recovery of Intracellular Proteins

T. R. Hopkins*

Phillips Petroleum Company
Bartlesville, Oklahoma

I. INTRODUCTION

This chapter surveys cell disruption devices and techniques for both the laboratory and the production plant. While host cells used for recombinant DNA protein production currently are either microorganisms or animal tissue culture cells, protein production in the future will most certainly include intact plants and animals. Therefore, the chapter covers disruption methods both for single-cell organisms and for animal and plant tissues. The subject has been reviewed previously (1–3).

Emphasis is on the practical aspects of the subject and generalizations are offered where possible. At the risk of dating the chapter, the names of commercial equipment, their manufacturers, and the approximate prices for the equipment are given. Many readers will find this information useful when considering the adoption of a cell disruption technique. The prices are for equipment purchased in the USA in 1990, and will undoubtedly require adjustment upward as time passes. It must be appreciated, of course, that the subject continues to evolve and that new applications, cell disruption methods, and equipment are likely to emerge in the future, especially in the "new" area of industrial biotechnology.

---

*Present affiliation: BioSpec Products, Inc., Bartlesville, Oklahoma.

II. METHODS OF DISRUPTION

A. Bead Mill Homogenizers

Bead mill homogenizing, a form of grinding, has been used for many years to disrupt microorganisms (4–8). The method works with tough-to-disrupt cells like yeast, spores, and microalgae where other techniques have failed and is considered by many to be the method of choice for the large-scale mechanical disruption of fungi. The method also has been used successfully for bacteria and, to a lesser extent, plant and animal tissue. In bead milling a large number of small glass beads (Ballotini) are rapidly agitated by shaking or stirring. Disruption is thought to occur by the shearing and crushing action of the minute glass beads as they collide with the cells.

There are several variables which must be studied in order to obtain efficient cell breakage. These are listed in Table 1. If these conditions are properly opitimized, cell disruption efficiencies of 90–95% can be obtained. The size of the glass bead is an important variable in cell disruption. Optimal bead size (mean diameter) is 0.1 mm for bacteria and spores; 0.5 mm for yeast, mycelia, microalgae, and unicellular animal cells such as leukocytes or trypsinized tissue culture cells; and 1.0–2.5 mm for tissues such as muscle, leaves, skin, etc. For small-volume bead mills, the speed of disruption is increased about 50% by using like-sized ceramic beads made of zirconium silicon oxide rather than glass (Hopkins, unpublished observations). Zirconium silicon oxide beads are more dense than glass beads and, therefore, collisional energy is increased. The loading of either glass or ceramic beads should be at least 50% of the total liquid-solid volume but can be up to 90% providing adequate agitation of the bead slurry is still possible. Generally, the higher the volume ratio of beads to cell suspension, the faster the rate of cell disruption. It is often best to completely fill the disruption vessel with beads and sample. While the efficiency of disruption is greater in only partially filled vessels, the presence of an air phase during disruption may denature sensitive proteins. After treatment, the beads settle by gravity in a few seconds and the cell homogenate is poured off. Alternatively, the homogenate can be separated from the beads by filtration or by suction through a tube fitted with fritted glass on

Table 1 Variables Influencing Efficiency of High-Speed Bead Mills

1. Composition and size of beads
2. Bead media load
3. Mix viscosity
4. Agitator speed and configuration
5. Disintegration time or throughput rate
6. Temperature

the end to exclude the beads. Beads washed in a good laboratory detergent and thoroughly rinsed can be reused many times.

1. Laboratory Bead Mills

In its simplest form, microorganisms can be disrupted by manually vortexing them in a test tube containing an excess of fine glass beads. A cell suspension the consistency of heavy cream is mixed with one to five volumes of grinding media in a test tube and the mixture is agitated at top speed on a vortex mixer for several minutes. The hand-held method is both slow and tedious. Therefore, mechanical bead mill agitators using a shaking action at 3000–6000 oscillations/min have been developed. Most shaking bead mills are restricted to sample sizes of 3.0 ml or less. Temperature control becomes a major concern in larger shaking bead mills and cooling is technically difficult to control. Bead-milling samples of 250 ml without external cooling, for example, increases sample temperature 10°C per minute of operation. The German-made Braun MSK (B. Braun Biotech, Bethlehem, PA) disrupts sample sizes up to 40 ml but must rely on somewhat undependable cooling with liquid CO_2 to keep the sample at an acceptable temperature. Smaller shaking bead mills have small breakage chambers with surface area-to-volume ratios high enough to permit adequate heat dissipation at ambient temperatures. Three commercial devices are the Mickle homogenizer (H. Mickle, Middlesex, England), the Micro-Dismembrator II (B. Braun Biotech, Bethlehem, PA), or the Mini-Bead Beater (BioSpec Products, Bartlesville, OK). Disruption of microorganisms takes about 1–5 min. There are two small-volume, shaking bead mill disrupters capable of handling multiple samples. One, called Shake It, Baby! (BioSpec Products, Bartlesville, OK), processes twelve 0.1- to 1.0-ml samples of biomass suspension in 2-ml microcentrifuge vials. Equipment to process up to 96 samples at a time using microtiter plates is under development. The other multiple-sample bead mill, the Mix-Tower Model A-14 (Taiyo Scientific Industrial Co., Tokyo, Japan), processes fourteen 1-ml samples in 18 × 110 mm test tubes. Cell disruption times are 10–25 min with the multiple-sample bead mills and are therefore considerably slower than the single-sample units. The price of shaking bead mills ranges from $400 to $5000.

Some laboratory bead mill cell disrupters agitate the beads with a rotor rather than by shaking. Equipped with cooling jackets, larger sample volumes can be processed without overheating. The most widely known is the Bead Beater (BioSpec Products, Bartlesville, OK). This unit will disrupt about 250 ml or, with small chamber attachments, 50-or 15-ml batches of cell suspension in 3–5 min. Cell concentrations as high as 50% (wet wt) can be used without appreciable degradation of the disruption efficiency. It is one of the least expensive powered cell disrupters available ($300). VirTis Company (Gardiner, NY) offers attachments for its line of high-speed rotary homogenizers which efficiently agitate glass beads in a test tube or fluted flask. The complete homogenizer unit costs about $700. Innomed (Stockholm, Sweden) makes a variable-speed, overhead-driven device called the SCP-100-MRE. It holds 15- to 80-ml samples and, because the rotor is isolated from the drive motor through a sealed magnetic coupling, it is said to be especially safe for use with biohazardous material. The device costs about $3000.

While the above small-volume cell disrupters are used primarily for the disruption of bacteria and yeast, they can also be used to homogenize plant and animal tissue. This newer application is suitable for both soft tissue and tough or fibrous samples such as skin, tendon, or leaves. Extraction yields of biomolecules such as nucleic acids, viruses, and receptor complexes are often superior to that of other methods. It is possible to selectively disrupt only the outer layers of whole leaves by using small charges of beads and short disruption times.

2. Pilot Plant and Large-Scale Bead Mills

There are several manufacturers of bead mill cell disrupters capable of operating in the continuous mode. The best known are Willy A. Bachofen Machinenfabrik (Basel, Switzerland), which makes the DYNO mill line, and Netzsch Incorporated (Exton, PA). The mills from both of these manufacturers are horizontal mills originally designed to wet-grind pigments for the paint industry. The machines have double-walled grinding chambers with a sieve plate or rotating gap at one end to retain the beads. The agitator consists of a shaft on which several slotted disks are mounted. Both commercial machines are well constructed and capable of continuous operation 24 hours a day, if desired. The smallest machine in the DYNO mill line is the model KDL. It is a pilot plant machine with 150-, 300-, and 600-ml glass or stainless steel chambers and can be operated in either a continuous or a batch mode.

Other lesser known manufacturers of continuous bead mills which have been demonstrated to disrupt microbial cells are BioSpec Products (Bartlesville, OK), Innomed (Stockholm, Sweden), Fryma (Edison, NJ), and Primier Mills (New York, NY). The Innomed Model SCP-011-IMC continuous disintegrator was expressly designed for production of bioactive products having pharmaceutical applications. The chamber of the SCP-011 is totally sealed. The internal horizontal rotor is connected to the drive motor by a magnetic coupling. Thus, the entire chamber assembly can be removed and sterilized without the operator making contact with parts exposed to the production microorganism or its product. The Fryma machines differ from the others because much smaller charges of beads are used for a given throughput capacity. The above pilot plant bead mill cell disrupters are priced in the $15,000–$25,000 range. An exception is the BioSpec Products machine which sells for approximately $5000. Being more of a laboratory machine, it is not designed for sustained usage in a production environment.

The KDL and the Innomed SCP-011-IMC are the only continuous bead mills capable of using 0.1-mm bead media. The agitator speed of the pilot plant bead mills is adjustable and good disruption is achieved in the 1500–5000 rpm range. Generally, higher agitator speeds give higher rates of cell disruption, but at the expense of increased heat generation. At flow rates of 20 liters/hr about 85% of microbial cells can be disrupted in a single pass through a pilot plant bead mill. Since feed concentrations within the range of 4–20% (dry wt) have little effect on the degree of cell disruption, these relatively small continuous mills have capacities of about 4 kg (dry wt) per hour.

Since heat generation during the operation of bead mills is considerable, all continuous bead mills used to disrupt microorganisms are equipped with cooling jackets and require aggressive external cooling. Even with well-chilled sample entering the bead mill and with cold tap water running through the water jacket, the sample can emerge from the machine at 40°C. Therefore, a refrigeration unit and pump to circulate coolant mixtures are essential accessories for the pilot plant and production plant bead mills. Some manufacturers of bead mills have addressed the cooling problem better than others. For example, the Netzsch bead mills not only have a jacketed chamber but also provide a cooling system in the agitator shafts and blades.

Exit temperatures of 40–50°C need not always be unacceptable because heat denaturation of proteins is dependent on time as well as temperature. Therefore, prompt cooling of the homogenate after passing through the bead mill may suffice. Heating also can be minimized by increasing the product flow rate through the bead mill, thus shortening the residence time. Fortunately, the percentage of cells disrupted in a pass through the mill is not very dependent on flow rate, providing that the bead size is properly matched to the cell type. For example, there is only a modest decrease in the efficiency of cell disruption from 90 to 70% upon increasing the flow rate four- to five fold. As an added bonus, the energy requirement per kilogram yeast disrupted decreases with increased flow rate. Thus, for both economic and technical reasons, high flow rates are attractive if the cellular raw material is plentiful.

Large production models of DYNO mill and Netzsch machines have been used to process microbial products at flow rates of 100–1200 liters/hr and cell concentrations up to 40% packed wet cell volume. This corresponds to a capacity of about 10–100 kg (dry wt) per hour. Agitator rotational speed is about 1000 rpm in these bigger machines, with higher speeds generating more heat but only marginal increases in cell disruption. Because these machines are unable to use 0.1-mm beads they are not as efficient for the disruption of bacteria as the laboratory and some of the pilot plant models.

Cell disruption with bead mills has been found to be equal and sometimes superior in energy efficiency to that of other large-scale mechanical breakage methods such as the Manton–Gaulin high-pressure homogenizer (see below).

With disruption efficiencies of 80–95% yielding 0.06–0.1 kg soluble protein/ kg (wet wt) yeast cells, the calculated energy requirements vary from 0.2 to 0.8 kWhr/kg (dry wt) disintegrated yeast (9,10). Capital investment in these industrial machines, including the essential cooling unit, is around $40,000–$100,000. An engineering study of the efficiency parameters given in Table 1 using a 20-liter industrial bead mill has been published (11), as have similar studies with 5-liter machines (9,12).

B. Rotor-Stator Homogenizers

Rotor-stator homogenizers (also called colloid mills or Willems homogenizers) generally outperform cutting blade-type blenders and are well suited for plant and animal tissue. Combined with glass beads, the rotor-stator homogenizer has been successfully used to disrupt microorganisms. However, the homogenized sample is contaminated with minute glass and stainless steel particles and the abrasive wear to the rotor-stator homogenizer is unacceptably high. Cell disruption with the rotor-stator homogenizer involves hydraulic and mechanical shear and to a lesser extent high-energy sonic and ultrasonic pressure gradients. First developed to make dispersions and emulsions, most biological tissues are quickly and thoroughly homogenized with the apparatus. Appropriately sized cellular material is drawn up into the bottom of the apparatus by a rapidly rotating rotor sited within a static head or tube containing slots or holes. There the material is centrifugally thrown outward in a pumplike fashion to exit through the slots or holes. Because the rotor turns at very high rpm, the tissue is rapidly reduced in size by a combination of extreme turbulence and scissorlike mechanical shearing occurring within the narrow gap between rotor and stator. Since most rotor-stator homogenizers have an open configuration, the product is repeatedly recirculated. The process is fast and, depending on the toughness of the tissue sample, desired results will usually be obtained in 15–120 sec. For the recovery of intracellular organelles or receptor site complexes, shorter times are used and the rotor speed is reduced.

The variables to be optimized for maximum efficiency are listed in Table 2. The size of the rotor-stator probe (also called the generator) can vary from the diameter of a pencil for 0.3–10-ml sample volumes to much larger units having batch capacities up to 19,000 liters or, for on-line units, capabilities of 68,000 liters/hr. Rotor speeds vary from 3000 rpm for large units to 8000–60,000 rpm for the smaller units. In principle, the rotor speed of the homogenizer should be doubled for each halving of the rotor diameter. It is not rpm per se but the tip velocity of the rotor that is the important operating parameter. Ten to twenty meters per second (2000–4000 fpm) is acceptable tip speed. Unfortunately, many of the smallest commercial laboratory-sized rotor-stator homogenizers are unable to meet this criterion. Other factors

CELL DISRUPTION AND PROTEIN RECOVERY

such as rotor-stator design (there are many), materials used in construction, and ease of cleaning are also important factors to consider in selecting a rotor-stator homogenizer.

1. Laboratory Rotor-Stator Homogenizers

Laboratory-sized rotor-stator homogenizers process liquid samples in the 0.3-ml to 20-liter range. Some manufactures are BioSpec Products (Bartlesville, OK), Brinkmann Instruments, (Westbury, NY), Charles Ross and Son (Hauppauge, NY), Craven Laboratories (Austin, TX), IKA Works (Cincinnati, OH), Omni International (Waterbury, CT), Silverson Machines (Bay Village, OH), and VirTis Company (Gardiner, NY). The cost of complete units (motor and rotor-stator head or generator) range from $500 to $5000. Some of these manufacturers offer several models operating at different speeds and having a variety of rotor-stator designs. Thus, the capacity of the rotor-stator can be matched to the viscosity and volume of the medium and with the type and amount of plant and animal tissue to be processed. The speed and efficiency of homogenization is greatly degraded by using too small a homogenizer, and the volume range over which a given homogenizer rotor-stator size will function efficiently is only about 10-fold. Also, most of the laboratory-sized homogenizers function properly only with liquid samples in the low- to medium-viscosity range (<10,000 cps). This must be balanced against the practical observation that concentrated samples, by colliding more frequently, are broken up more rapidly. Higher viscosity samples can be processed but require specially shaped homogenization vessels or unique rotor-stator configurations. The size of the sample prior to processing with the homogenizer must be small enough to be drawn inside the stator. Therefore, samples often must be pre-chopped, -cut, or -fragmented. Brinkmann and IKA manufacture special rotor-stator heads having exposed cutting knives projecting from the rotors or sawtooth edges on the stators which nibble away at larger samples.

Foaming and aerosols can be a problem with rotor-stator homogenizers. Keeping the tip of the homogenizer well submerged in the media and the use of properly sized vessels helps with the first problem. Square-shaped

Table 2 Variables Influencing the Efficiency of Rotor-Stator Homogenizers

1. Design and size of rotor-stator (generator)
2. Rotor tip speed
3. Initial size of sample
4. Viscosity of medium
5. Time of processing or flow rate
6. Volume of medium and concentration of sample
7. Shape of vessel and positioning of rotor-stator

homogenization vessels give better results than round vessels and it is also beneficial to hold the immersed tip off-center. Aerosols can be minimized by using covered vessels (VirTis, Brinkmann, and Omni). There are no aerosols with in-line homogenizers (Brinkmann and IKA). Even though a number of the laboratory rotor-stator homogenizers use sealed motors, none of them are truely explosion-proof. Due caution should be followed when using flammable organic solvents by conducting the homogenization in a well-ventilated hood. On the positive side, rotor-stator homogenizers generate minimal heat during operation and this can be easily dissipated by cooling the homogenization vessel in ice water during processing.

2. Pilot Plant and Large-Scale Rotor-Stator Homogenizers

The larger rotor-stator homogenizers are either scaled up versions of the laboratory models or in-line homogenizers. The latter contain teeth on the edge of a horizontally oriented, multibladed, high-speed impeller aligned in close tolerance to matching teeth in a static liner. Some manufacturers of pilot plant and production scale machines are Arde Barinco (Mahwah, NJ), BioSpec Products (Bartlesville, OK), Brinkmann Instruments (Westbury, NY), Charles Ross and Son (Hauppauge, NY), Eastern Rotostat (Clinton, CT), IKA Works (Cincinnati, OH), Kinematica GmbH (Lucerne, Switzerland), and Silverson Machines (Bay Village, OH). With the exception of one $400 pilot plant unit designed to make course homogenates (BioSpec Products), these machines range from $6000 to over $80,000. At least two manufacturers offer homogenizers capable of handling solution viscosities as high as 500,000 cps (Eastern Rotostat and Charles Ross).

3. Blade Homogenizers

Although less efficient than rotor-stator homogenizers, blade homogenizers (also called blenders) have been used for many years to produce fine brie and extracts from plant and animal tissue. The cutting blades on this class of homogenizer are either bottom- or top-driven and rotate at speeds of 6000–50,000 rpm. Blenders are not suitable for disruption of microorganisms unless glass beads or other abrasives are added to the media and then one encounters the same problems as were mentioned above for rotor-stator homogenizers. Many plant tissue homogenates undergo enzymatic browning–a biochemical oxidation process which can complicate subsequent separation procedures. Enzymatic browning is minimized by carrying out the extraction in the absence of oxygen or in the presence of thiol compounds such as mercaptoethanol. Sometimes addition of polyethylene imine, metal chelators, or detergents such as Triton X-100 or Tween-80 also helps.

Blade homogenizers are available for a range of liquid sample sizes from 0.2 ml to 1 gal. Some of the higher rpm homogenizers can reduce tissue samples to a consistent particulate size with distributions as small as 4 μm as determined by flow cytometric analysis. Accessories for some blenders include cooling jackets to control temperature and closed containers to minimize aerosol formation and entrainment of air. Manufactures of a scientific line of blenders include BioSpec Products (Bartlesville, OK), British Medical Enterprises (London, England), Omni International (Waterbury, CT), VirTis Company (Gardiner, NY), and Waring Products Division (New Hartford, CT). A variety of accessories for the Waring brand blenders are manufactured by Eberbach Corporation (Ann Arbor, MI). Prices for blade homogenizers range from about $100 to $2000.

C. High-Pressure Homogenizers

High-pressure homogenizers have been used to disrupt microbial cells for many years (13–16). With the exception of highly filamentous microorganisms, the method has been found to be generally suitable for a variety of bacteria, yeast, and mycelia.

This type of homogenizer works by forcing cell suspensions through a very narrow channel or orifice under high pressure. Subsequently, and depending on the type of high-pressure homogenizer, they may or may not impinge at high velocity on a hard-impact ring or against another high-velocity stream of cells coming from the opposite direction. Machines which include the impingement design are more effective than those which do not. Disruption of the cell wall occurs by a combination of the large pressure drop, highly focused turbulent eddies, and strong shearing forces. The rate of cell disruption is proportional to approximately the third power of the turbulent velocity of the product flowing through the homogenizer channel, which in turn is directly proportional to the applied pressure. Thus, the higher the pressure, the higher the efficiency of disruption per pass through the machine. The operating parameters which effect the efficiency of high-pressure homogenizers are listed in Table 3. There are fewer parameters to consider than for the bead mill. Like the bead mill, disruption is almost independent of cell concentration in the feed.

High-pressure homogenizers have long been the best available means to mechanically disrupt nonfilamentous microorganisms on a large scale. Animal tissue also can be processed but the tissue must be pretreated with

Table 3 Variables Influencing the Efficiency of High-Pressure Homogenizers

1. Pressure
2. Temperature
3. Number of passes
4. Valve and impingement design
5. Flow rate

a blade blender, rotor-stator homogenizer, or paddle blender (see below). The supremacy of high-pressure homogenizers for disruption of microorganisms is now being challenged by bead mill homogenizers. Still, in terms of throughput, the largest industrial models of high-pressure homogenizer outperform bead mills. The maximum volume of microbial suspension per hour that can be treated by the larger commercial machines is 4500 liters for high-pressure homogenizers versus about 1200 liters for bead mills. Even larger capacity high-pressure homogenizers are available but their efficiency in disrupting microbial cells has not been documented. This throughput advantage is diminished somewhat by the fact that most high-pressure homogenizers require several passes of the cell suspension to achieve high levels of cell disruption whereas bead mills frequently need only one.

1. Laboratory High-Pressure Homogenizers

A familiar commercial high-pressure homogenizer for the laboratory is the French press (SLM Instruments, Urbana, IL) which uses a motor-driven piston inside a steel cylinder to develop pressures up to 40,000 psi. Pressurized sample suspensions up to 35 ml are bled through a needle valve at a rate of about 1 ml/min. Because the process generates heat, the sample, piston, and cylinder are usually precooled. Typical pressures used to disrupt yeast are 8000–10,000 psi and several passes through the press may be required for high efficiency of disruption. Generally, the higher the pressure, the fewer the passes. Pressure cells rated at 20,000 psi maximum come in capacities of 3.7 and 35 ml and there is also a 35-ml capacity cell rated at 40,000 psi. The cost for a motor-driven lab press and a piston-cylinder pair is about $8000.

APV Gaulin (Everett, MA) recently introduced a high-pressure homogenizer which they claim can operate at pressures up to 23,000 psi. Called the Micron Lab 40, batch samples of 20–40 ml can be processed. The sample cylinder is filled with product, a homogenizing valve is stacked on top, and the system is hydraulically clamped closed. The homogenization pressure is selected and a piston drives the product upward through the valve and into a collection chamber. The homogenization process is said to be performed in under 1 min. No tools are required to operate or clean the Micron Lab. It is expected that at under these very high pressures little if any product recycling will be required. The preliminary literature of the manufacturer did not address how they deal with a probable temperature increase of 30–40°C during such a high-pressure pass. The unit is tentatively priced at $25,000.

A rather unique high-pressure homogenizer originally developed in England is the Microfluidizer (Microfluidics Corp., Newton, MA). The sample is pressurized with a pneumatically powered pump to about 20,000 psi, split into two streams, and the streams are fired at each other inside an impingement chamber at velocities up to 450 m/sec. Batch mode samples as small as 20 ml can be homogenized and in the continuous mode the laboratory model has flow rates of 50–800 ml/min. Because this machine operates in the 20,000 psi range, high cell breakage is possible in only one or two passes. As with other high-pressure cell disrupters, considerable heat is developed in the process (about a 2°C increase in the exiting product per 1000 psi of pressure drop) but can be removed with a properly designed heat exchanger. The price of the laboratory unit is about $18,000.

2. Pilot Plant and Large-Scale High-Pressure Homogenizers

Most high-pressure homogenizers used for homogenization were adapted from commercial equipment designed to produce emulsions and homogenates in the food and pharmaceutical industries. They combine high pressure with an impingement valve. The most documented large-scale homogenizer used for cell disruption is the Manton Gaulin-APV homogenizer (APV Gaulin, Everett, MA). Several models and valve designs are available. Those with a maximum pressure rating of 10,000 psi rupture about 40% of the cells on a single pass, 60% on the second pass, and 85% after four passes. At 110 liters/hr capacity and 15,000 psi the model 30CD pilot plant machine (model 30CD) is reported to rupture 85% on a single pass. Capacities of continuous homogenizers vary from 55 to 4500 liters/hr at 10–17% w/v cell concentrations. With the larger capacity machines several passes are needed to achieve high yields of disruption. Considerable heat can be generated during operation of these homogenizers and therefore a heat exchanger attached to the outlet port is essential. Another manufacturer of homogenizers with specifications similar to that of APV-Gaulin is APV Rannie AS (Albertslund, Denmark). The cost of a pilot plant machine is about $7500 and production machines are $20,000–$80,000.

As discussed earlier, the operation of the Microfluidizer differs from other high-pressure homogenizers by impinging two high-velocity streams of cells against each other. Although there is little literature on the use of this type of device for large-scale microbial cell disruption, pilot plant and production equipment are available. Power consumption for these machines is approximately 11 hp/liter/hr. The price of the pilot plant machine is in the $50,000 range and a production machine would be 2–20 times that amount, depending on the performance requirements.

3. Freeze-pressing

The Hughes press is a variation of high-pressure homogenizers. Here material is pressed through an narrow orifice in the "solid" frozen state at -25 to -30°C(2). A pressure-induced transition in the crystal structure of water from ice I to liquid or ice III starts the material flowing. Shear stress, eddying, and explosive decompression during flow disrupt the cells. In addition, the abrasive action of ice crystals may aid in disrupting the cells as they pass through the orifice. More popular in Europe than in the USA, this method is effective for the disruption of a wide variety of cell types. It is especially suitable where heat sensitivity could be a problem but, of course, it cannot be used for materials that are sensitive to freezing and thawing. Commercial equipment called the X-Press is available with sample capacities of 5 or 25 ml (AB Biox, Jarfalla, Sweden). Unfortunately, commercial large-scale equipment is not available. Using pressures of 30,000–50,000 psi, most microorganisms are 90% disrupted in three to seven passes through the orifice of the X-Press and animal tissue in one or two passes. The price of the X-Press disintegrator, cooling bath, and 25-ton hydraulic press is about $4000.

4. Nitrogen Decompression

High-pressure vessels which disrupt cells and tissues by rapid decompression are called disruption bombs. The sample is placed in a stainless steel vessel, sealed, and pressurized to around 2000 psi with a tank of compressed nitrogen gas. The gas is allowed to equilibrate 5–30 min, thus dissolving into the aqueous media and the intracellular volume of the cells. Then a valve connected to the cell suspension is opened and the sample suspension allowed to flow out of the pressure vessel. As the high-pressure sample passes through the valve to atmospheric pressure, the dissolved gas in the sample turns to gas and the individual cells explosively decompress. The cells are puffed and ruptured like so many heated kernels of popping corn. Disruption also occurs by shear as the sample passes through the outlet valve. Best suited for treating mammalian and some plant cells, this method gives variable results with bacteria and does not work with yeast, mycelia, and spores. Animal and plant tissue samples require pretreatment by some mechanical chopping or dispersing method to reduce them to a brie capable of passing through the discharge valve of the disruption bomb. Because of adiabatic expansion of the gas, this is one of the few cell disruption techniques which actually cools a sample during operation. Parr Instruments (Moline, IL) manufactures a series of disruption bombs sized from 45 ml to 2 gal (7.6 liters) costing from $800 to $7000. Cell suspension can occupy up to two-thirds of the total vessel capacity. A similar cell disruption bomb is manufactured by Kontes (Vineland, NJ). In this case, sample sizes from 1.0 to 15.0 ml can be processed. Selling for about $1000, the small size permits easy temperature control by immersion in a constant-temperature or ice bath.

5. Miscellaneous Pressure Homogenizers

Two devices which are said to be useful for the breakup of cellular materials are the Kirkland emulsifier available from Brinkmann Instruments (Westbury, NY) and the Chase-Logeman hand homogenizer (Hicksville, NY). Working with small volumes of sample (2–1000 ml), a hand-operated plunger forces the sample under hydraulic pressure of up to 1000 psi past a valve where instantaneous pressure drop and high-shear forces occur. These units, which cost about $700 and $200, respectively, are not in the same performance league as the high-pressure homogenizers. Nevertheless, they may prove adequate for some easy-to-break cellular materials.

D. Ultrasonic Disintegrators

One widely used method to disrupt cells is ultrasonic disruption. These devices work by generating intense sonic pressure waves in a liquid media. The pressure waves cause streaming in the liquid and, under the right conditions, rapid formation of microbubbles which grow and coalesce until they reach their resonant size, vibrate violently, and eventually collapse. This phenomenon is called cavitation. The implosion of the vapor phase bubbles generates a shock wave with sufficient energy to break covalent bonds. Shear from the imploding cavitation bubbles as well as from eddying induced by the vibrating sonic transducer disrupt the cells.

There are several external variables which must be optimized to achieve efficient cell disruption. These are listed in Table 4. Modern ultrasonic processors use piezoelectric generators made of lead zirconate titanate crystals. The vibrations are transmitted down a titanium metal horn or probe tuned to make the processor unit resonate at 15–25 kHz. The rated power output of ultrasonic processors vary from 10 to 375 W. Low power output does not necessarily mean that the cell disintegrator is less powerful because lower power transducers are generally matched to probes having smaller tips. It is the power density at the tip that counts. Higher output power is required to maintain the desired amplitude and intensity under conditions of increased load such as high viscosity or pressure. The larger the horn, the more power is required to drive it and the larger the volume of sample that can be processed. On the other hand, larger ultrasonic disintegrators generate considerable heat during operation and will necessitate aggressive external cooling of the sample. Typical maximum tip amplitudes are 30–250 μm and resultant output intensities are in the range of 200–2000 W/cm^2. Some manufacturers of ultrasonic disintegrators are Artek Systems (Farmingdale, NY), Branson Sonic Power Company (Danbury, CT), B. Braun Biotech (Bethlehem, PA), RIA

Table 4 Variables Influencing the Efficiency of Ultrasonic Disintegrators

1. Tip amplitude and intensity
2. Temperature
3. Cell concentration
4. Pressure
5. Vessel capacity and shape

Research Corp. (Hauppauge, NY), Sonic Systems (Newton, PA), and VirTis Company (Gardiner, NY). Laboratory-sized ultrasonic processors for cell disruption are priced from $1000 to $4,000.

The temperature of the sample suspension should be as low as possible. In addition to addressing the usual concerns about temperature lability of proteins, low media temperatures promote high-intensity shock front propagation. So ideally, the temperature of the ultrasonicated fluid should be kept just above its freezing point. The ultrasonic disintegrator generates considerable heat during precessing and this complicates matters. Disruption can also be enhanced by increased hydrostatic pressure (typically 15–60 psi) and increased viscosity, providing the ultrasonic processor has sufficient power to overcome the increased load demand and the associated sample heating problems can be solved. For microorganisms the addition of glass beads in the 0.05- to 0.5-mm size range enhances cell disruption by focusing energy released by the bubble implosions and by physical crushing. Beads are almost essential for disruption of spores and yeast. A good ratio is one volume beads to two volumes liquid. Tough tissues such as skin and muscle should be macerated first in a blender or the like and confined to a small vessel during ultrasonic treatment. The tip should not be placed so shallow in the vessel as to allow foaming. Antifoam agents or other materials which lower surface tension should be avoided. Finally, one must keep in mind that free radicals are formed in ultrasonic processes and that they are capable of reacting with biological material such as proteins, polysaccharides, or nucleic acids. Damage by oxidatire free radicals can be minimized by including scavengers like cysteine, dithiothreitol, or other SH compounds in the media or by saturating the sample with a protective atmosphere of helium or hydrogen gas.

For practical reasons, the tip diameter of ultrasonic horns cannot exceed about 3 in. This sets a limit on the scale-up of these devices. While standard-sized ultrasonic disrupters have been adapted to continuous operation by placing the probe tip in a chamber through which a stream of cells flow, cooling and free-radical release present problems. In one study (17), protein released from yeast being disrupted in a continuous ultrasonic process was independent of cell concentration up to 60 g wet w/100 ml. Maximum throughput at 60% cell disruption was 1.4 kg (wet wt) of yeast per hour. Continuous flow accessories can be obtained from several manufacturers of ultrasonic disrupters (Artek, Braun, Branson, and Sonic Systems). Chamber designs vary somewhat as do efficiencies of cell disruption and sample cooling.

E. Other Physical Methods

1. Grinding

Grinding biological material in a mortar or tube with fine sand, alumina, or glass powder is roughly the equivalent of bead milling. The method works reasonably well with all types of biomass but is strictly small scale and is quite labor-intensive. Cell pastes or solid mass with a minimum volume of buffer are mixed with 0.5–1 volume of grinding media and ground with a mortar and pestle. Disruption efficiency is poor if lower cell densities or charges of grinding media are used.

2. Pestle and Tube Homogenizers

Laboratory pestle and tube homogenizers (also called tissue grinders) are used to disrupt animal tissue. While modifications of the pestle and tube homogenizer have names like Potter, Potter–Elvehjem, Dounce, and Ten Broeck, as a group they consist of test tubes made of glass, inert plastic, or stainless steel into which is inserted a tight-fitting pestle (clearance about 0.1–0.2 mm) made of a like material. The walls of the test tube and pestle can be smooth or have a ground finish. Most tissues must be cut or chopped into small pieces (1–5 mm) with a pair of scissors or a blade before being suspended in a 4- to 10-volume excess of medium in the test tube. The pestle is manually worked to the bottom of the tube, thus tearing and fragmenting the tissue as it is forced to pass between the sides of the pestle and the wall of the tube. The grinding action occurs again as the pestle is withdrawn. Five to thirty repetitions of this low-shear method homogenizes the tissue. Rotation of the pestle at about 500–1000 rpm with an electric motor while the test tube is manually raised and lowered speeds up the process.

While pestle and tube homogenization is simple and the equipment is usually inexpensive, it is both labor-intensive and, in the case of fragile glass homogenizers, potentially dangerous. Even so, this homogenizer will continue to be popular because of its extremely gentle action. It often is the method of choice for the preparation of small quantities of subcellular organelles from soft animal tissues such as brain or liver. Microorganisms cannot be disrupted with pestle homogenizers.

Commercially available pestle homogenizers with batch capacities of 0.1–50 ml generally cost $15–$100 and are available from many manufacturers including Ace Glass (Vineland, NJ), Bellco Glass (Vineland, NJ), Biomedix (Pinner, England), Kontes (Vineland, NJ), Thomas Scientific (Swedesboro, NJ), Tri-R Instruments (Rockville Centre, NY), and Wheaton Industries (Milville, NJ). A Rolls Royce version costing about $3000 has a variable-speed motor, cooling jacket, and hand-operated lever to rise and lower the pestle (B. Braun Biotech, Bethlehem, PA). A continuous pestle homogenizer is available from Yamato USA (Northbrook, IL). Grooves machined on the upper one-third of the pestle catch and guide tissue through the close tolerance region of the lower two-thirds of the cylinder pestle. The resultant homogenate exits from the bottom of the cylinder. Flow rates are from 1 to 10 liters/hr, depending on the tissue type and premincing treatment. Recycling is usually necessary. The machine comes in two sizes and costs $2000–$3000.

3. Paddle Blender

This blender uses two reciprocating paddles to repeatedly pound a tightly sealed plastic bag containing soft animal tissue and media. The current use of the paddle blender is for the release of microorganisms from tissue and food samples for subsequent microbiological analysis. The tissue is effectively dispersed in the media in about 1 min but usually the cells themselves are not disrupted. It is mentioned here because of its potential to be combined with other physical or chemical disruption methods which have been shown to work well on cellular suspensions. The paddle blender manufactured by Seward Laboratory (London, England) is called the Stomacher. It is manufactured in three sizes having batch capacities from as small as 5 ml up to 3500 ml and costs $2000–$4000.

4. Freeze-fracturing

Although impractical for large samples, both microbial pastes and plant and animal tissue can be frozen in liquid nitrogen and then ground with a mortar and pestle at the same low temperature. Presumably the hard frozen cells are fractured under the mortar because of their brittle nature. Also, ice crystals at these low temperatures may act as an abrasive.

A little known but promising freeze-fracture cell disruption method is rasping. Bleeg and Christensen (18) obtained a 65–70% yield of disrupted yeast by feeding frozen 50 g blocks of yeast cells into a rapidly rotating grating disk of a Braun Multipress MP 50. The overall feed rate was about 2 kg yeast/hr. Using a 50% suspension of baker's yeast as starting material, the homogenate became sufficiently fluid during cell disruption to drain continuously. The enzyme activity of galactono-γ-lactone oxidase in the homogenate was comparable to that obtained by other yeast cell disruption methods.

A freeze-fracturing device called the Bessman or percussion tissue pulverizer is useful for preparing small quantities of tissue such as skin or cartilage for subsequent homogenization or extraction by other methods. Looking somewhat like a tablet press, the pulverizer consists of a hole machined into a stainless steel base into which fits a pestle. Depending on the mortar size, 10 mg to 1 g of animal or plant tissue hard-frozen in liquid nitrogen is placed in the prechilled mortar, the piston is placed in the hole, and the brittle material is given a sharp blow with a hammer. The still frozen, powdered material is poured into a beaker for further processing. The pulverizer costs about $250 and is manufactured by Biomedix (Pinner, England), BioSpec Products (Bartlesville, OK), and Spectrum Medical Industries (Carson, CA).

5. Freezing and Thawing

Cells having rigid walls such as plant material can sometimes be ruptured by several cycles of freezing to -20 to -30°C and thawing. It is listed as a physical method of disruption because it is believed that the formation of large ice crystals ruptures intracellular membrane structure. Freezing must be done slowly to get large ice crystals. While the method can be scaled up and is inexpensive, it is a relatively slow process and can also initiate intracellular metabolic degradation (see below). For this reason the disruption method may not be suitable when the recovery molecule is susceptible to enzymatic degradation. Plant material usually requires a large number of freeze-thaw cycles and the yields of soluble protein are not very high. Animal and yeast cells respond poorly to freezing and thawing, if at all, and bacteria fall somewhere in between the extremes.

6. Meat Mincer

The section on physical cell disruption is concluded by recalling that the household meat grinder or mincer has been used for many years for the preparation of animal tissue extracts. Tissue is mechanically pressed through holes in a metal sieve plate while rotating blades slowly sweep across the face of the plate cutting the meat in 0.3 to 0.5-mm fragments. While it is not an effective way to disrupt cells per se, it is useful as a preliminary step for complete homogenization using other physical or chemical methods. Meat grinders cut flexible tissue like muscle better if the tissue is processed slightly frozen. For small tissue samples, Biomedix (Pinner, England) and EDCO Scientific (Chapel Hill, NC) manufacture a stainless steel, hand-operated screw press for the preparation of tissue extracts. Capable of considerable force, sample sizes up to 30 ml are pushed through a sieve plate having 0.25 to 1.5-mm holes, much like the action of a kitchen garlic press. The unit costs $400. A simple cup sieve having a choice of fine-mesh stainless steel screens is sold by Sigma Chemical (St. Louis, MO) for $70. Up to 50 cm^3 of tissue culture cells or other soft animal tissue like brain or spleen is manually worked through the screen with a glass pestle.

F. Autolysis

In industry, autolysis is the primary method for disrupting yeast on a very large scale. For the biochemical engineer who is concerned about capital and production costs and wants a cell disruption process which can easily be scaled up, autolysis should be considered. Unfortunately, the details of autolytic processes are not well understood. Specific methods of cell autolysis will need to be developed empirically for each application.

Autolysis involves shock, i.e., a chemical or physical treatment which assaults the cell to such an extent that irreversible intracellular catabolic events are induced. Key to the method, the treatment conditions are mild enough to permit many endogenous enzymes to function for extended periods of time. Shock methods involve exposure to a nonpolar solvent (solvent shock), a rapid change in salt or sugar concentration (osmotic shock), acidification to lower pH or addition of alkali to increase pH (pH shock), or elevation of the temperature (heat shock). In some bacterial systems, sudden substrate or oxygen depletion will also induce autolytic events. The end result is the same: the integrity of the cell membranes is compromised, the porosity of the cell increases, and lysis occurs, releasing soluble intracellular constituents into the media.

Chemical and physical shock methods have two clear advantages. They are easily scaled up and use approximately 1% of the power required for an equivalent amount of disruption using mechanical methods. Furthermore, because much of the cell wall stays intact during autolysis, the remaining solids are easily removed by centrifugation or filtration to form clear extracts. In contrast, mechanical methods often fragment cell walls severely and yield opaque supernatants which can complicate the downstream recovery process. The disadvantages of autolysis methods are (a) much longer times for disruption than mechanical methods, (b) considerable holding tank space is needed, and (c) there is the possibility that the product will be modified or destroyed during the autolytic process—especially if the product is a peptide or resides within the cell as a nonnative inclusion. With respect to the later problem, it is known that individual cellular proteins are degraded at vastly different rates. As we gain more information on the correlation of the structure of proteins with their differential stabilities within the cell (19), it may be possible to design proteins to resist degradation.

1. Solvent-Induced Autolysis

Solvent shock is initiated by adding low concentrations of an appropriate organic solvent to concentrated cell suspensions (50–90% packed weight per volume) and incubating at 20–40°C for periods of several hours up to 3 days. The cells lose their integrity and the contents leak into the media. Typical solvents which have been used to induce cell autolysis are diethyl ether, toluene, ethyl acetate, chloroform, and methylene dichloride. Successful solvents have three common properties. They are nonpolar, lipophilic, and at least partially soluble in water. Polar solvents such as acetone or ethanol used alone at low concentrations do not work. They are, however, sometimes used as cosolvents to increase the solubility of the autolytic solvent. Solvents with extremely low water solubilities such as silicone or fluorocarbon solvents do not work. Supercritical solvents have etherlike properties and should induce autolysis of yeast. However, in the case of supercritical CO_2, the successfully lysed cell suspension was unacceptably acidified (Hopkins, unpublished observations). Other supercritical solvents should be investigated. There is a major problem with many of the common autolytic solvents. Solvents such as chloroform, methylene dichloride, toluene, or benzene are carcinogens or suspected carcinogens, and their use likely would be unacceptable in a production protocol for pharmaceuticals or food materials.

Some variables which influence the success of the method are listed in Table 5. Not all microbial cells are susceptible to solvent autolysis and, even when successful, one solvent may work while others fail. The initial events which trigger solvent-induced cell autolysis are not known but it is probable that the nonpolar solvent partitions into lipophilic patches in cell membranes, with the rate of partitioning being determined to some extent by the physical and chemical characteristics of the cell wall. Diffusion will also depend on the temperature and dissolved solvent concentrations surrounding the cell. With respect to the latter, frequent stirring of the cell suspension being treated effects better penetration of the solvent. Once the autolytic process is underway the surrounding cellular environment must be controlled to optimize rates of autolysis. It is important to determine and maintain the correct pH, ionic strength, temperature, and oxygen tension. High cell concentrations are often desirable for successful application of the method. This again may reflect the importance of maintaining a controlled cellular environment. Perhaps certain intracellular factors which leak out early such as divalent ions, cofactors, or coenzymes are needed at sufficiently high concentration for the autolytic process to continue on to completion. At low cell concentrations the essential factors would be diluted away as they leaked from the cells.

2. Heat, pH, and Osmotic Shock

Heat shock has been used by the yeast industry to prepare autolysed yeast and yeast hydrolysates for several decades. Many details of the art were proprietary then and still are now. In general, autolysis is initiated by slowly warming a stirred yeast suspension at a rate of about 2°C/min to 55–60°C. The onset of the autolytic process is characterized by the rapid evolution of CO_2 as stored carbohydrates and other molecules are rapidly metabolized by endoenzymes. The suspension is held at the elevated temperature for 12–36 hr while maintaining the suspension at the proper pH (usually near neutrality). Most of the variables listed in Table 5 apply equally as well to heat shock and other shock methods. While temperatures lower than 55°C but above the maximum growth temperature of the microorganism also support autolysis, in the

Table 5 Variables Which Affect the Efficiency of Solvent Autolysis

1. Type solvent and its concentration
2. Temperature
3. pH and ionic strength
4. Dissolved oxygen concentration
5. Cell concentration
6. Incubation time

past the aseptic technique of commercial processes were not reliable enough to avoid putrefaction.

Osmotic shock can be initiated by equilibrating cells in a medium of high osmotic pressure such as concentrated sucrose, glycerol, or salt and then rapidly decreasing the concentration of the medium by dilution. It is very gentle and has been used successfully to disrupt animal and blood cells on a large scale. The method does not lyse microorganisms very well but has been reported to release periplasmic enzymes (20). Osmotic shock has been used in combination with other lytic methods such as enzyme lysis. Combined shock methods often achieve enhanced results. For example, the heat shock autolytic process described above is greatly speeded up by addition of "autolytic precursors" such as chloroform, ether, or sodium chloride.

Many interesting combinations of temperature, pH, salt, and/or solvent have never been seriously explored because of the purification problem. Today, however, equipment design and industrial operations meet higher standards of good manufacturing practice (GMP) and new opportunities exist to develop methods of cell autolysis at more moderate pH values and temperatures under sterile conditions, as illustrated by the following examples (Hopkins, unpublished observations). In the first example, a thermophilic strain of *Bacillus* sp. was grown to cell densities of 80 g dry wt/liter in a custom-built foam fermentor maintained at pH 8.5. The cooling water to the fermentor was shut off and the heat from the rapidly metabolizing, high cell density culture quickly drove the temperature up to 76°C, some 6° above the usual growth temperature. The bacteria, which were nonviable at that point, were left overnight in the sterile fermentor without further pH or temperature adjustment. The pH slowly dropped to 6.5 and the temperature cooled to room temperature. Autolysis was complete by the next morning. After removing the solids by centrifugation, a crude preparation of alkaline protease was recovered. In the second example, 15 liters of *Pichia pastoris* yeast grown on methanol to a cell density of 140 g dry wt/liter was harvested from a foam fermentor in an aseptic manner. The cell cream was adjusted to pH 8.5 with ammonium hydroxide, made 3% in methylene dichloride, and slowly stirred at 37°C for 2 days. The solids were removed by centrifugation. Twenty percent of the soluble protein in the supernatant was alcohol oxidase. An estimated 85% of this intracellular enzyme was recovered. In this last example, temperature, solvent type, solvent concentration, incubation time, pH, and cell concentration were optimized for the success of the process. The experimental approach was largely empirical. A more rational approach will first require a better understanding of the underlying autolytic events initiated by chemical and physical shock.

G. Enzymatic Lysis

Enzymatic lysis of microbial cells is attractive because it is gentle and selective. The method is becoming more commonly used, especially at the laboratory level, as commercial lytic enzymes become available. Many of the current enzyme preparations offered are not pure but, as will be discussed below, treatment with a pure enzyme may not be desirable. Of the lytic enzymes, the most commonly known is lysozyme. While this enzyme is low cost (even raw egg white contains ample levels of the enzyme for the purposes of lysing cells), only a limited number of bacteria are lysed by lysozyme.

Yeast cells, which are difficult to disrupt by most mechanical methods, can be disrupted in low ionic strength media by glucanases (zymolyases) obtained from microbial sources. The pure lytic glucanases are not active on viable yeast cells without the aid of thiol reagents or lytic proteases to disrupt the crosslinked mannan-protein layer making up the outer surface of the cell wall (21). Ordinary proteases such as trypsin, pronase, and chymotrypsin may not substitute for the lytic protease. Once the outer wall structure is opened up, the exposed glucan surface can be attacked by one or more lytic glucanases. The attack takes place at several sites on the cell simultaneously and eventually the inner wall is opened enough (about one-third of the cell surface) to release the protoplast or spheroplast. Providing that the media buffer is dilute enough, the osmotic gradient quickly ruptures the plasma membrane releasing cytoplasmic proteins, storage carbohydrates, and membrane-associated subcellular structures into the medium. Under appropriate media conditions the subcellular organelles may also lyse.

The best strategy for breaching the defenses of microbial cell walls may be to use, either sequentially or simultaneously, a variety of hydrolytic and/or oxidative enzymes. The different enzymes act synergistically to enhance the overall lytic reaction rate. Consider, for example, that (a) invading bacteria are destroyed by white blood cells using several enzymes to generate highly oxidative species such as hydrogen peroxide and superoxide ion, (b) fungi attack the walls of plants using a variety of hydrolytic and oxidative enzymes, and (c) microorganisms which lyse yeast generally have several enzymatic activities which may include proteases, glucanases, mannanase, and chitinase. Successful application of a multienzymatic treatment requires attention to the same physical and chemical variables as autolysis (see Table 5). Indeed, autolysis might be viewed as a special case of enzymatic lysis where the lytic enzymes are endogenous. Anticipating the usefulness of the multienzyme strategy, Hunter and Asenjo (22) constructed a model for the kinetics of yeast cell lysis by microbial cell lytic enzymes and consider how physical and chemical variables might be adjusted to achieve selective product recovery.

There are several potential problems in using enzymatic lysis: (a) the product to be recovered from the cell might be destroyed or modified during lysis, (b) the added lytic enzymes in the cell lysate may complicate downstream product purification steps, and (c) lytic enzymes may be too expensive to permit scale-up to an industrial process, especially if the lytic enzymes are used only once.

With respect to the third point, there are ways to use less of or to reuse lytic enzymes. One approach is to prepare a soluble enzyme derivative having a molecular weight in the millions by chemically coupling it to a large, soluble polymer such as dextran (23). A reactor equipped with an ultrafiltration membrane retains the derivatized enzyme while lower molecular weight products released from the autolysed cells filter through the membrane and exit the reactor. Another approach relies on the property of lytic enzymes to bind strongly to their insoluble cell wall substrates (24). A fluidized bed or countercurrent flow reactor design would favor rebinding of the enzyme to incoming cells, thus reusing the enzyme. A third approach to overcoming the high cost of lytic enzymes is to use less enzyme. Instead of attempting to rupture the cell wall with enzyme, only enough is used to weaken the cell wall. Disruption of the pretreated cells is then quickly accomplished with a mechanical method. Rokem and Zomer (25), for example, obtained excellent yields of xylitol and alcohol dehydrogenase from the yeast *Pachysolen tannophilus* with a combination of glucanase (EC 3.2.1.21) and a commercial high-pressure homogenizer operating at a modest pressure. To get comparable yields with the homogenizer alone required twice the pressure and took five rather than two passes.

H. Dehydration

Drying animal and plant material can be a preliminary step to protein extraction. It is a long established and convenient way to preserve many tissues and the dried material can be ground up to hasten subsequent extraction procedures. The simplest and least expensive method is air drying. This relatively slow process can alter cell structure enough to enhance subsequent protein extraction by buffer. It has been applied mostly to plant tissue. Other direct drying methods such as spray drying and lyophilization preserve intracellular proteins but do not tend to permeabilize cells.

Dehydration with acetone or ethanol has been the starting point for extraction of proteins and other constituents for many years. The solvent dehydration method is fast and prepares the tissue in a powder form. In this method, a large excess of cold acetone or 100% ethanol is slowly added to tissue or microbial paste while agitating the mixture in an explosion-proof blade homogenizer (blender). The solvent is replaced with fresh solvent and blending is continued. The powdered, dehydrated tissue is recovered by filtration and, after evaporating off residual solvent, placed in a tightly sealed container.

I. Other Chemical Agents

Nonionic detergents, digitonin, butanol, isopentanol, and other agents which disrupt and disorganize membrane structures have been used to extract intracellular constituents such as receptor sites and mitochondrial particles from easily breached cells such as erythrocytes, leukocytes, and tissue culture cells. The disrupting agent is in contact with the cell for the shortest possible time and is often followed by a physical or mechanical homogenization method. Brief exposure to these agents can also induce autolysis.

High concentrations of chaotropic chemicals such as urea, guanidinium salts, or ionic detergents can break up the cell and solubilize (and denature) many proteins within the cell. When disulfide bond-reducing agents are present in addition to the chaotropic chemical, almost all the protein can be solubilized. The contaminant proteins then must be separated from the product protein in the presence of the chaotropic chemical before renaturation is attempted. This has been done successfully on a microgram to milligram scale using gel electrophoresis or chromatography. Renaturation of larger quantities of protein product has been done at the laboratory scale with relatively sturdy proteins such hormones, receptors, antigens, and viral proteins. Successful renaturation of denatured proteins is not always possible.

J. Cell Permeabilization

Most chemically induced cell lytic events are preceded by permeabilization of the cell, i.e., the plasma membrane is damaged to such an extent that the cell becomes leaky but the morphology of the cell remains intact. Permeabilized cells are particularly interesting because protein release can be selective. For example, a given treatment might remove only periplasmic proteins or only the low molecular weight cytoplasmic proteins. The choice of a permeabilization method depends on the composition of the cell wall and the cell membrane. A widely scattered literature on cell permeabilization has been reviewed by Felix (26). He found that toluene, chloroform, and diethyl ether were the most frequently used agents to permeabilize cells. The concentration of the solvent is usually higher than that used in solvent autolysis (see above) and the effects of the solvent are seen within minutes rather than tens of hours. Other permeabilizing agents discussed were antibiotics, basic proteins, chitosan, lysolecithin, detergents, chelating agents, and limited enzymatic digestion of the cell wall. A few physical methods which also permeabilize cells included freeze-thaw and ultrasonic treatment.

Take chitosan as an illustrative example. Incubation with chitosan, a soluble, deacetylated derivative of chitin, causes yeast to become leaky (27). The permeabilization method involves incubating a dilute, washed yeast suspension in the presence of 0.01% (w/v) chitosan at 30°C for 1 hr. In a recent application of this method about 40% of the alcohol oxidase, a 600,000-Da intracellular enzyme in methanol grown *Pichia pastoris* yeast leaked out of the cell (Hopkins, unpublished observations). Unfortunately, the method failed when concentrations of yeast higher than 2% (wet wt) were tried, even if the yeast were water-washed and higher concentrations of chitosan were used. It is not known how chitosan makes yeast cell leaky. However, because the method works only in low ionic strength media, it is likely that ionic binding of positively charged chitosan to the cell wall is an important early event. It is reasoned that at higher cell concentrations intracellular salts are released into the media at sufficiently high concentrations to interfere with the complex permeabilization process.

There appears to be no documented examples of cell permeabilization being used to recover intracellular proteins at the industrial level. This technique should be investigated further for its potential to recover intracellular proteins. Like autolysis, cell permeabilization is relatively simple, can be scaled up, and has attractive economics.

K. Programmed Self-Destruction

Cell walls of microorganisms can be weakened just before harvesting by dosing the growing cells with an antibiotic such as penicillin, which blocks cell wall synthesis (Hughes et al., 1971). In a similar vein, the cells can be disrupted with a lysogenic bacteriophage (28). And it may be possible to select production strains having wall-deficient mutations which are triggered to lyse by an increase in temperature, shift in pH, or the like. For example, Boudrant et al. (29) studied the temperature-programmed lysis of several mutants of *Saccharomyces cerevisiae*. The lytic event was initiated by an increase from 27 to 37°C and was sensitive to cell density and the physiological state of the microorganism.

III. SUMMARY

There are many ways to disrupt microorganisms and plant and animal tissue. Selecting the best cell disruption method depends on the factors listed in Table 6. The kind or type of cells is an important consideration. For example, some disruption methods which work well for animal tissue do not work at all for

Table 6 General Considerations in Selecting Cell Disruption Methods

1. Kind of cells and their history
2. Sample volume and number of samples
3. Disruption time
4. Availability of homogenization equipment
5. Possible scale-up potential
6. Effect on downstream purification processes
7. Economics of disruption

CELL DISRUPTION AND PROTEIN RECOVERY

Table 7 Selection Guide for Cell Disruption Methods

Method	Animal	Bacteria	Yeast	Mycelia	Plant
Bead mill[a]	Good	Good	Good	Good	Good
Rotor-stator[a]	Good	Poor	Poor	Fair	Good
Blade[a]	Good	Poor	Poor	Poor	Good
High-pressure[a]	?	Good	Good	Fair	?
Freeze-pressing	Good	Good	Good	Good	Good
N_2 decompression	Good	Fair	Poor	Poor	Fair
Ultrasonication	Good	Good	Fair	Fair	Good
Grinding	Good	Fair	Fair	Fair	Good
Pestle and tube	Good	Poor	Poor	Poor	Good
Paddle blender	Good[b]	Poor	Poor	Poor	Poor
Meat mincer[a]	Good[b]	Poor	Poor	Poor	Poor
Freeze fracturing	Good[b]	Fair	Fair	Fair	Good[b]
Freezing and thawing[a]	Poor	Fair	Poor	Poor	Good
Autolysis[a]	?	Good	Good	?	?
Enzyme lysis[a]	?	Good	Good	?	?
Dehydration[a]	Good	Fair	Poor	Poor	Fair
Chaotropic agents	?	?	?	?	?
Permeabilization	?	?	?	?	?
Programmed destruction	?	?	?	?	?

[a]Production scale available.
[b]Disperses but does not disrupt cells.

microorganisms. A guideline for the suitability of a given disruption method for some cell types is given in Table 7. The ratings in this table are not incontestable and, as mentioned earlier, combinations of methods can sometimes produce satisfactory results whereas one method alone fails.

The disruptibility of cells can be influenced by their growth and storage history. For microorganisms, cells in log phase growth tend to produce thinner cell walls which are more easy to disrupt. This and other conditions which can influence microbial cell disruptiability are listed in Table 8.

The cell disruption method selected will depend on its capability to process samples of a certain size or to be able to process multiple samples in a

Table 8 Fermentation Conditions Affecting the Disruption of Microbial Cells

1. Carbon source
2. Micronutrients and media richness
3. Phase of growth (batch fermentation)
4. Retention time (continuous fermentation)
5. Strain of microorganism

reasonable period of time. Other considerations are the availability, cost, and general utility of the disruption equipment. Thus, in a research environment the purchase of an expensive cell disrupter which processes a wide variety of cell types may be more easy to justify than a specialized disrupter. And if the long-term goal is to scale up, the choice of disruption methods narrow considerably. Indeed, several of the most successful laboratory cell disruption methods have no possibility of being scaled up.

Despite possible scale-up difficulties, in the case of many bioactive recombinant products expressed at high levels in microorganisms, this concern may be irrelevant. Few of these products are likely to be manufactured in really large amounts and current laboratory scale or pilot plant scale production equipment may be entirely adequate. For instance, active human TNF (tissue necrosis factor) can be expressed in *Pichia pastoris* yeast at levels of 100 g/kg of yeast (dry weight) (30). At this level of expression, only a few kilograms of r-DNA yeast needs be disrupted to meet the worldwide demand for this research material.

Finally, the operating and energy requirements which affect the economics of the disruption process (batch versus continuous, disruption yield, cell fragment size, effect of added enzymes on downstream separation, etc.) are important considerations in the selection of production equipment.

REFERENCES

1. Edebo L. Disintegration of cells. In: Perlman D, ed. Fermentation advances. New York: Academic Press, 1969:249–271.
2. Hughes DE, Wimpenny JWT, Lloyd D. The disintegration of microorganisms. In: Norris JR, Ribbons DW, eds. Methods in microbiology. Vol. 5B. New York: Academic Press, 1971:1–54.
3. Engler CR. Disruption of microbial cells, In: Cooney CL, Humphrey AE. Comprehensive biotechnology. Vol 2. Toronto: Pergamon Press, 1985; 305–324.
4. Curran HR, Evans ER. The accelerating effect of sublethal heat on spore germination in mesophilic aerobic bacteria. J Bacteriol 1943; 46:513–523.
5. Foulkes EC. The occurrence of the tricarboxylic acid cycle in yeast. Biochem J 1951; 48:378–383.
6. Mickle H. Tissue disintegrator. J Roy Microsc Soc 1948; 68:10–12.
7. Northcote DH, Thorne RW. The chemical composition and structure of the yeast cell wall. Biochem J 1952; 51:232–236.
8. Nossal PM. A mechanical cell disintegrator. Aust J Exp Biol 1953; 31:583–590.
9. Morgen H, Lindblom M, Hedenskog G. Mechanical disintegration of microorganisms in an industrial homogenizer. Biotechnol Bioeng 1974; 16:261–274.
10. Rehacek J, Schaefer J. Disintegration of microorganisms in an industrial horizontal mill of novel design. Biotechnol Bioeng 1977; 19:1523–1534.
11. Schutte H, Kroner KH, Husstedt H, Kula M-R. Experiences with a 20 liter industrial bead mill for the disruption of microorganisms. Enzyme Microb Technol 1983; 5:143–148.

12. Limon-Lason J, Hoare H, Orsborn CB, Doyle DJ, Dunnill P. Reactor properties of a high-speed bead mill for microbial cell rupture. Biotechnol Bioeng 1979; 21:745–774.
13. Milner HW, Lawrence NS, French CS. Colloidal dispersion of chloroplast material. Science 1950; 111:633–634.
14. Ribi E, Perrine T, List R, Brown W, Goode G. Use of pressure cell to prepare cell walls from mycobacteria. Proc Soc Exp Biol Med 1959; 100:647–649.
15. Edebo L. A new press for the disruption of micro-organisms and other cells. J Biochem Microbiol Technol Eng 1960; 2:453–479.
16. Hetherington P, Follows M, Dunnill P, Lilly MD. Release of protein from baker's yeast (*Saccharomyces cerevisiae*) by disruption in an industrial homogenizer. Trans Inst Chem Eng 1971; 49:142–148.
17. Lilly MD, Dunnill P. Isolation of intracellular enzymes from microorganisms: The development of a continuous process. In: Perlman D, ed. Fermentation advances. New York: Academic Press, 1969:225–247.
18. Bleeg HS, Christensen F. Facile, large scale yeast disintegration. In: Phaff HJ, ed. Yeast: A news letter. Vol. 28(1). Davis, CA: University of California, 1979:32–33.
19. Rechsteiner M, Rogers S, Rote K. Protein structure and intracellular stability. Trends Biochem Sci 1987; 12:390–394.
20. Neu HC, Heppel LA. The release of enzymes from Escherichia coli by osmotic shock and during the formation of speroplasts. J Biol Chem 1965; 240:3685–3692.
21. Kitamura K. A protease that participates in yeast cell lysis during zymolyase digestion. Agr Biol Chem 1982; 46:2093–2099.
22. Hunter JB, Asenjo JA. A structured mechanistic model of the kinetics of enzymatic lysis and disruption of yeast cells. Biotechnol Bioeng 1988; 31:929–943.
23. Asenjo JA, Dunnill P. The isolation of lytic enzymes from Cytophaga and their application to the rupture of yeast cells. Biotechnol Bioeng 1981; 23:1045–1056.
24. Kitamura K. preparation of yeast cell wall lytic enzyme form *Arthrobacter luteus* by its adsorption on b-glucan. J Ferment Technol 1982; 50:257–260.
25. Rokem SJ, Zomer E. Scale up of the production of xylitol dehydrogenase and alcohol dehydrogenase from *Pachysolen tannophilus* grown on xylose. Appl Microbiol Biotechnol 1987; 26:231–233.
26. Felix H. Permeabilized cells. Anal Biochem 1982; 120:211–234.
27. Jaspers HTA, Christianse K, Van Stevenick J. Improved method for the preparation of yeast enzymes in situ. Biochem Biophys Res Commun 1975; 65:1434–1439.
28. Bird I, Porter FHK, Stocking CR. Intracellular localization of enzymes associated with sucrose synthesis in leaves. Biochem Biophys 1965; 100:366–375.
29. Boudrant J, DeAngelo J, Sinskey AJ, Tannenbaum SR. Process characteristics of cell lysis mutants of *Saccharomyces cerevisiae*. Biotechnol Bioeng 1979; 21:659–670.
30. Sreekrishna K, Potenz RH, Cruze JA, McCombie WR, Parker KA, Nelles L, Mazzaferro PK, Holden KA, Harrison RG, Wood PJ, Phelps DA, Hubbard CE Fuke M. High level expression of heterologous proteins in methylotrophic yeast *Pichia pastoris*. J Basic Microbiol 1988; 28:265–278.

4

Proteases During Purification

Georg-B. Kresze

Biochemical Research Center
Boehringer Mannheim GmbH
Penzberg, Federal Republic of Germany

I. PROTEOLYSIS OF NATIVE AND DENATURED PROTEINS

A. Biological Functions of Proteases

In living organisms, proteases serve a large number of different functions, e.g.:

—Digestion of protein foodstuffs
—Intracellular protein turnover
—Processing of precursor forms of proteins (preproteins), e.g., secretory, organelle or viral proteins, and peptides, such as peptide hormones
—Activation of zymogens (proproteins)
—Maturation of newly synthesized proteins, e.g., removal of amino terminal f-Met residues
—Participation in diverse biological processes requiring degradation of peptide bonds, such as fibrinolysis, inflammation, tissue disintegration and remodeling, tumor invasion, fertilization, control of blood pressure, cell division, or sporulation

Due to the wealth of their essential functions, proteases are present in all living cells and organisms. Several reviews on protease functions are available (1–5).

This chapter will not deal with the use of proteases for processing, maturation, or modification of recombinant proteins, but will focus on the question of how to avoid undesirable proteolysis during purification of recombinant proteins.

B. Classes of Proteases and Protease Mechanisms

Proteases are unique among enzymes in that they are classified depending on their catalytic mechanism (or, more precisely, on the essential catalytic residues at their active sites) rather than their substrate specificity, which is decisive for classification of most other enzymes. Generally, proteases are grouped in four categories; for details, see (6–9).

Serine proteases, such as trypsin, chymotrypsin, or subtilisin, possess a highly reactive serine residue at their active center which during catalysis forms a covalent bond with the carbonyl group of the protein substrate, and a sterically adjacent histidine residue which enhances the nucleophilicity of the serine hydroxyl group by forming a hydrogen bond. Formerly, an essential role for a neighboring aspartic residue had been widely accepted ("charge relay system" or "catalytic triad") but its importance has been challenged by the fact that submaxillary tonin lacks this residue (10).

Cysteine proteases (formerly called thiol proteases), such as papain, calpain, or cathepsins B, H, and L, contain an essential cysteine residue that is involved in a covalent intermediate complex with the substrate (11,12).

Aspartic proteases (formerly designated as acid or carboxyl proteases) such as pepsin, cathepsin D, renin, or chymosin have been found as yet only in eukaryotes including eukaryotic microorganisms, but not in prokaryotes. These enzymes contain at their active sites two aspartic residues. Although their catalytic mechanism has not yet been definitively elucidated, it appears that it involves general acid–base catalysis rather than a covalent enzyme–substrate intermediate (13,14).

Metalloproteases, e.g., thermolysin, carboxypeptidase A, and bacterial and mammalian collagenases, contain a metal ion, usually zinc, at their active center. It probably assists in catalysis by enhancing the nucleophilicity of the attacking water molecule and polarizing the peptide bond to be cleaved (15,16).

Proteases generally are classified in one of these groups depending on their reaction with group-specific inhibitors (see Sec. IV.c) and, more recently, on information obtained through X-ray crystallography and amino acid sequence homologies. There are a number of proteases which have not (yet) been assigned to one of the four classes. It is unclear whether some of them, e.g., human insulin-degrading protease (216), may represent new types of catalytic mechanism [the protease from the crayfish *Astacus*, formerly thought to belong to none of the known classes of proteases (17,18), has been identified as a zinc metalloenzyme (217)].

C. Protease Specificity and Susceptibility of Protein Substrates to Degradation

Protease specificities usually are described in terms of the particular amino acid residues of the protein (or peptide) substrate whose participation in a peptide bond renders that bond most sensitive to the protease. In some cases, this primary specificity is directed toward only one of the residues involved [P_1 or P'_1 residue according to the nomenclature introduced by Schechter and Berger (19)], e.g., in cleavage of Lys-X or Arg-X bonds by trypsin. However, analysis of the structures of several proteases by X-ray crystallography has shown that they possess extended catalytic sites in a cleft on the surface of the enzyme molecule which are provided with a number of specificity subsites. Therefore, many proteases possess cleavage specificities directed toward somewhat longer amino acid sequences, e.g., preferential cleavage at the sequence Ile-Glu-Gly-Arg-X by factor Xa (20,21).

These primary and secondary interactions clearly are very important in the proteolytic cleavage of oligopeptide substrates and denatured proteins. However, they are only one of several factors controlling degradation of native proteins by proteases (22). In this process, the sites susceptible to the initial proteolytic attack are dictated by the steric accessibility of potentially sensitive peptide bonds to the active site of the protease. Thus, they are located preferentially at "hinges and fringes," i.e., exposed polypeptide chain segments (23) which may also possess enhanced segmental flexibility (24,25). The intermediate (N^*) formed during the first proteolytic event may then be more or less susceptible to further proteolysis. Depending on the relative magnitudes of the rate constants of the initial and the subsequent proteolytic steps, N^* may either accumulate as is the case in limited proteolysis (e.g., in zymogen activation), exist transiently, or be cleared away by rapid further proteolysis as in intracellular protein catabolism (26).

When recombinant proteins are expressed within heterologous host cells, the situation is even more complex. Prokaryotic (see II.A) and eukaryotic (see II.C and II.D) cells contain systems for proteolytic degradation of their own intracellular proteins, which are degraded with widely varying rates, as well as of denatured or abnormal proteins (27). Degradation is accelerated when the protein is damaged by oxidation (28,29) or free radicals (30). Clearly, recombinant heterologous proteins fulfill excellently the criteria of being "foreign" and therefore should be prime targets for proteolytic breakdown.

Degradation of proteins generally is accelerated at elevated temperatures— on the one hand, through the normal thermodynamic increase of reaction rate, and on the other hand, due to the loss of structural stability of the protein substrates at higher temperatures with a resulting increase in their susceptibility to proteases. Therefore, as usual in enzymological work, as a rule it is advisable to work at low temperatures (e.g., 4°C) during protein purification.

II. PROTEASES OF EUKARYOTIC AND PROKARYOTIC ORGANISMS USED AS HOSTS

In this chapter the proteolytic systems of several organisms frequently used as hosts for the expression of recombinant proteins will be concisely described.

A. Escherichia coli

Although more has been learned about the biochemistry of *Escherichia coli* than any other organism, the proteases of this gram-negative bacterium have received surprisingly little attention for a long time. Mainly due to the work of Goldberg and his coworkers, a considerable number of *E. coli* proteases have been described (31): about 10 soluble proteases, at least five membrane-bound proteases (32–34, 218–222) and numerous peptidases. The properties of the soluble proteases are summarized in Table 1.

Although the functions of the individual proteases are not known, it is clear that *E. coli* contains systems for the rapid degradation of polypeptides with abnormal structures, such as those resulting from nonsense mutations, deletions, or incorporation of amino acid analogs (35,36). However, not all recombinant proteins are unstable in *E. coli* even if they are produced as insoluble

Table 1 Soluble Proteases of *Escherichia coli*

Protease	Localization	Mol. mass 10^3 Da	pH optimum	Inhibitors[a]
Do	Cytosol	~ 500 or 300	6.0–8.5	DFP, PMSF, pMB
Re[b]	Cytosol	82	7.0–8.5	DFP, PMSF, TPCK
Mi	Periplasm	110	n.d.	DFP, EDTA, ophen
Fa	Cytosol	110	n.d.	DFP, EDTA, ophen, TPCK
So	Cytosol	140	6.5–8.0	DFP, PMSF, TPCK, pentamidine
La[c]	Cytosol (and membrane?)	450 (subunit 87–92)	~ 8.0	DFP, NEM, IAM, EDTA, vanadate
Pi (= III)	Periplasm	108	7.5	EDTA, ophen
Ci	Cytosol	125	7.5	ophen, pMB
Clp or Ti[c]	Cytosol	370 (subunits 23 + 80–81)	n.d.	DFP, NEM, pMB

[a]For abbreviations, see Table 4.
[b]Probably identical with the protease cleaving oxidatively damaged glutamine synthetase (190).
[c]ATP-dependent.
Source: Data from Refs. 31,51,187–189,223–225,254.

"inclusion bodies" which would no doubt represent an abnormal conformation (37). This has been attributed to the cell's degradation apparatus being saturated, under these conditions, by the large amounts of abnormal proteins (38).

One important feature of protein degradation in *E. coli* is its requirement for metabolic energy in form of ATP (38). It seems to result essentially from the ATP needed for activity of protease La. This enzyme was shown to be the product of the *lon* gene locus (39, 40, 223). A number of observations suggest that it catalyzes the rate-limiting step in the breakdown of most abnormal proteins to acid-soluble peptides with molecular weights >1500 Da which then are degraded further, probably by protease Ci and soluble exopeptidases (38). The in vivo synthesis of protease La is induced when large amounts of aberrant polypeptides (e.g., containing the arginine analog canavanine, or incomplete due to incorporation of puromycin) or of cloned heterologous polypeptides (e.g., tissue plasminogen activator) are produced in the cell (41). An increased content of protease La, on the other hand, increases the rates of degradation of abnormal proteins (42). This may represent an important protective mechanism for the cell that helps to prevent continuing accumulation of potentially toxic polypeptides (38).

Overall protein breakdown in *E. coli* rises sharply at elevated temperatures, i.e., above 42°C (43). Under these conditions, the content of protease La increases about twofold (44). Thus, under conditions widely used to cause induction of temperature-inducible promoters (e.g., the λP_L promoter), degradation of the cloned product may be induced.

When *E. coli* is shifted to high temperatures, a limited number of polypeptides is induced which are known as heat shock proteins. The transcription of their genes is regulated by the *htpR* gene product (45) which functions as an alternative σ subunit of RNA polymerase (46). Protease La was found to be one of the heat shock proteins (44, 47); therefore, the increase of protease La activity at 42°C or in the presence of abnormal proteins does not occur in *htpR* mutants (41,44) which, as predicted, show a lower rate of degradation of abnormal, incomplete, or mutant proteins (44,48).

Although these results clearly indicate an important function of protease La in the degradation of various types of abnormal proteins in *E. coli*, in *lon*⁻ cells such proteins are still degraded relatively quickly compared to the bulk of cell proteins in a process also requiring ATP (44,49). Recently it was found that *E. coli* contains, in addition to protease La, a second soluble ATP-dependent protease (named Clp protease or protease Ti) degrading proteins such as [^{3}H]methyl casein in an ATP-dependent manner (50,51,224,225). The importance of this enzyme for intracellular protein degradation in *E. coli* is not yet clear.

B. *Bacillus subtilis*

Bacillus subtilis, a gram-positive bacterium, has the potential to be a highly useful system for producing recombinant proteins and peptides since it is amendable to genetic manipulation, can be adapted to various nutritional and physical growth conditions, is not pathogenic to humans, does not form endotoxins, and has the capacity to secrete proteins efficiently into the medium (52).

However, one main disadvantage is the presence of high levels of intracellular and extracellular proteases which may degrade foreign proteins. *Bacilli* produce quite a number of various proteases (53). Particularly in *B. subtilis* two major and one minor extracellular proteases, at least two intracellular proteases, one membrane-bound protease, and a pyrrolidonyl peptidase have been described (Table 2).

A neutral metalloprotease and an alkaline serine protease (subtilisin) are secreted into the medium (54). Both are produced after the exponential growth phase, at the beginning of the stationary phase. Their physiological role is unclear. The genes of both proteases have been cloned, and in vitro-derived deletion mutants have been described (55,56). Since the only detectable phenotypic effect of the deletion of both proteases was the loss of protease activity (56), a function in sporulation or in the regulation of cell morphology seems improbable. It rather appears that the proteases might have a role as

Table 2 Proteases of *Bacillus subtilis*

Protease	Localization	Mol. mass 10^3 Da	pH optimum	Inhibitors[a]
Neutral protease	Extracellular	33	7.3	EDTA
Alkaline protease (subtilisin)	Extracellular	27.5	7.0–8.0	DFP, PMSF
Bacillo-peptidase F	Extracellular	33 + 50	7.5–8.0	DFP, PMSF
ISP-I	Intracellular	56–60	7.5–10.0; 8.0–8.4	PMSF, EDTA, pMB, IAA
ISP-II	Intracellular	47	6.6–7.0	PMSF, ovomucoid, TPCK
Serine protease	Membrane fraction	540	11.0	PMSF, chymostatin
Pyrrolidonyl peptidase	Intracellular	n.d.	8.0–9.0	IAA, IAM, Hg^{2+}, Cu^{2+}

[a] For abbreviations, see Table 4.
Source: Data from Refs. 53,57–61,191–194,215.

scavengers (54,56). A further extracellular serine protease from *B. subtilis* was reported (57); however, this enzyme appears to be of minor importance.

At least two intracellular serine proteases are produced in *B. subtilis* (58–60). The major intracellular protease, ISP-I, shows considerable amino acid sequence homology with subtilisin and is characterized by its sensitivity to EDTA due to an absolute requirement for Ca^{2+} for stability and activity (59). It has been proposed that ISP-I [as well as ISP-II, a minor serine protease with trypsinlike specificity (60)], is involved in the process of sporulation. The gene coding for ISP-I has been cloned, and a deletion mutant indeed was reported to show decreased sporulation in a synthetic medium but to sporulate normally in a nutritionally rich medium (61). A frameshift mutant, on the other hand, sporulated at the same frequency as the wild type in nutrient broth as well as in minimal medium (120). ISP-I therefore seems not to be essential in sporulation although it evidently accounts for the majority of intracellular protease activity. This is confirmed by the result that ISP-I activity can be detected in sporulation-deficient strains, and its activation from an enzymatically inactive precursor form is not dependent on a sporulation-specific gene product (226).

C. Yeast *(Saccharomyces cerevisiae)*

Whereas in 1970, only three proteases were known to occur in *Saccharomyces cerevisiae*, this number had increased to eight by 1979, and more than 30 discernible proteolytic activities have been described as of today (Table 3) (62). In studies with yeast mutants defective in the principal proteases, a multitude of new proteolytic enzymes have been detected which, in some cases, apparently can substitute for the missing activities [for review, see (62–64)]. Furthermore, proteases of highly restricted and specific functions (such as pheromone precursor processing) have been found, e.g., (65,66); review: (227).

The principal protein-degrading proteases of *S. cerevisiae* are the proteases yscA (an aspartic protease), yscB (a serine protease), the carboxypeptidases yscS (a metalloprotease), and yscY (a serine protease). They are all located within the vacuole of the yeast cell, which appears to be the equivalent to the mammalian digestive intracellular compartment, the lysosome (67). yscA and yscB are highly nonspecific proteases (64). They, as well as carboxypeptidase yscY (and probably yscS), are synthesized as higher molecular weight precursors which undergo proteolytic processing to yield the mature enzymes. The maturation and activation mechanism has been intensively studied by Wolf's group: whereas maturation of yscA occurs autocatalytically upon transfer from the Golgi apparatus to the vacuole, yscA triggers processing of pro-yscB, and yscA as well as yscB are involved in maturation of pro-yscY (68). All four

Table 3 Proteases of *Saccharomyces cerevisiae*

Protease	Localization	Mol. mass 10^3 Da	pH optimum	Inhibitors[a]	Ref.
yscA	Vacuolar	41.5	2.4–6[b]	I^A, pepstatin, diazoacetyl-norleucin-methyl ester	197, 198
yscB	Vacuolar	33	7.0	I^B, Hg^{2+}, PMSF, chymostatin	195
yscD	Nonvacuolar	83	5.75–7.0	EDTA, Hg^{2+}	201
yscE	Nonvacuolar	600	8.2–8.6	Chymostatin, Hg^{2+}	204
yscF	Membrane fr.	n.d.	7.2	Hg^{2+}, EDTA	206
yscG	Membrane fr.	n.d.	n.d.	EDTA, Dithiothreitol	62
yscH	Membrane fr.	n.d.	n.d.	EDTA	62
yscK	n.d.	>600	n.d.	n.d.	62
Propheromone Convertase Y	Soluble fr.	43	7.5	DFP, PMSF	213
Calcium-dependent protease	Membrane fr.	100–120	7.0	IAM, IAA, pMB, EDTA, EGTA, Hg^{2+}, Cu^{2+}, Zn^{2+}	200
Mitochondrial protease	Mitoch. matrix	115	7.5	ophen, EDTA	209
Mitochondrial protease[c]	Mitoch. matrix	550	~8.0	DFP, PMSF, vandate	208
Mitochondrial protease	Mitoch. inter-membrane space	n.d.	n.d.	n.d.	209
Three mitochondrial proteases	Mito. membrane	17	n.d.	PMSF, pMBS, leupeptin	205
Leulysin	Secreted	750	4.0	n.d.	202
CPase yscY	Vacuolar	61	6.5	I^C, DFP, PMSF	196
CPase yscS	Vacuolar	n.d.	n.d.	EDTA	211
CPase yscα	Membrane fr.	n.d.	7.5	n.d.	212
CPase yscγ	n.d.	n.d.	n.d.	PMSF, antipain	168
CPase yscδ	n.d.	n.d.	n.d.	EDTA	168
CPase yscε	n.d.	n.d.	n.d.	PMSF	168
APase yscI	Vacuolar	640	n.d.	EDTA, bestatin	210
APase yscII	Periplasm	85–140	7.5	EDTA, Hg^{2+}, bestatin	203
APase yscIII	n.d.	30	n.d.	n.d.	210

(*Continued*)

Table 3 (Continued)

Protease	Localization	Mol. mass 10^3 Da	pH optimum	Inhibitors[a]	Ref.
APases IV–XIV	n.d.	—	7.0–7.5	Bestatin	203
APase yscCo	Vacuolar	100	8.5	EDTA, Zn^{2+}	207
APase yscP	Membrane fr.	n.d.	7.5	EDTA, bestatin	203
DPAP yscI	n.d.	n.d.	7.0	n.d.	203
DPAP yscII	n.d.	78	6.5	EDTA	203
DPAP yscIII	n.d.	n.d.	n.d.	n.d.	62
DPAP yscIV	Membrane fr.	n.d.	7.5	PMSF	203
DPAP yscV	Vacuolar membrane fr.	40	7.0–7.5	PMSF, Hg^{2+}, Ni^{2+}, Cu^{2+}, Zn^{2+}	199

[a] For abbreviations, see Table 4.
[b] Depending on substrate.
[c] ATP-dependent
Abbreviations: CPase, carboxypeptidase; APase, aminopeptidase; DPAP, dipeptidyl aminopeptidase.

enzymes apparently participate in the degradation of proteins to acid-soluble material and, thus, apparently play important roles in such processes as starvation, sporulation, and protein degradation under growing conditions.

In addition to these proteases, the yeast cell contains inhibitory proteins to some of these proteases, termed inhibitors I^A (inhibitor of yscA), I^B (inhibitor of yscB), and I^C (inhibitor of yscY). In contrast to the respective proteases, these inhibitors are located in the cytoplasm of the cell. Their function is not known (review: 63,69).

Furthermore, yeast contains a ubiquitin-dependent protein degradation system. This topic will be discussed in the next section.

D. Mammalian Cells

Mammalian cells, tissues, and body fluids contain a large number of different proteases which are located in various compartments of the cells or extracellularly. A comprehensive bibliography (70,71) as well as excellent reviews (72,73) have been published.

In the context of this chapter, the most interesting among the intracellular proteolytic systems (74,75,228) is the soluble ATP-dependent pathway responsible for the selective elimination of many short-lived enzymes and of proteins with highly abnormal structures, as it may result from mutations, biosynthetic errors, or postsynthetic damage (35,76). This system identified in reticulocytes (77) and in a wide variation of other cells (78) apparently uses ubiquitin, a small protein found in all eukaryotic cells, as a marker to identify proteins for rapid degradation (79). Ubiquitin is activated in an ATP-requiring process and then conjugated to substrate proteins to form (iso)peptide bonds between the carboxy terminus of ubiquitin and α- and ϵ-amino groups of lysine residues within the protein substrate (78). The ubiquitin-protein conjugate may then be de-ubiquinated by multiple isopeptidases, or degraded to short peptides and amino acids by a protease which also requires ATP (80). According to (81), ubiquitin itself also has intrinsic proteolytic activity.

It is not yet clear which molecular determinants are decisive for intracellular protein half-lives. Apparently, an unblocked α-amino terminus is important since blocking the amino terminus affects protein degradation by the ubiquitin-dependent system (82) as well as by other cytosolic proteases (83).

Two "rules" were recently proposed for explanation of factors dictating the rates of catabolism of proteins within the cell (84). The "N-end" rule (85) states that proteins with "destabilizing" amino terminal residues—i.e., Asp, Glu, Phe, Ile, Lys, Leu, Gln, Arg, or Tyr—are rapidly degraded whereas proteins with "stabilizing" amino termini (Ala, Gly, Met, Ser, Thr, or Val) are metabolically more stable. The rule was established by measurement of half-lives of ubiquitin-β-galactosidase fusion proteins within yeast cells where the N-terminus of β-galactosidase had been altered by site-directed mutagenesis, and was supported by a survey of amino termini of long-lived eukaryotic and prokaryotic intracellular proteins from the literature (85).

On the other hand, the "PEST" hypothesis was developed by computer analysis of the amino acid sequences of 12 short-lived and 35 long-lived proteins (86). It suggests that sequences rich in proline (P), glutamic acid (E), serine (S), and threonine (T) residues are present in proteins with short half-lives, but not in long-lived proteins. These regions may be sites of cleavage for proline endopeptidases (86).

Although the validity of both of these hypotheses is not yet clear and their general applicability awaits proof, there may be underlying unifying principles, such as preferential cleavage at proline residues with PEST regions to give destabilizing Glu termini (30).

There are additional ATP-dependent cytosolic pathways of protein degradation that are ubiquitin-independent (229,230). Furthermore, another cytosolic protease which attacks protein substrates is calcium-dependent neutral protease (CANP, calpain) (87,88). Its role in intracellular protein degradation is not yet clear (74,228).

Another important pathway of protein degradation occurs through lysosomes which are responsible for degradation of many long-lived proteins as well as for the increased protein degradation in tissues of animals and in cultured cells under starvation conditions (231–233).

Mammalian cells are able to secrete recombinant proteins into the culture medium. If secretion occurs cotranslationally, intracellular proteolytic systems should be expected not to be harmful for these proteins. However, they are exposed to proteolytic attack either by extracellular proteases secreted by the host cells or by proteases introduced with components of the growth medium. This may not only apply when peptones or tryptones are used which are not always protease-free, but specially when mammalian cells are grown in the presence of serum (e.g., fetal calf serum). Serum inevitably contains a number of proteases and protease zymogens such as plasminogen (but also protease inhibitors) which may effect damage of the target protein.

III. METHODS TO DETECT PROTEASES

The activity of proteinases can be assayed either with natural substrates, i.e., native or denatured proteins such as hemoglobin, casein, or fibrin, or with synthetic substrates. Most of the latter are composed of a small peptide portion located on the amino terminal side of the bond to be cleaved and a leaving group which can be measured (usually photometrically or fluorometrically). Although the convenience of assays using synthetic chromogenic substrates is higher, they are in most cases unsuitable when the precise nature and specificity of the protease(s) are unknown (89).

A survey of protease assay methods using proteins as substrates is given in (90). These assays can be based on the following:

Measuring the loss of turbidity as a protein is solubilized by proteolysis, e.g., plate tests with fibrin (91,92) or casein agar (93)

Measuring the loss of activity of a protease-sensitive enzyme such as luciferase (94)

Measuring released amino groups with ninhydrin (95), fluorescamine (96), or trinitrobenzenesulfonic acid (97)

Measuring of the amount of acid-soluble peptides released, e.g., from casein or hemoglobin (98)

Measuring of dye-stained acid-soluble peptides, e.g., assay with azocoll (99)

Measuring of fluorescence-labeled acid-soluble peptides with substrates such as fluorescamine-coupled hemoglobin (100) or casein coupled with fluorescamine (101) or with resorufin (234) which can be measured either photometrically or fluorometrically

Measuring of radiolabeled peptides released from proteins labeled with ^{125}I or ^{3}H (102,103)

Measuring the amount of active (protease-resistant) "reporter" enzyme released from a protein–enzyme conjugate (104)

The merits and disadvantages as well as the detection limits of these methods are discussed in (90). These assays are useful for estimation of the total proteolytic activity present in cell homogenates and fractions obtained during purification of a recombinant protein as well as for following the removal of proteases (cf. IV.B). However, they are not suitable to establish the absence of proteases since many proteases may work only under special conditions (concerning, e.g., pH values, presence of cofactors, etc.) and may not be active toward all substrates (depending on the protease specificity). Furthermore, even very small amounts of proteases may be sufficient to cause limited proteolysis of highly susceptible recombinant proteins. So the only definite proof of the absence of undesirable proteases is the control of the target protein's intactness, after each step of the purification procedure, by gel electrophoresis, HPLC, or even partial amino acid sequencing.

IV. STRATEGIES TO PREVENT PROTEOLYSIS OF RECOMBINANT PROTEINS

From Chapter 2 it should be clear that in the purification of (recombinant or nonrecombinant) proteins, one always has to be on one's guard against undesired proteolytic degradation of the target protein. In the special case of recombinant proteins, proteolysis may occur

Inside the host cell where the protein may be recognized as abnormal.

In the culture medium (in case of secreted proteins) due to attack by extracellular proteases. These may be produced by the host cell itself or may be introduced with components of the growth medium.

In the homogenate after cell disintegration, which may lead to breakdown of cell organelles, e.g., lysosomes or vacuoles, and release of proteases contained in these compartments. However, even lysozyme, which is often used for enzymatic digestion of the cell wall of gram-negative bacteria, has been reported to possess intrinsic proteolytic activity and may cause proteolytic artifacts (105).

At later stages of the protein purification if proteases have not been sufficiently removed and if conditions are suitable to proteolysis (e.g., incubation at elevated temperatures, suitable pH, presence of cofactors such as metal ions or ATP, etc.). This may even be possible if initially no protease activity had been detectable since proteases (e.g., yeast proteases yscA or yscB) may have been complexed with inhibitors in the homogenate, but later purification steps may have lead to dissociation of the protease–inhibitor complex and release of active protease.

It has also been found that dead cells, such as yeast (106) or mammalian cells, release proteases upon lysis which might degrade secreted recombinant proteins. To avoid this, it is essential to control precisely the conditions of fermentation and cell harvest.

Various approaches are conceivable to prevent proteolytic damage of recombinant proteins:

Strategies to *avoid* proteolysis by genetic manipulation of the host cell or the target protein

Strategies to *remove* proteases from the solution of the recombinant protein, and

Strategies to *inhibit* proteases by the use of protease inhibitors

These will be discussed in the following chapters. It appears that the effectiveness of all these methods individually is limited so that a combination of the different approaches is recommended.

A. Genetic Strategies to Avoid Proteolysis

1. Use of Protease-Deficient Host Mutants

E. coli. Although the cell's life is impossible without proteases, an obvious concept to avoid proteolysis would be to use host strains genetically deficient in those proteases responsible for degradation of the recombinant target protein. Therefore, in the case of *E. coli, lon* mutants which contain a defective but partially active protease La have been studied (39,107–109). In these cells, degradation of abnormal proteins occurs two to four times more slowly than in the wild type. Unfortunately, these *lon* mutants show a number of undesirable phenotypes such as the overproduction and accumulation of polysaccharides (referred to as mucoidy), defective cell division and filament formation, and increased sensitivity to ultraviolet radiation or other DNA-damaging agents.

Protease La is a heat shock protein whose synthesis is controlled by the *htpR* gene product (cf. II.A). *htpR* mutants have indeed been found to stabilize the expression of normally short-lived endogenous proteins (110), of analog-containing, mutant, or puromycyl polypeptides (44,48) and of cloned foreign polypeptides (44,111,112,234). The *htpR* mutants do not show the adverse properties of *lon* mutants (44,111) and are therefore more viable than *lon* mutants under a variety of conditions. An alternative approach may be the use of vectors allowing thermoinducible synthesis of *htpR* gene antisense RNA which inactivates the normal *htpR* transcripts (236).

It has been reported that a double *htpR lon* mutant constructed by P1-mediated transduction produces a labile protease in reduced amounts and exhibits even less capability to degrade abnormal proteins than strains carrying either mutation alone (48,111). At present, however, it is not clear whether these mutations also impair activity of the other ATP-dependent *E. coli* protease, protease Ti, and whether other proteases present in *E. coli* can substitute in degrading recombinant polypeptides.

Use of protease-deficient mutants has allowed large improvements in the yields of several cloned proteins otherwise subject to rapid degradation, such as somatomedin (115), α-tubulin, and C5a (38).

Alternatively, *E. coli* host strains which, in contrast to *E. coli* K12 wild types, do not contain protease La could be used, such as *E. coli* B (116,117).

Another *E. coli* protease which may destroy recombinant proteins after cell lysis is the membrane-bound protease VII (or ompT) (220,222). Use of *ompT* deletion strains has allowed production of T7 RNA polymerase, which is very susceptible to proteolysis, from its cloned gene (222).

Additionally, proteases Re (237) [probably identical with the protease described in (190)] and So (238) were reported to cleave oxidatively damaged glutamine synthetase and thus may play a role in the degradation of modified or abnormal proteins in *E. coli*.

Other *E. coli* mutants have been described which are defective in the degradation of abnormal proteins, such as canavanyl proteins (113) or recombinant human interferon-γ (114). Whether they are different from the *lon* and *htpR* mutants remains to be clarified.

An alternative general approach is to enhance expression, e.g., by increasing the copy number of the expression plasmid, to such an extent that the cellular proteolytic systems are saturated with the protein substrate so that a large part of the protein product will remain intact (118,119). Sometimes it may not be easy to remove fragments formed by limited proteolysis from the undegraded target protein. Furthermore, a relatively rapid burst of expression extending over no more than one or two generations is necessary rather than continuous low expression over an extended period.

B. subtilis. Strains have been described which are defective in the intracellular serine protease ISP-I (120), the extracellular alkaline (55) and neutral (56) proteases, or both extracellular enzymes (121–123). These have proven valuable in the expression and secretion of proteins such as β-lactamase (214) or human growth hormone (123) which are partially degraded when secreted from wild-type *B. subtilis*, and of mutant subtilisins (121,124) which can be easily studied in the absence of a wild-type subtilisin background. It has been reported that most of the residual protease activity present in cultures of a double-mutant *B. subtilis*, which is deficient in neutral and alkaline proteases, is dependent on the presence of calcium ions, and can be inhibited by adding EGTA (255). Novel strains of *Bacillus brevis* which produce a large amount of recombinant proteins (such as α-amylase from *Bacillus stearothermophilus*) but no extracellular proteases were recently claimed (239).

S. cerevisiae. A number of protease-deficient yeast mutants have been reported (62,125–127,168). These have been found to be useful as hosts in the intracellular expression of recombinant polypeptides, e.g., α-glucosidase

(240), α_1-proteinase inhibitor (241,242), an α-interferon analog (243), or β-lactamase precursor (244).

2. Coexpression of Protease Inhibitor Genes

The protease inhibitor (*pin*) gene of bacteriophage T4 has been shown (128,245) to stabilize incomplete, abnormal, and recombinant eukaryotic proteins when expressed in the same *E. coli* cell as the T4 *pin* gene by inhibiting the Lon protease (246). This technique requires to maintain two different plasmids in the *E. coli* host cell. This may be difficult because of the danger of alterations or loss of compatibility. An improvement of the method has been described which avoids the issue of plasmid incompatibility by the use of a double-chimeric plasmid from which a foreign DNA sequence is recombined into a T4 bacteriophage, which is then used to infect the *E. coli* host (129). It appears, however, that proteolysis cannot be prevented but only partially inhibited by this approach (128).

3. Compartmentation

A conceptually plausible idea would be to translocate recombinant proteins, during or as soon as possible after their synthesis, to a (preferentially extracellular, to facilitate purification) compartment where they are protected from proteolytic attack. Of course, this does not apply to host systems secreting proteases, such as *Bacillus* wild-type strains.

Methods have been described to direct recombinant polypeptides to the periplasm of the *E. coli* cell (130,133,136,247,248) [in this case, the use of periplasmic protease mutants of *E. coli* may be advantageous (249,250)] as well as to secrete them from *E. coli* (251) or yeast (131,132,252). The latter system is especially interesting not only in view of the glycosylation occurring during secretion from eukaryotes, but also due to the potentially attractive production costs. Unfortunately, efficiency of secretion is as yet rather low in most cases. This might be overcome by the use of "supersecreting" yeast strains (131).

4. Genetic Modification of the Target Protein

We have seen that the initial proteolytic attack on native proteins usually occurs at regions of high accessibility (cf. I.D.). It therefore might seem attractive to remove potential cleavage sites within the recombinant protein by site-specific mutation of the peculiar amino acid residues involved in forming the sensitive peptide bond. This approach has been used, e.g., for preparing mutants of urokinase (134) and tissue plasminogen activator (135) which are resistant to processing to the two-chain forms by plasmin or related proteases, and to prevent cleavage of α-interferon by a yeast protease specific for dibasic sequences (243). Clearly this method can only be used when the cleavage sites are exactly known, and is therefore not generally applicable.

It has also been claimed that the metabolic stability of proteins can be altered by modifying the amino terminus according to the N-end rule (cf. II.D) (85,253).

Low molecular weight proteins have repeatedly been found to be unstable in *E. coli*. For example, the half-life of rat proinsulin in *E. coli* has been described to be only 2 min (136). This instability can in many cases be overcome by generating gene fusions that increase the size of the protein, e.g., with β-galactosidase (137). This now classical approach has proven useful for small peptides such as somatostatin (138) and others (139,140), proteins such as insulin (141), antigens (142,143), and viral envelope proteins (144–146). In the latter case (146), clever use was made of the cytoplasmic aggregation commonly observed when certain fusion proteins or heterologous proteins are produced at high levels in *E. coli*: the vector used for this strategy has a cloning site which is an open reading frame except for the presence of a single nonsense codon. When a construct containing the gene of interest is placed in a suppressor host, a fusion protein is made which is found as an aggregate in the cytoplasm. Since the efficiency of suppression is only about 50%, a rather large amount of the unfused protein of interest is also made which, under these conditions, is not degraded since it coprecipitates with the fusion protein.

Fusion proteins with β-galactosidase can be in principle easily purified, either by conventional methods developed for β-galactosidase itself (147) or by affinity chromatography (148,149).

A disadvantage of this approach is that the target protein is not obtained in free form but as a fusion protein which in most cases is not acceptable for the intended use. A number of methods have been described to introduce amino acid sequences suitable for specific processing, e.g., by collagenase (150), CNBr (138), acid (151), or factor Xa (152), but none of them is completely satisfactory for larger proteins, which indeed would not be expected to be stabilized by gene fusion.

As an alternative method, the expression of a multimeric form of the protein coding sequence was reported producing a multidomain polypeptide which stabilized itself and could be converted to a monomeric unit by specific cleavage (153).

B. Strategies to Remove Proteases

When purifying recombinant proteins, it would be highly desirable to have purification steps available for the selective removal of proteolytic activities. Unfortunately, one usually has to deal with mixtures of various proteases with different and unknown properties, so that in most cases a rational approach for complete protease removal by a general method is not possible. It should be recalled that even reducing the protease content of a solution to 1% of the initial value will not be sufficient to solve a potential protease problem but will just delay the rate of proteolytic attack by a factor of 100. It generally appears more attractive to remove the target protein selectively from the protease-containing solution, e.g., by binding to a specific adsorbent.

Proteases are characterized by their affinity to proteins. Therefore, it should be possible to prepare a general protease affinity adsorbent by immobilizing a suitable protein to an insoluble carrier. In fact, binding of proteases to protein matrices such a hemoglobin-Sepharose has been reported (154). If an insoluble protein substrate is used, e.g., collagen (155) or thermally modified casein (156), immobilization is not necessary. This approach, although plausible, suffers from the disadvantages that binding usually is far from complete, and the solution may become contaminated with degradation products of the protein substrate.

Alternatively, affinity chromatography on immobilized protease inhibitors can be tried. A rather large number of specific inhibitor matrices have been described, several of which may be of value not only for particular proteases but at least for groups of proteases. Examples are carrier-fixed p-aminobenzamidin (157,158), aprotinin, ovomucoid, soy bean trypsin inhibitor (159), or argininal (160), which have been used to bind proteases possessing trypsin-like specificity.

Attempts have also been made to construct general protease-binding inhibitor matrices. Bacitracin, an antibiotic cyclopeptide produced by *Bacillus licheniformis*, is a weak inhibitor of several proteases. Bound to Sepharose it has been described to allow the binding and isolation of a number of proteases of all four classes (161) and may therefore also be useful for the removal of proteases. Another protease inhibitor which interacts with a large number of serine, cysteine, aspartic, and metalloproteases is α_2-macroglobulin (162). When bound to zinc chelate-Sepharose, this adsorbent (which is commercially available) can be used for the removal of many endoproteinases from biological fluids (163).

Binding of proteases to their substrates usually involves hydrophobic interactions. It therefore has been tried to remove proteases selectively by binding to hydrophobic adsorbents such as octyl- or phenyl-Sepharose (164). These methods may be successful in particular cases but require optimization for the particular system.

Conceptually fast and simple systems for the extraction of proteins are the aqueous two-phase systems formed from polymers such as polyethyleneglycol (PEG) or dextran and salt solutions (165–167). These may be applied for separation of recombinant proteins from proteases provided that the partition coefficients are sufficiently different. Recently, use of a two-phase system formed from PEG and potassium phosphate for the rapid extraction of a fusion protein formed from staphylococcal protein A and *E. coli* β-galactosidase has been reported (169). Since β-galactosidase dissolves preferentially in the PEG phase, dragging the protein A with it, the latter escapes instant cleavage by the *E. coli* proteases. This system may be generally applicable for the extraction of β-galactosidase fusions.

C. Strategies to Inhibit Proteases

Due to the potentially deleterious effect of proteases, there is most likely no protease without an in vivo acting endogenous inhibitor. Additionally, there is a wealth of protease inhibitors formed by microorganisms as well as synthetic protease inhibitors.

The field of protease inhibitors has been thoroughly described in a number of monographs and conference proceedings (e.g., 170,171). The reader is particularly referred to a recent comprehensive volume (172).

Protease inhibitors can be classified according to the class of proteases which they inhibit (inhibitors of serine, cysteine, aspartic, or metalloproteases); they may be discriminated as reversible and irreversible inhibitors (173); or they may be distinguished according to their chemical structure by classification as low molecular weight inhibitors, oligopeptide and related inhibitors, and protein-protease inhibitors.

1. Synthetic Low Molecular Weight Inhibitors

The rational design of synthetic protease inhibitors requires an understanding both of the catalytic mechanism and the substrate specificity of the enzyme under study (174) (cf.I.C.). The effect of low-M_r inhibitors of serine and cysteine proteases in most cases depends on covalent modification of the active serine or cysteine residue (or of the adjacent histidine residue) whereas aspartic proteases are inhibited by diazoacetyl compounds + Cu^{2+} (175), and metalloproteases by complexating agents such as EDTA or *ortho*-phenanthroline. A list of synthetic protease inhibitors is shown in Table 4; like the following Tables 5 and 6, it should be considered representative rather than exhaustive, and is confined to inhibitors which are easily obtained commercially. More detailed information can be found in (172).

2. Oligopeptide and Related Inhibitors

Many protease inhibitors of this class have either been isolated from culture filtrates of microorganisms, especially by Umezawa and his coworkers (176,177), or chemically synthesized. They either comprise substratelike peptide sequences coupled to reactive groups such as aldehyde, chloromethyl ketone, or epoxy groups, or contain analogs of the usual naturally occurring amino acid residues (such as D-amino acid residues or exotic residues like statine, (3S,4S)-4-amino-3-hydroxy-6-methyl heptanoic acid) which mimic transition states of proteolytic cleavage or form bonds uncleavable by pro-

Table 4 Synthetic Low Molecular Weight Protease Inhibitors

Protease inhibitor	Abbr.	Proteases Inhibited (examples)
Diisopropylfluorophosphate	DFP	Nearly all serine proteases
Phenylmethanesulfonyl fluoride	PMSF	Most serine proteases
(p-Amidinophenyl)methanesulfonyl fluoride	pAPMSF	Several serine proteases
Tosylphenylalanylchloromethyl ketone	TPCK	Chymomotrypsin, chymase I, bromelain, ficin, papain (*not* trypsin)
Tosyllysylchloromethyl ketone	TLCK	Trypsin, thrombin, enterokinase, plasmin, acrosin, C1r, endoproteinase Lys-C, endoproteinase Arg-C (*not* chymotrypsin)
Benzamidine, p-aminobenzamidine	PAB	Trypsin and many trypsinlike proteases
3,4-Dichloroisocoumarin	DCI	Many serine proteases
Iodoacetate	IAA	Many cysteine proteases
Iodoacetamide	IAM	Most cysteine proteases
N-Ethyl maleimide	NEM	Most cysteine proteases
p-Hydroxymercuribenzoate	pMB	Most cysteine proteases
p-Hydroxymercuriphenylsulfonate	pMBS	Most cysteine proteases
Ethylenediamine tetraacetic acid	EDTA	Most metalloproteases
Ethyleneglycol bis (β-aminoethyl ether)-N,N,N',N'-tetraacetic acid	EGTA	Ca^{2+}-dependent proteases
1,10-Phenanthroline	ophen	Many metalloproteases
Diazoacetyl-DL-norleucine methyl ester + Cu^{2+}	—	Many aspartic proteases

teases. Unless these inhibitors contain groups reacting irreversibly with the active site of the protease, e.g., the chloromethyl ketones, they usually act as reversible inhibitors. A representative list of commercially available oligopeptide inhibitors is given in Table 5.

3. Protein-Proteinase Inhibitors

Protease-inhibiting proteins are widely distributed and are present in animal tissues and body fluids, plants, and microorganisms (178,179). Since they are indeed proteins, they should be substrates rather than inhibitors of proteases. However, it has been established that many such inhibitors work according to a standard mechanism (179): the inhibitor is bound very tightly to the protease. On the surface of the inhibitor protein, there is at least one peptide bond specifically interacting with the active center of the protease. The reactive bond is cleaved with high catalytic efficiency (k_{cat}/K_m is large), but the individual values of k_{cat} and K_m are very low, so that hydrolysis occurs very slowly.

Table 5 Oligopeptide and Related Inhibitors

Protease inhibitor	Proteases inhibited (examples)	Solubility
Antipain	Papain, trypsin, cathepsin B *not* chymotrypsin, elastase, pepsin	Soluble in water, methanol, DMSO
Bestatin	Many aminopeptidases	Soluble to 20 mg/ml in HCl (1 mol/liter), to 5 mg/ml in methanol, and 1 mg/ml in 0.15 mol/liter NaCl
Calpain inhibitor I (*N*-acetyl-Leu-Leu-norleucinal)	Calcium-dependent neutral protease (calpain)	Soluble in water
Calpain inhibitor II (*N*-acetyl-Leu-Leu-Leu-norleucinal)	Calcium-dependent neutral protease (calpain)	Soluble in water
Chymostatin	Chymotrypsin, papain, cathepsin B, *not* trypsin, plasmin, thrombin, kallikrein	Soluble in glacial acetic acid and DMSO, sparingly soluble in water, methanol
Elastatinal	Elastase, *not* trypsin, chymotrypsin, papain, pepsin, thermolysin	Soluble in ethanol and acetone
E-64	Many cysteine proteases	Soluble to 20 mg/ml in a 1:1 (v/v) mixture of ethanol and water
Leupeptin	Many serine and cysteine proteases (e.g. trypsin, plasmin, endoproteinase Lys-C, kallikrein, papain, cathepsins B and L, calpain), *not* chymotrypsin, elastase, pepsin, thrombin	Well soluble in water
Pepstatin A	Aspartic proteases (e.g., pepsin, cathepsin D, renin, penicillopepsin), *not* serine, cysteine, or metalloproteases	Soluble in methanol to approx. 1 mg/ml
Phosphoramidon	Metalloproteases (e.g., thermolysin, collagenase), *not* serine, cysteine, or aspartic proteases	Salts of phosphoramidon are soluble to 20 mg/ml in water; also soluble in methanol and DMSO

Furthermore, cleavage of the inhibitor's reactive site does not occur completely, but an equilibrium between native ("virgin") and modified inhibitor is reached, both of which can bind to the protease.

Protein-protease inhibitors mostly possess important functions in the regulation of biological processes involving proteolysis, such as coagulation, complement activation, or inflammation (172). A list of some protease-inhibiting proteins is shown in Table 6.

Since most of the protease inhibitors only inhibit proteases of one (or at most two) classes, it is generally necessary to use mixtures (or "cocktails") of a number of protease inhibitors with different specificities unless the class of the disturbing protease(s) has been determined by the use of group-specific inhibitors before. When working with prokaryotic expression systems, the use of inhibitors of aspartic proteases should not be necessary since these proteases have as yet only been found in eukaryotes and viruses.

Table 6 Protein-Protease Inhibitors

Protease inhibitor	Source	Mol. mass 10^3 Da	Proteases inhibited (examples)
α_1-Proteinase inhibitor	Serum (2.9 g/liter)[a]	51	Many serine proteases (trypsin, elastase, chymotrypsin, etc.)
α_2-Macroglobulin	Serum (2.6 g/liter)	725	Nearly all endoproteinases
Antithrombin III	Serum (0.235 g/liter)	65	Trypsin, kallikrein, plasmin, coagulation factors IXa, Xa, XIa, XIIa, Thrombin
α_2-Antiplasmin	Serum (0.07 g/liter)	65	Plasmin, kallikrein, coagulation factors XIa, XIIa, thrombin, trypsin
Cystatins A,B,C	Serum, liver, chicken egg	11–13	Many cysteine proteases
Eglin C	Leeches	8.1	Elastase, cathepsin G, chymotrypsin, subtilisin
Hirudin	Leeches	7	Thrombin, factor IXa
Trypsin inhibitor (aprotinin)	Bovine lung	6.5	Trypsin, plasmin, endoproteinase Lys-C, kallikrein, factor XIa, chymotrypsin
Trypsin inhibitor (ovomucoid)	Chicken egg white	28	Trypsin, chymases I and II
Trypsin inhibitor (STI)	Soy beans	20.1	Trypsin, plasmin, factors Xa, XIa, chymases I and II, cathepsin G

[a] Normal values in human serum.

Examples for protease inhibitor cocktails, which have been successfully used in the purification of proteins from various sources, are the following (the final concentrations in the solution containing the target protein to be purified are given):

1. TLCK, 1 mmol/liter; TPCK, 1 mmol/liter; PMSF, 1 mmol/liter; EDTA, 1.5 mmol/liter; EGTA, 1.5 mmol/liter [used for isolation of precursors of mitochondrially-synthesized yeast proteins (184)]
2. EDTA, 1 mmol/liter; leupeptin, 0.5 mg/liter; pepstatin, 0.68 mg/liter; PMSF, 0.2 mmol/liter [used in the purification of a highly protease-sensitive multienzyme complex from yeast (185)]
3. EGTA, 5 mmol/liter; PMSF, 1 mmol/liter; aprotinin, 1 mmol/liter; leupeptin, 1 mmol/liter; phosphoramidon, 1 mmol/liter; pepstatin, 1 mmol/liter; antipain, 1 mmol/liter; bacitracin, 0.1% [used to isolate atrial natriuretic hormone receptor from rabbit plasma membranes (186)].

REFERENCES

1. Reich E, Rifkin DB, Shaw E. Proteases and biological control. New York: Cold Spring Harbor Laboratory, 1975.
2. Neurath H, Walsh KA. Role of proteolytic enzymes in biological regulation. Proc Natl Acad Sci USA 1976; 73:3825–3832.
3. Holzer H, Tschesche H. Biological functions of proteases. Berlin: Springer-Verlag, 1979.
4. Hörl WH, Heidland A. Proteases: Potential role in health and disease. New York: Plenum Press, 1984.
5. Mullins DE, Rifkin DB. Proteases in metastasis and angiogenesis. In: Developments in cell biology: Secretory processes. Dean RT, Stahl P. eds. London: Butterworths, 1985:159–177.
6. Hartley BS. Proteolytic enzymes. Ann Rev Biochem 1960; 29:45–72.
7. Mihalyi E. Application of proteolytic enzymes to protein structure studies. Boca Raton: CRC Press, 1978:43–113.
8. Barrett AJ. The many forms and functions of cellular proteinases. Fed Proc 1980; 39:9–14.
9. Barrett AJ, Salvesen G. Proteinase inhibitors. Amsterdam: Elsevier, 1986.
10. Lazure C, Leduc R, Seidah NG, Thibault G, Genest J, Chrétien M. Amino acid sequence of rat submaxillary tonin reveals similarities to serine proteases. Nature 1984; 307:555–558.
11. Polgar L, Halasz P. Current problems in mechanistic studies of serine and cysteine proteinases. Biochem J 1982; 207:1–10.
12. Willenbrock F, Brocklehurst K. A general framework of cysteine-proteinase mechanism deduced from studies on enzymes with structurally different analogous catalytic-site residues Asp-158 and -161 (papain and actinidin), Gly-196 (cathepsin B) and Asn-165 (cathepsin H). Biochem J 1985; 227:521–528.

13. Pearl LH. The catalytic mechanism of aspartic proteinases. FEBS Lett 1987; 214:8–12.
14. Polgár L. The machanism of action of aspartic proteases involves "push-pull" catalysis. FEBS Lett 1987; 219:1–4.
15. Kester WR, Matthews BJ. Crystallographic study of the binding of dipeptide inhibitors to thermolysin: Implications for the mechanism of catalysis. Biochemistry 1977; 16:2506–2516.
16. Vallee BL, Galdes A, Auld DS, Riordan JF. Carboxypeptidase A. In: Zinc proteins. Spiro TG ed. Vol. 5. New York: John Wiley and Sons, 1983:25–75.
17. Zwilling R, Dörsam H, Torff H-J, Rödl J. Low molecular mass protease: evidence for a new family of proteolytic enzymes. FEBS Lett 1981; 127:75–78.
18. Titani K, Torff H-J, Hormel S, Kumar S, Walsh KA, Rödl J, Neurath H, Zwilling R. Amino acid sequence of a unique protease from the crayfish *Astacus fluvialis*. Biochemistry 1987; 26:222–226.
19. Schechter I, Berger A. On the size of the active site in proteases. I. Papain. Biochem. Biophys. Res. Commun. 1967; 27:157–162.
20. Magnusson S, Petersen TE, Sottrup-Jensen L, Claeys H. Complete primary structure of prothrombin: Isolation, structure and reactivity of ten carboxylated glutamic acid residues and regulation of prothrombin activation by thrombin. In: Reich E, Rifkin D.B., Shaw E. eds. Proteases and biological control. New York: Cold Spring Harbor Laboratory, 1975:123–149.
21. Lottenberg R, Hall JA, Pautler E, Zupan A, Christensen U, Jackson CM. The action of factor Xa on peptide p-nitroanilide substrates: Substrate selectivity and examination of hydrolysis with different reaction conditions. Biochim. Biophys. Acta 1986; 874:326–336.
22. Fruton JS. The specificity of proteinases toward protein substrates. In: Reich E, Rifkin DB, Shaw E. Proteases and biological control. New York: Cold Spring Harbor Laboratory, 1975:33–50.
23. Neurath H. Limited proteolysis, protein folding and physiological regulation. In: Protein folding. Jaenicke R, ed. Amsterdam: Elsevier, 1980:501–523.
24. Fontana A, Fassina G, Vita C, Dalzoppo D, Zamai M, Zambonin M. Correlation between sites of limited proteolysis and segmental mobility in thermolysin. Biochemistry 1986; 25:1847–1851.
25. Novotny J, Bruccoleri RE. Correlation among sites of limited proteolysis, enzyme accessibility and segmental mobility. FEBS Lett 1987; 211:185–189.
26. Beynon RJ, Place GA, Butler PE. Limited proteolysis of enzymes: The generation of functionally modified derivatives in vitro and in vivo. Biochem. Soc. Trans. 1985; 13:306–308.
27. Mayer RJ, Doherty F. Intracellular protein catabolism: State of the art. FEBS Lett 1986; 198:181–193.
28. Horowitz PM, Bowman S. Oxidation increases the proteolytic susceptibility of a localized region in rhodanese. J Biol Chem 1987; 262:14544–14548.
29. Rivett AJ, Levine RL. Enhanced proteolytic susceptibility of oxidized proteins. Biochem. Soc. Trans. 1987; 15:816–818.
30. Dean RT. A mechanism for accelerated degradation of intracellular proteins after limited damage by free radicals. FEBS Lett 1987; 220:278–282.

31. Goldberg AL, Swamy KHS, Chung CH, Larimore FS. Proteases in *Escherichia coli*. Meth Enzymol 1981; 80:680–702.
32. Pacaud M. Purification and characterization of two novel proteolytic enzymes in membranes of *Escherichia coli*. Protease IV and Protease V. J Biol Chem 1982; 257:4333–4339.
33. Wolfe PB, Silver P, Wickner W. The isolation of homogenous leader peptidase from a strain of *Escherichia coli* which overproduces the enzyme. J Biol Chem 1982; 257:7898–7902.
34. Suzuki T, Itoh A, Ichihara S, Mizushima S. Characterization of the sppA gene coding for Protease IV, a signal peptide peptidase of *Escherichia coli*. J Bacteriol 1987; 169:2523–2528.
35. Goldberg AL, St. John AC. Intracellular protein degradation in mammalian and bacterial cells, Part 2. Ann Rev Biochem 1976; 45:747–803.
36. Mount DW. The genetics of protein degradation in bacteria. Annu Rev Genet 1980; 14:279–294.
37. Chang YSE, Kwok DY, Kwok TJ, Solvest BC, Zipser D. Gene 1981; 14:121–130.
38. Goldberg AL, Goff SA. The selective degradation of abnormal proteins in bacteria. In: Reznikoff W, Gold L, eds. Maximizing gene expression. Boston: Butterworths, 1986: 287–314.
39. Chung CH, Goldberg AL. The product of the *lon* (*capR*) gene in *Escherichia coli* is the ATP-dependent protease, protease La. Proc Natl Acad Sci USA 1981; 78:4931–4935.
40. Charette MF, Henderson GW, Marowitz A. ATP hydrolysis-dependent protease activity of the *lon* (*capR*) protein of *Escherichia coli* K-12. Proc Natl Acad Sci USA 1981; 78:4728–4732.
41. Goff SA, Goldberg AL. Production of abnormal proteins in *E. coli* stimulates transcription of *lon* and other heat shock genes. Cell 1985; 41:587–595.
42. Goff SA, Goldberg AL. An increased content of protease La, the lon gene product, increases protein degradation and blocks growth in *Escherichia coli*. J Biol Chem 1987; 262:4508–4515.
43. St. John AC, Conklin K, Rosenthal E, Goldberg AL. Further evidence for the involvement of charged tRNA and Guanosine tetraphosphate in the control of protein degradation in *Escherichia coli*. J Biol Chem 1978; 253:3945–3951.
44. Goff SA, Casson LP, Goldberg AL. Heat shock regulatory gene *htpR* influences rates of protein degradation and expression of the *lon* gene in *Escherichia coli*. Proc Natl Acad Sci USA 1984; 81:6647–6651.
45. Neidhardt FC, VanBogelen RA, Vaughn V. The genetics and regulation of heat-shock proteins. Ann Rev Genet 1984; 18:295–330.
46. Grossman AD, Erickson JW, Gross CA. The *htpR* gene product of *E. coli* is a sigma factor for heat-shock promoters. Cell 1984; 38:383–390.
47. Philipps TA, VanBogelen RA, Neidhardt FC. *lon* gene product of *Escherichia coli* is a heat-shock protein. J Bacteriol 1984; 159:283–287.
48. Baker TA, Grossman AD, Cross CA. A gene regulating the heat shock response in *Escherichia coli* also affects proteolysis. Proc Natl Acad Sci. USA 1984; 81:6779–6783.

49. Maurizi MR, Trisler P, Gottesman S. Insertional mutagenesis of the *lon* gene in *Escherichia coli*: *lon* is dispensable. J Bacteriol 1985; 164:1124–1135.
50. Katayama-Fujimura Y, Gottesman S, Maurizi MR. A multiple-component, ATP-dependent protease from *Escherichia coli*. J Biol Chem 1987; 262:4477–4485.
51. Hwang BJ, Park WJ, Chung CH, Goldberg AL. *Escherichia coli* contains a soluble ATP-dependent protease (Ti) distinct from protease La. Proc Natl Acad Sci USA 1987; 84:5550–5554.
52. Doi RH, Wong S-L, Kawamura F. Potential use of *Bacillus subtilis* for secretion and production of foreign proteins. Trends Biotechnol 1986; 4:232–235.
53. Keay L. Proteases of the genus *Bacillus*. Proc. IV. IFS: Ferment. Techol. Today, 1972; 289–298.
54. Priest FG. Extracellular enzyme synthesis in the genus *Bacillus*. Bacteriol Rev 1977; 41:711–753.
55. Stahl ML, Ferrari E. Replacement of the *Bacillus subtilis* subtilisin structural gene with an in vitro-derived deletion mutation. J Bacteriol 1984; 158:411–418.
56. Yang MY, Ferrari E, Henner DJ. Cloning of the neutral protease gene of *Bacillus subtilis* and the use of the cloned gene to create an in vitro-derived deletion mutation. J Bacteriol 1984; 160:15–21.
57. Roitsch CA, Hageman JH. Bacillopeptidase F: Two forms of a glycoprotein serine protease from *Bacillus subtilis* 168. J Bacteriol 1983; 155:145–152.
58. Hiroishi S, Kadota H. Intracellular proteases of *Bacillus subtilis*. Agric Biol Chem 1976; 40:1047–1049.
59. Strongin AY, Izotova LS, Abramov ZT, Gorodetsky DI, Ermakova LH, Baratova LA, Belyanova LP, Stepanov VM. Intracellular serine protease of *Bacillus subtilis*: Sequence homology with extracellular subtilisins. J Bacteriol 1978; 133:1401–1411.
60. Srivastava OP, Aronson AI. Isolation and characterization of a unique protease from sporulating cells of *Bacillus subtilis*. Arch Microbiol 1981; 129:227–232.
61. Koide Y, Nakamura A, Uozumi T, Beppu T. Cloning and sequencing of the major intracellular serine protease gene of *Bacillus subtilis*. J Bacteriol 1986; 167:110–116.
62. Wolf DH. Proteinases, proteolysis and regulation in yeast. Biochem Soc Trans 1984; 13:279–283.
63. Wolf DH. Control of metabolism in yeast and other lower eukaryotes through action of proteinases. Adv Microb Physiol 1980; 21:267–338.
64. Achstetter T, Wolf DH. Proteinases, proteolysis and biological control in the yeast *Saccharomyces cerevisiae*. Yeast 1985; 1:139–157.
65. Dmochowska A, Dignard D, Henning D, Thomas, D.Y., Bussey H. Yeast KEX1 gene encodes a putative protease with a carboxypeptidase B-like function involved in killer toxin and α-factor precursor processing. Cell 1987; 50:573–584.
66. Wagner J-C, Escher C, Wolf DH. Some characteristics of hormone (pheromone) processing enzymes in yeast. FEBS Lett 1987; 218:31–34.
67. Wiemken A, Schellenberg M, Urech K. Vacuoles: The sole compartment of digestive enzymes in yeast (*Saccharomyces cerevisiae*)? Arch Microbiol 1979; 123:23–35.

68. Teichert U, Mechler B, Müller H, Wolf DH. Protein degradation in yeast. Biochem Soc Trans 1987; 15:811–815.
69. Wolf DH, Holzer H. Proteolysis in yeast. In: Payne JW, ed. Transport and utilization of amino acids, peptides and proteins by microorganisms. Chichester: John Wiley and Sons, 1980: 431–458.
70. Barrett AJ, McDonald JK. Mammalian proteases: A glossary and bibliography. Vol.1: Endopeptidases. London: Academic Press, 1980.
71. McDonald JK, Barrett AJ. Mammalian proteases: A glossary and bibliography. Vol.2: Exopeptidases. London: Academic Press, 1986.
72. Barrett AJ, Dingle JT. Tissue proteinases. Amsterdam: North-Holland, 1971.
73. Barrett AJ (ed.), Proteinases in mammalian cells and tissues. Amsterdam: North-Holland, 1977.
74. Bond JS, Butler PE. Intracellular proteases. Ann Rev Biochem 1987; 56: 333–364.
75. Dice JF. Molecular determinants of protein half-lives in eukaryotic cells. FASEB J 1987; 1:349–357.
76. Hershko A, Ciechanover A. Mechanisms of intracellular protein breakdown. Ann Rev Biochem 1982; 51:335–364.
77. Ciechanover A, Finley D, Varshavsky A. The ubiquitin-mediated proteolytic pathway and mechanisms of energy-dependent intracellular protein degradation. J Cell Biochem 1984; 24:27–53.
78. Finley D, Varshavsky A. The ubiquitin system: Functions and mechanism. Trends Biochem Sci 1985; 10:343–346.
79. Hershko A, Ciechanover A. The ubiquitin pathway for the degradation of intracellular proteins. Prog Nucl Acids Res 1986; 33:19–56.
80. Waxman L, Fagan J, Goldberg A. Demonstration of two distinct high molecular weight proteases in rabbit reticulocytes, one of which degrades ubiquitin conjugates. J Biol Chem 1987; 262:2451–2457.
81. Fried VA, Smith HT, Hildebrandt E, Weiner K. Ubiquitin has intrinsic proteolytic activity: Implications for cellular regulation. Proc Natl Acad Sci USA 1987; 84:3685–3689.
82. Hershko A, Heller H, Eytan E, Kaklu G, Rose I. Role of the α-amino group of protein in ubiquitin-mediated protein breakdown. Proc Natl Acad Sci USA 1984; 81:7021–7025.
83. Murakami K, Etlinger JD. Degradation of proteins with blocked amino groups by cytoplasmic proteases. Biochem Biophys Res Commun 1987; 146: 1249–1255.
84. Rechsteiner M, Rogers S, Rote K. Protein structure and intracellular stability. Trends Biochem Sci 1987; 90–94.
85. Bachmair A, Finley D, Varshavsky A. In vivo half-life of a protein is a function of its amino-terminal residue. Science 1986; 234:179–186.
86. Rogers S, Wells R, Rechsteiner M. Amino acid sequences common to rapidly degraded proteins: The PEST hypothesis. Science 1986; 234:364–368.
87. Murachi T, Tanaka K, Hatanaka M, Murakami T. Intracellular Ca^{2+}-dependent protease (calpain) and its high-molecular-weight endogenous inhibitor (calpastatin). Adv Enzyme Regul 1981; 19:407–424.

88. Murachi T. The proteolytic system involving calpains. Bio Soc Trans 1985; 13:1015-1018.
89. Fritz H, Seemüller U, Tschesche H, Proteinases and their inhibitors: general review. In: Bergmeyer HU, Grassl M, Bergmeyer J, eds. Methods of enzymatic analysis, 3rd ed, Vol 5. Weinheim: Verlag Chemie, 1984:74-98.
90. Wunderwald, P. Proteinases (proteins as substrates). In: Bergmeyer HU, Grassl M, Bergmeyer J, eds. Methods of enzymatic analysis, 3rd ed, Vol 5. Weinheim: Verlag Chemie, 1984:258-270.
91. Wunderwald P, Schrenk J. Method with fibrin as substrate (fibrin plate assay). In: Bergmeyer HU, Grassl M, Bergmeyer J, eds. Methods of enzymatic analysis, 3rd ed, Vol 5. Weinheim: Verlag Chemie, 1984:285-289.
92. Granelli-Piperno A, Reich E. A study of proteases and protease-inhibitor complexes in biological fluids. J Exp Med 1978; 148:223-234.
93. Schumacher GBF, Schill WB. Radial diffusion in gel for micro determination of enzymes. Anal Biochem 1972; 48:9-26.
94. Njus D, Baldwin TO, Hastings JW. A sensitive assay for proteolytic enzymes using bacterial luciferase as a substrate. Anal Biochem 1974; 61:280-287.
95. Reimerdes EH, Klostermeyer H. Determination of proteolytic activities on casein substrates. In: Wood AW, ed. Methods in Enzymology, Vol XLV, Part B. New York: Academic Press, 1976:26-28.
96. Coleman PL, Latham HG Jr, Shaw EN. Some sensitive methods for the assay of trypsin-like enzymes. In: Hirs CHW, Timasheff SN, eds. Methods in Enzymology, Vol XLV, Part B. New York: Academic Press, 1976:12-26.
97. Habeeb AFSA. Determination of free amino groups in proteins by trinitrobenzenesulfonic acid. Anal Biochem 1966; 14:328-336.
98. Walter H-E. Method with haemoglobin, casein and azocoll as substrate. In: Bergmeyer HU, Grassl M, Bergmeyer J, eds. Methods of enzymatic analysis, 3rd ed, Vol 5. Weinheim: Verlag Chemie, 1984:270-277.
99. Chavira R Jr, Burnett TJ, Hageman JH. Assaying proteinases with azocoll. Anal Biochem 1984; 136:446-450.
100. De Lumen BO, Tappel AL. Fluorescein-hemoglobin as a substrate for cathepsin D and other proteases. Anal Biochem 1970; 36:22-29.
101. Twining SS. Fluorescein isothiocyanate-labeled casein assay for proteolytic enzymes. Anal Biochem 1984; 143:30-34.
102. Varani J, Johnson K, Kaplan J. Development of a solid-phase assay for measurement of proteolytic enzyme activity. Anal Biochem 1980; 107:377-384.
103. Bosmann HB. Elevated glycosidases and proteolytic enzymes in cells transformed by RNA tumor virus. Biochim Biophys Acta 1972; 264:339-343.
104. Andrews AT. Method with immobilized enzyme-labelled proteins as substrates. In: Bergmeyer HU, Grassl M, Bergmeyer J, eds. Methods of enzymatic analysis, 3rd ed. Vol 5. Weinheim: Verlag Chemie, 1984:289-297.
105. Oliver CN, Stadtman ER. A proteolytic artifact associated with the lysis of bacteria by egg white lysozyme. Proc Natl Acad Sci USA 1983; 80:2156-2160.
106. Sato M, Morimoto H, Oohashi T. Increase in intracellular proteinase activities in extinct cells of *Saccharomyces cerevisiae*. Agric Biol Chem 1987; 51: 2609-2610.

107. Gottesman S, Zipser D. *deg* phenotype of *E. coli lon* mutants. J Bacteriol 1978; 133:844–851.
108. Kowit JD, Goldberg AL. Intermediate steps in the degradation of a specific abnormal protein in *Escherichia coli*. J Biol Chem 1977; 252:8350–8357.
109. Chung CH, Waxman L, Goldberg AL. Studies on the protein encoded by the *lon* mutation capR9 in *Escherichia coli*: A labile form of the ATP-dependent protease La that inhibits the wild-type protease. J Biol Chem 1983; 258:215–221.
110. Grossman AD, Zhou Y-N, Gross C, Heilig J, Christie GE, Calendar R. Mutations in the *rpoH (htpR)* gene of *Escherichia coli* K-12 phenotypically suppress a temperature-sensitive mutant defective in the σ^{70} subunit of RNA polymerase. J Bacteriol 1985; 161:939–943.
111. Goldberg AL, Goff SA, Casson LP. Hosts and methods for producing recombinant proteins in high yields. PCT patent application no. WO 85/03949, 1984.
112. Debouck CM, Ho YS, Rosenberg M. Proteolytic deficient *E. coli*. Eur. patent application no. 0216747, 1985.
113. Carr AJ, Rosenberger RF, Hipkiss AR. *Escherichia coli* mutants defective in the degradation of abnormal proteins. Biochem Soc Trans 1985; 13:337.
114. Sugimura K, Sugimoto S, Shirasawa H. Novel host *E. coli* and use thereof. Eur patent application no. 0230670, 1985.
115. Buell GN, Schulz MF, Selzer G, Chollet A, Movva NR, Semon D, Escanez S, Kawashima E. Optimizing the expression in *Escherichia coli* of a synthetic gene encoding somatomedin-C (IGF-I). Nucl Acids Res 1985; 13:1923–1938.
116. Donch K, Greenberg J. Ultraviolet sensitivity genes in *Escherichia coli*. J Bacteriol 1986; 95:1555–1559.
117. Donch J, Chung YS, Greenberg J. Locus for radiation resistance in *Escherichia coli* strain B/r. Genetics 1969; 61:363–370.
118. Sharma SK. On the recovery of genetically engineered proteins from *Escherichia coli*. Sep Sci Technol 1986; 21:701–726.
119. Carrier MJ, Nugent ME, Tacon WCA, Primrose SB. High expression of cloned genes in *Escherichia coli*. Trends Biotechnol 1983; 1:109–113.
120. Band L, Henner DJ, Ruppen M. Construction and properties of an intracellular serine protease mutant of *Bacillus subtilis*. J Bacteriol 1987; 169:444–446.
121. Bott RB, Ferrari E, Wells JA, Estell DA, Henner DJ. Bacillus incapable of excreting subtilisin or neutral protease. Eur Patent Application no. 0246678, 1984.
122. Kawamura F, Doi, RH. Construction of a *Bacillus subtilis* double mutant deficient in extracellular akaline and neutral proteases. J Bacteriol 1984; 160: 442–444.
123. Honjo M, Nakayama A, Iio A, Mita I, Kaweamura K, Sawakura A, Furutani Y. Construction of a highly efficient host-vector system for secretion of heterologous protein in *Bacillus subtilis*. J Biotechnol 1987; 6:191–204.
124. Cunningham BC, Wells JA. Improvement of the alkaline stability of subtilisin using an efficient random mutagenesis and screening procedure. Prot Eng 1987; 1:319–325.
125. Hemmings BA, Zubenko GS, Jones EW. Proteolytic inactivation of the NADP-dependent glutamate dehydrogenase in proteinase-deficient mutants of *Saccharomyces cerevisiae*. Arch Biochem Biophys 1980; 202:657–660.

126. Achstetter T, Ehmann C, Wolf DH. New proteolytic enzymes in yeast. Arch Biochem Biophys 1981; 207:445–454.
127. Achstetter T, Emter O, Ehmann C, Wolf DH. Proteolysis in eukaryotic cells: Identification of multiple proteolytic enzymes in yeast. J Biol Chem 1984; 259:13334–13343.
128. Simon LD, Randolph B, Irwin N, Binkowski G. Stabilization of proteins by a bacteriophage T4 gene cloned in *Escherichia coli*. Proc Natl Acad Sci USA 1983; 80:2059–2062.
129. Shub DA, Casna NJ. Stabilising foreign proteins, and plasmids for use therein. Eur. patent application no. 0133044, 1983.
130. Bochner BR, Olson KC, Pai R-C. Periplasmic protein recovery. U.S. patent no. 4,680,262, 1984.
131. Smith RA, Duncan MJ, Moir DT. Heterologous protein secretion from yeast. Science 1985; 229:1219–1224.
132. Kingsman SM, Kingsman AJ, Mellor J. The production of mammalian proteins in *Saccharomyces cerevisiae*. Trends Biotechnol 1987; 5:53–57.
133. Chang SJ, Weiss J, Konrad M, White T, Bahl C, Yu S-D, Mark D, Steiner DF. Biosynthesis and periplasmic segregation of human proinsulin in *Escherichia coli*. Proc Natl Acad Sci USA 1981; 78:5401–5405.
134. Piérard L, Jacobs P, Gheysen D, Hoylaerts M, André B, Topisirovic L, Cravador A, de Foresta F, Herzog A, Collen D, De Wilde M, Bollen A. Mutant and chimeric recombinant plasminogen activators. Production in eukaryotic cells and preliminary characterization. J Biol Chem 1987; 262:11771–11778.
135. Tate KM, Higgins DL, Holmes WE, Winkler ME, Heynecker HL, Vehar GA. Functional role of proteolytic cleavage at arginine-275 of human tissue plasminogen activator as assessed by site-directed mutagenesis. Biochemistry 1987; 26:338–343.
136. Talmadge K, Gilbert W. Cellular location affects protein stability in *Escherichia coli*. Proc Natl Acad Sci USA 1982; 79:1830–1833.
137. Silhavy TJ, Beckwith JR. Use of *lac* fusions for the study of biological problems. Microbiol Rev 1985; 49:398–418.
138. Itakura K, Hirose T, Crea R, Riggs AD, Heynecker HL, Bolivar F, Boyer HW. Expression in *Escherichia coli* of a chemically synthesized gene for the hormone somatostatin. Science 1977; 198:1056–1063.
139. Shine J, Fettes I, Lan NCY, Roberts JL, Baxter JD. Expression of cloned β-endorphin gene sequences by *E. coli*. Nature 1980; 285:456–461.
140. Wetzel R, Heynecker H, Goeddel DV, Jhurani P, Shapiro J, Crea R, Low TLK, McClure JE, Thurman GB, Goldstein AL. Production of biologically active N-desacetylthymosin α1 in *E. coli* through expression of a chemically synthesized gene. Biochemistry 1980; 19:6096–6104.
141. Goeddel DV, Kleid DG, Bolivar F, Heynecker HL, Yansura DG, Crea R, Hirose T, Kraszewsky A, Itakura K, Riggs AD. Expression in *E. coli* of chemically synthesized genes for human insulin. Proc Natl Acad Sci USA 1979; 76:106–110.
142. Ellis J, Ozaki LS, Gwadz RW, Cochrane AH, Nussenzweig RS, Godson GN. Cloning and expression in *E. coli* of the malarial sporozoite surface antigen gene from *Plasmodium knowlesi*. Nature 1983; 302:536–538.

143. Leonardo MJ, Brentano ST, Donelson JD. Expression of antigenic regions of a trypanosome variable surface glycoprotein in *E. coli* using Bal-31 nuclease digestion. Nucl Acids Res 1984; 12:4637–4652.
144. Van der Werf S, Dreano M, Bruneau P, Kopecka H, Girard M. Expression of poliovirus capsid polypeptide VP1 in *E. coli*. Gene 1983; 23:85–93.
145. Kiyokawa T, Yoshikura H, Hattori S, Seiki M, Yoshida M. Envelope proteins of human T-cell leukemia virus: Expression in *E. coli* and its application to env gene functions. Proc Natl Acad Sci USA 1984; 81:6202–6206.
146. Watson RJ, Weis JH, Salstrom JS, Enquist LW. Bacterial synthesis of herpes simplex virus types 1 and 2 glycoprotein D antigens. J Invest Dermat 1984; 83:102s–111s.
147. Fowler AV, Zabin I. Purification, structure, and properties of hybrid β-galactosidase proteins. J Biol Chem 1983; 258:14354–14358.
148. Germino JJ, Gary JG, Charbonneau H, Vanaman T, Bastia D. Use of gene fusions and protein-protein interaction in the isolation of a biologically active regulatory protein: The replication initiator protein of plasmid R6K. Proc Natl Acad Sci USA 1983; 80:6848–6852.
149. Ullman A. One-step purification of hybrid proteins which have β-galactosidase activity. Gene 1984; 29:27–31.
150. Germino J, Bastia D. Rapid purification of a cloned gene product by genetic fusion and site-specific proteolysis. Proc Natl Acad Sci USA 1984; 81:4692–4696.
151. Szoka PR, Schreiber AB, Chan H, Murthy J. A general method for retrieving the components of a genetically engineered fusion protein. DNA 1986; 5:11–20.
152. Nagai K, Thøgersen HC. Generation of β-globin by sequence-specific proteolysis of a hybrid protein produced in *Escherichia coli*. Nature 1984; 309:810–812.
153. Shen S-H. Multiple joined genes prevent product degradation in *Escherichia coli*. Proc Natl Acad Sci USA 1984; 81:4627–4631.
154. Nakayama T, Munoz L, Doi RH. A procedure to remove protease activities from *Bacillus subtilis* sporulating cells and their crude extracts. Anal Biochem 1977; 78:165–170.
155. Gallop TM, Seifter S, Meilman E. Studies on collagen. I. The partial purification, assay, and mode of activation of bacterial collagenase. J Biol chem 1957; 227:891.
156. Safarik I. Affinity chromatography of trypsin on thermally modified casein. J Chromatogr 1983; 261:138–141.
157. Hixson HF, Nishikawa A. Affinity chromatography of trypsin and thrombin. Arch Biochem Biophys 1973; 154:501–509.
158. Varady A, Patthy L. Kringle 5 of human plasminogen carries a benzamidine-binding site. Biochem Biophys Res Commun 1981; 103:97–102.
159. Büttner W, Kraeft U, Etzold G, Verfahren zur Herstellung proteasefreier Glucoseoxidase. GDR patent no. 204944, 1982.
160. Patel AH, Schultz RM. Transition-state affinity purification of proteases; the preparation of an argininal affinity resin for the selective binding of trypsin-like proteases. Biochem Biophys Res Commun 1982; 104:181–186.

161. Stepanov VM, Rudenskaya GN. Proteinase affinity chromatography on bacitracin-Sepharose. J Appl Biochem 1983; 5:420–428.
162. Barrett AJ, Starkey PM. The interaction of α_2-macroglobulin with proteinases. Characteristics and specificity of the reaction, and a hypothesis concerning its molecular mechanism. Biochem J 1983; 13:709–724.
163. Wunderwald P, Schrenk WJ, Port H, Kresze G-B. Removal of endoproteinases from biological fluids by "sandwich affinity chromatography" with α_2-macroglobulin bound to zinc chelate-Sepharose. J Appl Biochem 1983; 5:31–42.
164. Hedman P, Gustafsson J-G. Verminderung der Proteaseaktivität in Zellhomogenaten aus Mikrooroganismen. Deutsche Pharmacia GmbH, Freiburg i.Br., 1984.
165. Kula M-R, Kroner KH, Hustedt H. Purification of enzymes by liquid-liquid extraction. In: Fiechter A, ed. Advances in biochemical engineering. Berlin: Springer-Verlag, 1982:73–118.
166. Mattiasson B. Applications of aqueous two-phase systems in biotechnology. Trends Biotechnol 1983; 1:16–20.
167. Johansson G. Aqueous two-phase systems in protein purificaton. J Biotechnol 1985; 3:11–18.
168. Wolf DH, Ehmann C. Carboxypeptidase S- and carboxypeptidase Y-deficient mutants of *Saccharomyces cerevisiae*. J Bacteriol 1981; 147:418–426.
169. Veide A, Strandberg L, Enfors S-O. Extraction of β-galactosidase fused protein A in aqueous two-phase systems. Enzyme Microb Technol 1987; 9:730–738.
170. Fritz H, Tschesche H, Greene LJ, Truscheit E. eds. Proteinase inhibitors. Berlin: Springer-Verlag, 1974.
171. Katunuma N, Umezawa H, Holzer H. Proteinase inhibitors: Medical and biological aspects. Tokyo: Japan Scientific Society Press, and Berlin: Springer-Verlag, 1983.
172. Barrett AJ, Salvesen G. Proteinase inhibitors. Amsterdam: Elsevier, 1986.
173. Knight CG. The characterization of enzyme inhibition. In: Barrett AJ, Salvesen G, eds. Proteinase inhibitors. Amsterdam: Elsevier, 1986:23–51.
174. Powers JC, Harper JW. Inhibitors of serine proteinases. In: Barrett AJ, Salvesen G. Proteinase inhibitors. Amsterdam: Elsevier, 1986:55–152.
175. Lundblad RL, Stein WH. On the reaction of diazoacetyl compounds with pepsin. J Biol Chem 1968; 244:154–160.
176. Umezawa H. Enzyme inhibitors of microbial origin. Tokyo: University of Tokyo Press, 1972.
177. Aoyagi T, Umezawa H. Structures and activities of protease inhibitors of microbial origin. In: Reich E, Rifkin DB, Shaw E, eds. Proteases and biological control. New York: Cold Spring Harbor Laboratory, 1975: 429–454.
178. Tschesche H. Biochemie natürlicher Proteinase-Inhibitoren. Angew Chemie 1974; 86:21–40.
179. Laskowski M Jr, Kato I. Protein inhibitors of proteinases. Annu Rev Biochem 1980; 49:593–626.
180. Laura R, Robison DJ, Bing DH. (*p*-Amidinophenyl) methanesulfonyl fluoride, an irreversible inhibitor of serine proteases. Biochemistry 1980; 19:4859–4864.

181. James GT. Inactivation of the proteinase inhibitor phenylmethanesulfonyl fluoride in buffers. Anal Biochem 1978; 86:574–579.
182. Harpel PC, Hayes MG, Hugli TE. Heat-induced fragmentation of human α_2-macroglobulin. J Biol Chem 1979; 254:8669–8678.
183. Howard JB, Zieske L, Clarkson J, Rathe L. Mechanism-based fragmentation of coenzyme A transferase. Comparison of α_2-macroglobulin and coenzyme A transferase thiol ester reactions. J Biol Chem 1986; 261:60–65.
184. Coté C, Solioz M, Schatz G. Biogenesis of the cytochrome bc_1 complex of yeast mitochondria. J Biol Chem 1979; 254:1437–1439.
185. Kresze G-B, Ronft H. Pyruvate dehydrogenase complex from baker's yeast. 1. Purification and some kinetic and regulatory properties. Eur J Biochem 1981; 119:573–579.
186. Budzik GP, Bush EN, Holleman WH. Solubilization and molecular size of atrial naturiuretic hormone (ANH) receptors from rabbit aorta, renal cortex and adrenal. Fed Proc 1986; 45:1737.
187. Waxman L, Goldberg AL, Protease La from *Escherichia coli* hydrolyzes ATP and proteins in a linked fashion. Proc Natl Acad Sci USA 1982; 79:4883–4887.
188. Swamy KHS, Chung CH, Goldberg AL. Isolation and characterization of protease Do from *Escherichia coli*, a large serine protease containing multiple subunits. Arch Biochem Biophys 1983; 224:543–554.
189. Chung CH, Goldberg AL. Purification and characterization of protease So, a cytoplasmic serine protease in *Escherichia coli*. J Bacteriol 1983; 154:231–238.
190. Roseman JE, Levine RL. Purification of a protease from *Escherichia coli* with specificity for oxidized glutamine synthetase. J Biol Chem 1987; 262: 2101–2110.
191. Yasunobu KT, McConn J. *Bacillus subtilis* neutral protease. Methods Enzymol 1970; 19:569–575.
192. Nedkov P, Oberthür W, Braunitzer G. Determination of the complete amino-acid sequence of subtilisin DY and its comparison with the primary structures of the subtilisins BPN', Carlsberg and Amylosacchariticus. Biol Chem Hoppe-Seyler 1985; 366:421–430.
193. Shimizu Y, Nishino T, Murao S. Purification and characterization of a membrane bound serine protease of *Bacillus subtilis* IFO 3027. Agric Biol Chem 1983; 47:1775–1782.
194. Szewczuk A, Mulczyk M. Pyrrolidonyl peptidase in bacteria. The enzyme from *Bacillus subtilis*. Eur J Biochem 1969; 8:63–67.
195. Fujishiro K, Sanada Y, Tanaka H, Katunuma N. Purification and characterization of yeast proteinase B. J Biochem 1980; 87:1321–1326.
196. Hayashi R, Bai Y, Hata T. Further confirmation of carboxypeptidase Y as a metal-free enzyme having a reactive serine residue. J Biochem 1975; 77: 1313–1318.
197. Magni G, Natalini P, Santarelli I, Vita A. Baker's yeast proteinase A purification and enzymatic and molecular properties. Arch Biochem Biophys 1982; 213:426–433.
198. Meussdoerffer F, Tortora P, Holzer H. Purification and properties of proteinase A from yeast. J Biol Chem 1980; 255:12087–12093.

199. Alvárez NG, Bordallo C, Gascón S, Rendueles PS. Purification and characterization of a thermosensitive X-prolyl dipeptidyl aminopeptidase (dipeptidyl aminopeptidase yscV) from *Saccharomyces cerevisiae*. Biochim Biophys Acta 1985; 832:119–125.
200. Mizuno K, Nakamura T, Takada K, Sakakibara S, Matsuo H. A membrane-bound, calcium-dependent protease in yeast α-cell cleaving on the carboxyl side of paired basic residues. Biochem Biophys Res Commun 187; 144:807–814.
201. Achstetter T, Ehmann C, Wolf DH. Proteinase yscD: Purification and characterization of a new yeast peptidase. J Biol Chem 1985; 260:4585–4590.
202. Okada T, Sonomoto K, Tanaka A. Novel Leu-Lys specific peptidase (leulysin) produced by gel-entrapped yeast cells. Biochem Biophys Res Commun 1987; 145:316–322.
203. Achstetter T, Ehmann C, Wolf DH. Proteolysis in eucaryotic cells: Aminopeptidases and dipeptidyl aminopeptidases of yeast revisited. Arch Biochem Biophys 1983; 226:292–305.
204. Achstetter T, Ehmann C, Osaki A, Wolf, DH. Proteolysis in yeast cells: Proteinase yscE, a new yeast peptidase. J Biol Chem 1984; 259:13344–13348.
205. Zubatov AS, Mikhailova AE, Luzikov VN. Detection, isolation and some properties of membrane proteinases from yeast mitochondria. Biochim Biophys Acta 1984; 787:188–195.
206. Achstetter T, Wolf DH. Hormone processing and membrane-bound proteinases in yeast. EMBO J 1985; 4:173–177.
207. Achstetter T, Ehmann C, Wolf DH. Aminopeptidase Co, a new yeast peptidase. Biochem Biophys Res Commun 1982; 109:341–347.
208. Desautels M, Goldberg AL. Demonstration of an ATP-dependent, vanadate-sensitive endoproteinase in the matrix of rat liver mitochondria. J Biol Chem 1982; 257:11673–11679.
209. Hay R, Böhni P, Gasser S. How mitochondria import proteins. Biochim Biophys Acta 1984; 779:65–87.
210. Frey J, Röhm KH. Subcellular localization and levels of aminopeptidases and dipeptidases in *Saccharomyces cerevisiae*. Biochim Biophys Acta 1978; 527:31–41.
211. Wolf DH, Weiser U. Studies on a carboxypeptidase Y mutant of yeast and evidence for a second carboxypeptidase activity. Eur J Biochem 1977; 73:553–556.
212. Wagner J-C, Escher C, Wolf DH. Some characteristics of hormone (pheromone) processing enzymes in yeast. FEBS Lett 1987; 218:31–34.
213. Mizuno K, Matsuo H. A novel protease from yeast with specificity towards paired basic residues. Nature 1984; 309:558–560.
214. Wong S-L, Kawamura F, Doi RH. Use of the *Bacillus subtilis* signal peptide for efficient secretion of TEM β-lactamase during growth. J Bacteriol 1986; 168:1005–1009.
215. Boyer HW, Carlton BC. Production of two proteolytic enzymes by a transformed strain of *Bacillus subtilis*. Arch Biochem Biophys 1968; 128:442–455.
216. Affholter JA, Fried VA, Roth RA. Human insulin-degrading enzyme shares structural and functional homologies with *E. coli* protease III. Science 1988; 242:1415–1418.

217. Stoĉker, W, Wolz RL, Zwilling R. *Astacus* protease, a zinc metalloenzyme. Biochemistry 1988; 27:5026–5032.
218. Palmer SM, St. John AC. Characterization of a membrane-associated serine protease in *Escherichia coli*. J Bacteriol 1987; 169:1474–1497.
219. Sugimura K, Higashi N. Novel outer membrane-associated protease in *Escherichia coli*. J Bacteriol 1988; 170:3650–3654.
220. Sugimura K, Nishihara T. Purification, characterization, and primary structure of *Escherichia coli* Protease VII with specificity for paired basic residues: Identity of protease VII and OmpT. J Bacteriol 1988; 170:5625–5632.
221. Sugimura K. Mutant isolation and cloning of the gene encoding protease VII from *Escherichia coli*. Biochem Biophys Res Commun 1988; 153:753–759.
222. Grodberg J, Dunn JJ. *ompT* encodes the *Escherichia coli* outer membrane protease that cleaves T7 RNA polymerase during purification. J Bacteriol 1988; 170:1245–1253.
223. Chin DT, Goff SA, Webster T, Smith T, Goldberg AL. Sequence of the *lon* gene in *Escherichia coli*: A heat-shock gene which encodes the ATP-dependent protease La. J Biol Chem 1988; 263:11718–11728.
224. Hwang BJ, Woo KM, Goldberg AL, Chung CH. Protease Ti, a new ATP-dependent protease in *Escherichia coli*, contains protein-activated ATPase and proteolytic functions in distinct subunits. J Biol Chem 1988; 263:8727–8734.
225. Katayama Y, Gottesman S, Pumphrey J, Rudikoff S, Clark WP, Maurizi MR. The two-component, ATP-dependent Clp protease of *Escherichia coli*: Purification, cloning, and mutational analysis of the ATP-binding component. J Biol Chem 1988; 263:15226–15236.
226. Ruppen M, van Alstine GL, Band L. Control of intracellular serine protease expression in *Bacillus subtilis*. J Bacteriol 1988; 170:136–140.
227. Bussey H. Proteases and the processing of precursors to secreted proteins in yeast. Yeast 1988; 4:17–26.
228. Pontremoli S, Melloni E. Extralysosomal protein degradation. Annu Rev Biochem 1986; 55:455–481.
229. Tanaka D, Waxman L, Goldberg A. Vanadate inhibits the ATP-dependent degradation of proteins in reticulocytes without affecting ubiquitin conjugation. J Biol Chem 1984; 259:2803–2809.
230. Waxman L, Fagan J, Tanaka K, Goldberg A. A soluble ATP-dependent system for protein degradation from murine erythroleukemia cells: Evidence for a protease which requires ATP hydrolysis but not ubiquitin. J Biol Chem 1985; 260:11994–12000.
231. Mortimore GE, Pösö AR. Lysosomal pathways in hepatic protein degradation: Regulatory role of amino acids. Fed Proc 1984; 43:1289–1294.
232. Ahlberg J, Berkenstam A, Henell F, Glaumann H. Degradation of short and long lived proteins in isolated rat liver lysosomes. J Biol Chem 1985; 260:5847–5854.
233. Amenta J, Hlivko T, McBee A, Shimozuka H, Brocher S. Specific inhibition by ammonium chloride of autophagy-associated proteolysis in cultured fibloblasts. Exp Cell Res 1978; 115:157–166.

234. Schickaneder E, Geuß U, Hösel W, vdEltz H. Casein-resorufin, a new substrate for a highly sensitive protease assay. Fresenius' Z Analyt Chem 1988; 330:360.
235. Kiselev VI. Use of cloned htpR gene of *Escherichia coli* to introduce htpR mutation into the chromosome. Biotechnol Appl Biochem 1988; 10:397–401.
236. Kiselev VI, Tarasova IM. Regulation of proteolysis in *Escherichia coli* cells by antisense RNA of htpR gene. Biotechnol Appl Biochem 1988; 10:59–62.
237. Park JH, Lee YS, Chung CH, Goldberg AL. Purification and characterization of protease Re, a cytoplasmic endoprotease in *Escherichia coli*. J Bacteriol 1988; 170:921–926.
238. Lee YS, Park SC, Goldberg AL, Chung CH. Protease So from *Escherichia coli* preferentially degrades oxidatively damaged glutamine synthetase. J Biol Chem 1988; 263:6643–6646.
239. Udaka S, Takagi H, Kadowaki K. Novel *Bacillus brevis* strains and application thereof. Eur patent application 0,257,189, 1986.
240. Kopetzki E, Buckel P, Schumacher G. Cloning and characterization of baker's yeast α-glucosidase: Over-expression in a yeast strain devoid of vacuolar proteinases. Yeast 1989; 5:11–24.
241. Cabezón T, De Wilde M, Herion P, Loriau R, Bollen A. Expression of human α_1-antitrypsin cDNA in the yeast *Saccharomyces cerevisiae*. Proc Natl Acad Sci USA 1984; 81:6594–6598.
242. Hyolaerts M, Weyens A, Bollen A, Harford N, Cabezón T. High-level production and isolation of human recombinant α_1-proteinase inhibitor in yeast. FEBS Lett 1986; 204:83–87.
243. O'Loughlin JT. An alpha interferon analogue. Eur. Patent Application 0,240,224, 1986.
244. Roggenkamp R, Dargatz H, Hollenberg CP. Precursor of β-lactamase is enzymatically inactive: Accumulation of the preprotein in *Saccharomyces cerevisiae*. J Biol Chem 1985; 260:1508–1512.
245. Simon LD, Fay RB. T4 DNA fragment as a stabilizer for proteins expressed by cloned DNA. Eur. Patent Application 0,072,925, 1981.
246. Skorupski K, Tomaschewski J, Rüger W, Simon LD. A bacteriophage T4 gene which functions to inhibit *Escherichia coli* Lon protease. J Bacteriol 1988; 170:3016–3024.
247. Oka T, Sakamoto S, Miyoshi K-I, Fuwa T, Yoda K, Yamasaki M, Tamura G, Miyake T. Synthesis and secretion of human epidermal growth factor by *Escherichia coli*. Proc Natl Acad Sci USA 1985; 82:7212–7217.
248. Barbero JL, Buesa JM, Penalva MA, Pérez-Aranda A, Garcia JL. Secretion of mature human interferon alpha 2 into the periplasmic space of *Escherichia coli*. J Biotechnol 1986; 4:255–267.
249. Strauch KL, Beckwith J. An *Escherichia coli* mutation preventing degradation of abnormal periplasmic proteins. Proc Natl Acad Sci USA 1988; 85:1576–1580.
250. Beckwith J, Strauch KL. Periplasmic protease mutants of *Escherichia coli*. PCT patent application WO 88/05821, 1987.
251. Luria SE, Suit JL, Jackson JA. A system for release of proteins from microbe cells. Eur. patent application 0,278,697, 1987.

252. Lancashire WE. Stabilised polypeptides and their expression and use in fermentation. Eur. patent application 0,262,805, 1986.
253. Bachmair A, Finley D, Varshavsky A. Methods of regulating metabolic stability of proteins. PCT patent application WO 88/02406, 1986.
254. Finch PW, Wilson RE, Brown K, Hickson ID, Emmerson PT. Complete nucleotide sequence of the *Escherichia coli ptr* gene encoding protease III. Nucl Acids Res 1986; 14:7695–7703.
255. Longo A, Snay JR, Zimmer L, Ataai MM. Calcium requirement of residual protease in *Bacillus subtilis* DB104. Biotechnol Lett 1988; 10:649–654.

5
Properties of Recombinant Protein-Containing Inclusion Bodies in *Escherichia coli*

James F. Kane

Monsanto Company
St. Louis, Missouri

Donna L. Hartley

Centre International de Recherche Daniel Carasso
Le Plessis-Robinson, France

I. INTRODUCTION

With the advent of recombinant DNA technology, it is possible to produce peptides and proteins with phenomenal therapeutic and diagnostic value for human and animal health care. As with any technology, the problems or difficulties shift as scientific discoveries reveal appropriate solutions. One of the many challenges was to isolate, clone, and efficiently express the gene product of interest. Although cloning techniques were available, it was no easy task to get the protein expressed at levels that could be considered commercially feasible (1–7). Now it is not surprising to read of host/vector systems that generate the recombinant product up to 40% of the total cell protein (8–11). Although this difficulty was solved, many hurdles remain, not the least of which is the production of the peptides and proteins by an economically viable process. For the most part entirely new technologies had to be developed. Once the host/vector was constructed the next step in the process was fermentation, a technology with the most historical data base on which to draw. Unfortunately, the microbial workhorses of this technology, namely, *Bacillus*, *Streptomyces*, and yeast, were not particularly well suited for production of recombinant proteins. The workhorse of the recombinant DNA technology was *Escherichia coli*, a gram-negative microorganism with a vast data base in physiology, biochemistry, and genetics but essentially no background in fermentation. In the past decade alone, there were over 85,000 papers published that dealt with *E. coli*. The sum total of papers on *Bacillus sp.*, *Streptomyces sp.*, and yeast is less than that. Thus, the challenge is either to bring *E. coli* into fermentation processes and make it produce the recombinant peptides and proteins at levels that are commercially viable, or to generate the necessary data base in physiology, biochemistry, and genetics with another microorganism. At this point it is easier to develop fermentation conditions around *E. coli* than to devote research efforts to develop another system. As our knowledge of these other hosts continues to expand, it is likely that these microorganisms will be used commercially to produce recombinant products. The purported problem of endotoxin associated with the gram-negative *E. coli* can be substantially reduced by simple washing procedures (12,13). Once fermentation conditions are established for the recombinant *E. coli*, a still greater challenge remains, i.e., the isolation and purification of large quantities, literally millions of grams, of highly purified and active products and the formulation of these products into a readily acceptable state for use by the consumer. In these latter areas totally new technologies need to be developed. Proteins for both human and animal use must be highly purified and contain no components that could cause adverse side effects, and this has to be done in an economical manner. It is therefore essential that the entire process be geared to make the job of protein purification as easy as possible. Since *E. coli* is the major microbial host for production of recombinant proteins, this chapter will focus primarily on *E. coli* host/vector systems.

II. PROTEIN INCLUSION BODIES: AN ASSET OR A LIABILITY?

Many recombinant proteins form proteinaceous aggregates inside the host cell. This represents a distinct advantage to purification because the product can be isolated in a highly purified state ($> = 50\%$) in a concentrated form by a simple chemical or physical disruption and differential centrifugation (14,15). But as Mother Nature is prone to do, she giveth and she taketh away. The major disadvantage is that the protein aggregate is essentially in a denatured state. This insoluble aggregate must be solubilized, usually with strong chaotropic agents such as guanidine-HCl, and allowed to refold into a native, active conformation (14,15). There is not much information on how to do this, although several recent reports on the refolding pathway of specific proteins have been published (16–20). So let us consider in more detail the advantages and disadvantages of having the product in intracellular inclusion bodies.

A. Quantitation of Product

The electron-dense inclusion bodies are readily visible with a light microscope equipped with either dark-field or phase contrast lens and condensers (13,21–

26). These inclusions appear refractile to light and are very similar to spores found in several gram-positive microorganisms. Several prokaryotic species produce inclusions (27) but only a few produce exclusively protein-containing particles with defined structures. On the other hand, recombinant protein aggregates appear to be unorganized. In most, if not all, cases the mass and number of inclusion bodies (IBs) have been found to be proportional to the level of product (23; Kane, this chapter). Thus, one can follow the formation of the product by quantitating these morphological changes using techniques such as light scattering (28) or computer-enhanced image analysis (CEIA). The advantages of the latter method are ease of operation, rapid analysis, and on-line potential for assessing the productivity of the fermentation. There are several problems that make it difficult to get absolute numbers. These include glass slides of varying thickness, so that intensity levels used to enhance the images vary with each slide; streaming of the cells; cells not being on the same plane. Nevertheless, analysis of the inclusion bodies provides a rapid means of following product formation.

1. CEIA System to Measure Bovine Somatotropin Formation

Although several CEIA systems are available, we have used the following components which we have found perform exceptionally well:

1. IBM-AT personal computer with IBM graphics card
2. IBM color monitor
3. Panasonic monitor
4. Houston HiPad digitizing tablet and mouse
5. Dage Newvicon camera
6. Modified PC vision frame-grabber card (Southern Micro Instruments, SMI)
7. Nikon Optiphot microscope

The microscope contained a 40× objective and a 2.5× projection lens was installed in the tube supporting the Newvicon camera. The image from the microscope was enhanced for contrast and clarity by the Newvicon camera and was transmitted to the frame-grabber card in the IBM-AT. The image was transmitted from the computer to the Panasonic monitor for viewing; the image analysis software package (SMI) freezes the frame and allows the viewer to perform the necessary editing before determining the area of each cell and the associated IBs. Intensity limits can be set to optimize the image and distinguish the IB from the rest of the cell's cytoplasm, and both cell and IB in the frame are counted. The results can be individually represented as the mean area or can be used to generate histograms or other statistical analyses. In addition, the images can be stored for analysis at a later date.

2. Production of BST in *E. coli*

The method described above was used to determine the levels of bovine somatotropin (BST) made by *E. coli* W3110G containing the plasmid pBGH1 (29).

In this plasmid (Fig. 1), the BST gene is controlled by the *E coli* tryptophan promoter/operator (trp p/o). High rates of BST synthesis are initiated by the addition of indole acrylic acid (IAA) and BST accumulates to approximately 20% of the total cell protein (29,30). The results from a typical fermentation are shown in Table 1. It is interesting to note that the size of the cells increased twofold during the fermentation. The increase in optical density (OD) is attributed to this increase in cell size and the presence and intensity of the

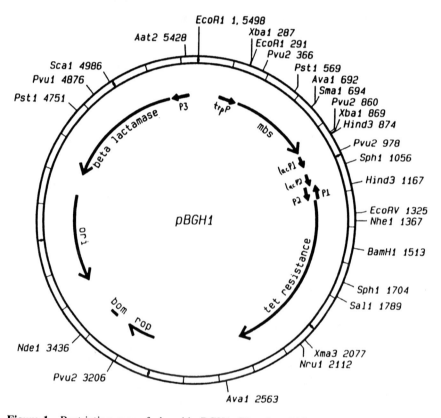

Figure 1 Restriction map of plasmid pBGH1. The plasmid is a pBR322-based vector that is 5498 bp. The restriction sites for each enzyme are indicated. The abbreviations are; *trp* p, tryptophan promoter; MBS, BST structural gene; *lacP1*, *lacP*, tandem *lacUV5* promoters; P1, P2, partial promoters found on pBR322; *tet* resistance tetracycline resistance gene; ropgene involved in the control of copy number; *bom*, gene involved in mobilization in the presence of ColE1 plasmids; *ori*, origin of plasmid replication; β-lactamase, gene for ampicillin resistance; P3, promoter from pBR322 for ampicillin resistance gene. (Reprinted from Calcott et al. J Ind Microbiol 1988; 29:257–266, with permission from Elsevier Publishing Co.)

Table 1 CEIA of 10-liter Fermentation Samples from I0 to I4[a]

Sample	Cells counted	Mean area (μm²)		
		Cells	IBs	OD
I0	85	3.38	0.06	22
I1	173	6.65	0.23	36
I2	126	5.98	0.62	58
I3	179	5.87	0.88	66
I4	131	7.79	1.00	74

[a]$E.\ coli$ W3110G/pBGH1 was grown in a 10-liter fermentor in a minimal salts medium at 37°C. The pH was maintained at 7.0 by the addition of NH^4OH and glucose was fed to the culture to maintain a concentration of 0.5%. At an optical density of 20 at 550 nm, indole acrylic acid was added to initiate the synthesis of BST, I0. Samples were taken every hour through I4 and analyzed for IBs.

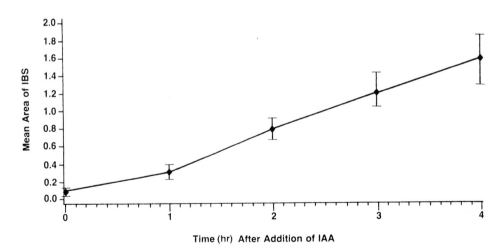

Figure 2 Mean area of IB at various hours of induction. The mean area of the IBs was determined with samples removed at hours 0 (i.e., the time of addition of IAA), 1, 2, 3, and 4. The plotted values are the average of 4- to 10-liter fermentations. Ranges are indicated by the bars.

IBs. The viable cell count and particle count (W. E. Workman, unpublished data) did not increase after the addition of the inducer IAA. In this particular experiment, the grams of BST per liter at I4 was 1.03 as measured by reverse phase HPLC. The area of the IBs was related to the expression level by the following equation:

$$\text{Factor} = \frac{(\text{mean IB area I*})(\text{OD I*})}{(\text{grams BST/liter at I*})}$$

For the results shown in Table 1, the factor would be determined by substituting the appropriate values as follows:

$$\text{Factor} = \frac{(1.0)(74)}{(1.03)}$$

to give a value of 71.8. When the product of the mean IB area multiplied by the OD for I0 through I3 was divided by the factor, the level of BST expression was determined. These results are shown in Table 2 and illustrate the relationship between the mean area of the IBs and the hour of induction. Samples were removed from four separate fermentations at I0–I4, and the mean IB areas were measured. As expected, the size of the IBs increased with time in the fermentation. It is important to note that the IBs contribute to the measured OD value. Therefore, the larger and more numerous the IBs in the cells the larger the apparent increase in the OD. This could result in an overestima-

Table 2 BST Expression as Measured by HPLC and CEIA

Sample	g BST/liter	
	HPLC[a]	CEIA[b]
I0	0.03	0.02
I1	0.20	0.12
I2	0.49	0.50
I3	0.90	0.81

[a]The grams of BST per liter were measured by reverse phase HPLC from whole-cell samples taken at the indicated times.

[b]The grams of BST per liter were calculated using the equation:

$$\text{g BST/liter} = \frac{(\text{IB area})(\text{OD})}{71.8}$$

The IB area and OD values are shown in Table 1.

tion of product expression as shown in Fig. 3 unless appropriate standard curves are used.

B. Stability of Cloned Protein

E. coli contains a set of proteolytic enzymes, both cytoplasmic and periplasmic (30,31), that function in the cell to degrade or turn over proteins. These are generally, although not exclusively, related to stress conditions in the cell, such as the heat shock response. An intriguing yet unanswered question is how do these proteolytic enzymes recognize the abnormal protein conformation. With proteins that require the formation of disulfide bonds, it seems reasonable to assume that these proteins are not in their native conformation in *E. coli* since disulfide bridge formation is an unlikely intracellular event (32). This is based on the fact that *E. coli* is geared to a reducing environment (33) and possesses enzymes (glutathione reductase, thioredoxin reductase, and dihydrolipoamide dehydrogenase) whose activities are the reduction of disulfide bonds. Reduced glutathione, for example, plays a crucial role in ensuring that other thiol groups remain in the reduced state inside the cell. However, disulfide bond formation, or a lack thereof, is not the only factor involved in determining an abnormal conformation and some *E. coli* proteins also can be

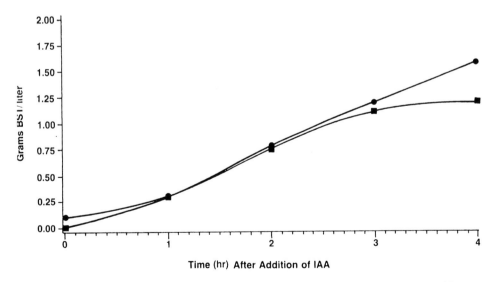

Figure 3 Increase in IB area and g BST/liter with time. The BST was measured by HPLC (■) and the IBs (●) were determined by CEIA. The results are plotted as a function of time after the addition of IAA to a 10-liter fermenter.

rapidly degraded. These include abnormally short proteins generated by nonsense mutations and mistranslated proteins that result from the addition of streptomycin or amino acid analogs (34,35). A number of studies indicate that the degradative system in *E. coli* can be overwhelmed. Presumably, under these conditions the concentration of the abnormal proteins increases and aggregation and IB formation follow. For example, when *E. coli* auxotrophs were fed ^{14}C-labeled amino acid analogs of the required amino acid (36,37), these labeled analogs were incorporated into proteins. These "abnormal" proteins became substrates for the intracellular proteases. When the rate of synthesis exceeded the rate of degradation and IBs were formed, the label was found in the acid-precipitable fraction. When the analog was removed and the appropriate L-amino acid was substituted, these inclusion bodies disappeared and the labeled amino acid analogs were found in the soluble fraction. The interpretation of these results is that the IBs are susceptible to proteolytic degradation once the synthesis of the abnormal proteins stops. The most susceptible form of the protein, however, appears to be soluble protein. These results are consistent with the idea that a key factor to stabilizing a protein is a high rate of synthesis such that the accumulation exceeds the proteolytic capacity and allows for the formation of IBs. When the inducer is removed or the

Table 3 Stability of Labeled BST in *E. coli* During High Rates of Product Synthesis[a]

Sample time after chase (min)	Integral of radioactive band	
	I1/2	I41/2
0	7244	9800
2	6922	—
4	6502	9136
8	5986	8160
16	4848	7730
32	4078	7088
64	3418	5928

[a]Five hundred μC^i of carrier-free ^{35}S-methionine were added to each of two shake flasks containing *E. coli* W3110G/pBGH1 in 50 ml of a minimal salts glucose medium at 37°C. After 3 min, a 6 log excess of cold methionine was added to each flask and samples were taken at the indicated times. The cells were collected by centrifugation and prepared for PAGE under denaturing conditions. The gels were overlaid with photographic film and the films were scanned to determine the intensity of the BST band.

cloned gene is otherwise repressed, the degradative enzymes eventually destroy the aggregated proteins.

1. Half-life of BST in *E. coli*

The host/vector system used to produce BST is *E. coli* K12 strain W3110G/pBGH1. In order to demonstrate the stabilizing effect of IB formation on BST, we measured the half-life ($t1/2$) at the BST at 1/2 hr (I1/2) and 4-1/2 hr (I4 1/2) after the addition of IAA (Table 3, Fig. 4). There appeared to be at least two different populations of BST present in the cell. At I1/2 approximately 66% of the molecules had a $t1/2$ of 100 min and 37% had a $t1/2$ of 8 min or less. By I41/2 over 85% of the material had a $t1/2$ of over 100 min. The remaining 15% of the molecules had a $t1/2$ of < 5 min. It was surprising that such a significant population of the BST molecules were so resistant at I1/2. Apparently, IB formation begins almost immediately after the initiation of high rates of product synthesis and sequesters the recombinant protein.

We observed that prototrophic *E. coli* formed IBs when grown on the amino acid analog azetidine-2-carboxylic acid (a proline analog). Therefore, we added this analog to *E. coli* W3110G/pBGH1 growing in minimal salts glucose medium, i.e., under noninducing conditions. If BST were being synthesized and rapidly degraded, then the formation of IBs by the analog should trap some of the BST and render it more stable. This hypothesis was verified when

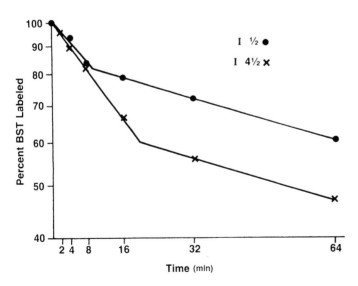

Figure 4 Half-Life of BST in *E. coli*. The results from Table 3 were plotted as the percentage of labeled BST as a function of time after the addition of the cold methionine.

we found BST in cells grown in the presence of the analog (Bogosian, unpublished observations; Fig. 5). These results lend further support to the idea that BST is produced but rapidly degraded in the absence of IB formation.

2. Half-lives of Other Cloned Proteins

A number of other studies indicate that labile proteins are made more stable by aggregation and IB formation. One interesting example is the defective form of β-galactosidase, the so-called X90 protomer. This protein lacks the final 12 amino acids and is rapidly degraded when present at low concentrations (38,39). When the gene for the X90 protomer was cloned on a plasmid, this

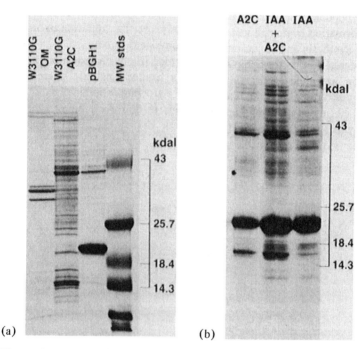

Figure 5 Accumulation of BST in cells treated with the proline analog azetidine-2-carboxylic acid. *E. coli* W3110G with and without pBGH1 was grown in the presence and absence of azetidine-2-carboxylic acid (A2C). IB-like preparations were isolated and subjected to PAGE analysis. (a) Lane 1, outer membrane preparation from *E. coli* W3110G; lane 2, *E. coli* W3110G grown in the presence of A2C; lane 3, *E. coli* W3110G/pBGH1 induced with IAA; lane 4, molecular weight markers. (b) Lane 1, *E. coli* W3110G/pBGH1 grown in the presence of A2C; lane 2, *E. coli* W3110G/pBGH1 grown in the presence of A2C and IAA; lane 3, *E. coli* W3110G/pBGH1 induced with IAA.

protein was produced at much higher levels. Under these conditions the protein formed IBs and accumulated in the cell.

An excellent example of designing the protein to form IB and thus become stabilized is illustrated in the experiments of Saito et al. (40,41). These investigators were trying to express somatomedin C and found that it had a short half-life and did not accumulate to high levels. They therefore fused the somatomedin C to a portion of γ-interferon making a protein with an isoelectric point of 7.64. They reasoned that such a protein would precipitate in the intracellular environment of *E. coli* (pH 6.8–7.2). When this fused protein was overexpressed in *E. coli*, IBs formed and the fused protein accumulated to over 20% of the total cell protein. Other workers (2,3,42–49) used similar approaches to stabilize rapidly degrading proteins by creating situations that favor the formation of IBs.

It is still not clear what factors control the proteolytic digestion. Several properties which may affect this have been proposed (50), including the conformation state (17,51). It is not unreasonable to expect cloned proteins to be present in an "abnormal" state inside *E. coli*. The question remains, however, as to what factors allow the cell proteases to recognize these conformations as "abnormal."

3. Protease-Deficient Strains

The in vivo stability of cloned proteins can be markedly improved with the use of protease-deficient strains. One interesting class of proteases requires metabolic energy (33,52). A well-studied example of such an enzyme in *E. coli* is the so-called La protease, the product of the *lon* gene (53). The *lon* gene is a member of a group of genes that constitute the heat shock response. Perhaps, it is more appropriately called the stress response, since a number of agents in addition to temperature elicit these cellular responses to an environmental stimulus. The genes that constitute the heat shock response are controlled by a positive regulator, the product of the *htpR* locus. This protein appears to code for a sigma-like transcription factor that is required for induction of the heat shock proteins (54). Accumulation of abnormal proteins, e.g., cloned aggregated proteins (34), induces the heat shock response not only in *E. coli* but also in mammalian cells (55). In *E. coli* the La protease is produced and the abnormal protein degraded. If this were true then one could increase the stability of such proteins by altering the *htpR* and/or *lon* genes. This hypothesis was tested with several cloned proteins. Expression of human complement C5a (56) was increased four- to 10-fold in an *hrpR* mutant and somatomedin C levels increased in a stepwise fashion in a *lon* mutant, *htpR* mutant, and a *lon* *htpR* double mutant (57). Similar increased levels of expression were seen with the T7 phage protein (gene 18) (58), tissue plasminogen activator (59), and polio virus 3C protease (23). Although the *htpR* locus controls some proteolytic functions, it is not the only regulatory factor involved. Similarly, not all proteins may have increased levels of stability in *htpR* or *lon* mutants, indicating that other as yet unidentified genes are involved in protein degradation. There is another side to the *lon* mutation, namely, the La protease is involved in the degradation of the *sulA* gene product which causes filamentation (60,61). Thus, in *lon* mutants the cells form long filaments and tend to be mucoid. Such hosts are very difficult to ferment because the viscosity is so high.

An alternative approach is to clone into the cells proteins which inhibit proteolytic function. Nature has already done this with the phage, specifically the T phages. Early after infection a phage protein is synthesized that has been called "pin" for proteolysis inhibition. This protein functions inside the cell to inhibit, albeit not completely, the degradation of abnormal proteins such as nonsense peptides (62) without affecting the turnover of normal proteins. One could view the presence of phage proteins as abnormal in *E. coli* and such a protein evolved to allow for phage maturation. When the phage T4 gene encoding the pin protein was cloned into *E. coli* abnormal proteins, including cloned eukaryotic gene, were stabilized (63). The utility of this was realized and a patent application has been submitted (64).

C. Purification Steps

The importance of reducing the costs of goods significantly impacts the number and types of steps employed in the purification scheme. Proteins that are needed at low levels (either per dose or number of recipients) can be made by using some sophisticated chromatographic steps assuming that the sale price of the product can support these costs. With proteins required in much larger doses, such as proteins to increase animal efficiency, the costs associated with column chromatography may significantly increase production costs to the point that there is little profit in the sale of the product. Thus, it is desirable for this latter type of product to purify with effective but cost-efficient means, such as heat and extractions with salt and/or organic solvents. Inclusion bodies represent a significant step toward such a purification scheme by presenting the product in both a relatively pure state and a highly concentrated form following disruption and centrifugation. Reports indicate that the product in the protein particle could be >50% of the total protein present. Thus, one needs only a twofold or less purification regimen to achieve the desired level of product quality. If the product were soluble and present at 10–40% of the total cell protein, at least two steps would be required to get to the same point in the purification scheme. In addition, one is generally working with larger volumes of soluble material than with the refractile bodies. However, even when there is a disparity between the amount of protein present in the inclusion particle and the soluble phase, there are advantages to working with the insoluble form of the protein. A good example of this was described with the purification of the cellulase from *Clostridium thermocellum* (65). The *celD* gene was cloned in *E. coli* and high-level expression (15%) was achieved (66). In this case, approximately 25% of the enzyme was in the pellet fraction and 75% was soluble. Yet the insoluble protein was purified more readily by a few simple procedures than was the soluble enzyme.

D. Renaturation

There is one major problem associated with IB formation. Since the protein is insoluble and very likely not in its native state either conformationally or with respect to disulfide bond formation, the protein must first be solubilized. This requires rather strong denaturing conditions such as high concentrations of urea or guanidine-HCl. These chaotropic agents solubilize the protein but two questions remain: will the protein refold to its native conformation when the agent is removed and will it do so in a highly efficient manner? Again, it is important to have this step occur with high efficiencies at high protein concentrations in order to develop a commercially viable purification scheme. There is obviously a great deal to be learned in this area in order to formulate some general principles and rules for protein folding. Several studies in this area indicate that there are discrete steps along the pathway to proper protein folding (16–20). The challenge is to be able to predict what this pathway would be for a given protein rather than empirically determining the rate-limiting steps of this routine.

III. COMPONENTS OF INCLUSION BODIES

Despite the fact that the major protein in the inclusion body is the cloned gene product, there are numerous other components that remain associated with the pellet after a wash with a neutral pH buffer. We have examined these components in bovine somatotropin (BST) containing IB from *E. coli*.

A. *E. coli* Proteins

We identified some of the proteins associated with these proteinaceous aggregates. The four subunits of RNA polymerase were detected in the BST-containing granules as well as in inclusion bodies produced by other plasmids and cloned gene products (Fig. 6). Presumably the transcription apparatus was closely associated with the formation of these granules. In addition, components of the translational machinery were also found tightly associated with these granules, i.e., ribosomal RNA and the L13 protein of the large ribosomal

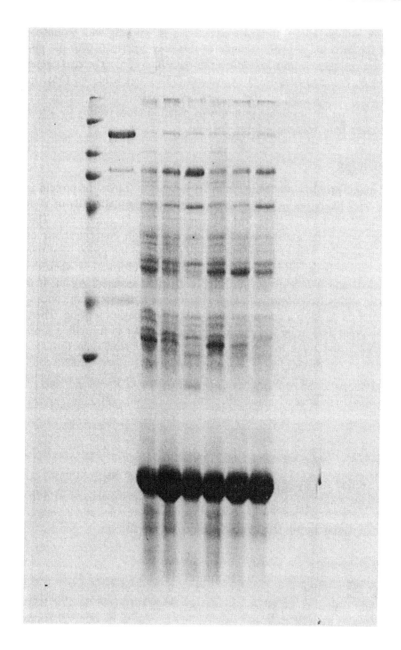

subunit were isolated. The L13 protein was purified by ion exchange chromatography and identified by its N-terminal sequence. That the ribosomes were somehow associated with the formation of IB was first suggested by Paul et al. (67). These authors speculated that the small, dense areas which were visible by electron microscopy in preparations of inclusion bodies were ribosomes that were trapped during aggregation of the overexpressed protein.

Other proteins associated with the inclusion body preparation were tentatively identified as the 35- and 38-kDa outer membrane proteins. Indeed, membrane fragments can be seen in the inclusion bodies using the electron microscope. It is clear, however, that the granules are not membrane-enclosed and that the fragments were coisolated with the granules during centrifugation.

The presence of *E. coli* proteins associated with the inclusion granules has been noted by other groups as well (12,13). It is clear that some of these proteins were attached to the particle in such a way that detergent washing removed these contaminating proteins without solubilizing the granules. One could think of these particles not as solid masses but rather as "whiffle balls" with holes to the interior. Proteins or other materials may adhere to the various surfaces of the particle by ionic or other attractions. Some of these attractions were so strong that the proteins remained attached to IB until it was solubilized. A good example of such a protein is kanamycin phosphotransferase (13), a plasmid-encoded protein (see below).

B. Nucleic Acids

BST-containing inclusion bodies were treated to extract nucleic acids, and both DNA and RNA were present (Fig. 7). We observed a significant portion of a DNase-sensitive ethidium bromide staining material which was less than 2 kb in length. This material hybridized to labeled pBR322, which is the vector portion of pBGH1. Apparently, these lower sized DNA fragments were pieces of the vector and not chromosomal DNA. Most of the larger sized DNA comigrated with nicked pBGH1 while a small portion appeared to be covalently closed circular DNA. We found that the plasmid DNA that is present in the IB can transform *E. coli* to tetracycline resistance (Bogosian, unpublished results).

Electrophoresis on formaldehyde gels demonstrated that both 16 and 23S RNA were present. Both of these bands disappeared when the preparation was treated with RNase. The implication of these results is that the N terminus of

---

Figure 6 Presence of RNA polymerase subunits in BST-containing IB preparations. Six preparations of IBs from BST fermentations were subjected to denaturing polyacrylamide gel electrophoresis (lanes 3–8). Lane 1 contains molecular weight standards and lane 2 contains the purified subunits of RNA polymerase.

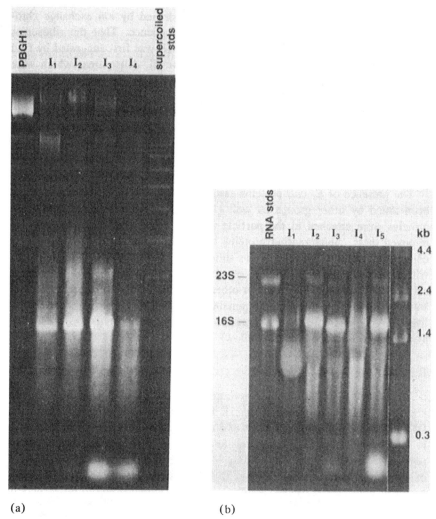

(a) (b)

Figure 7 Bulk nucleic acids present in BST-containing IBs. Nucleic acids were isolated by a modification of the procedure of Sarmientos et al. (83), and subjected to electrophoresis on nondenaturing (a) and denaturing (b) gels. Nondenaturing gels were composed of 0.8% agarose in 90 mM Tris-borate buffer, pH 8.0, containing 2.5 mM EDTA. The denaturing gels contained 1% agarose with 2.2 M formaldehyde in 50 mM 3-(N-morpholino) propanesulfonic acid (MOPS) buffer, pH7.0, containing 1 mM EDTA. After electrophoresis, the gels were stained with a 0.1% ethidium bromide solution. (Reprinted from Ref. 24, with permission from Biochemical Society and Portland Press.)

the nascent polypeptide begins to aggregate immediately after it is translated, trapping the plasmid(s), mRNA, and the enzymes involved in transcription and translation (24,26). This model is consistent with the demonstration of the various contaminants, both nucleic acid and proteins, that we have found to be associated with IBs. Other researchers (12) also observed the association of nucleic acids with aggregated protein granules. It has been reported that more than 95% of the nucleic acids could be removed by a simple detergent wash, but the types of nucleic acids that were present before and after the wash were not described.

Since these inclusion bodies entrap nucleic acids, there could be some adverse consequences on product expression, namely, plasmid instability (23) and decreased level of translation resulting from loss of RNA polymerase and ribosomes. If the plasmid(s) were sequestered in the IBs as suggested (23,68), then cell divisions after IBs are formed would result in plasmidless cells by preventing partitioning during cell division. Kenealy et al. (23) suggested that microscopic observation can be used to detect such an event by noting the increase in the number of IB-less cells. Although we have noted some cells lacking IBs at the end of the fermentation, we have been unable to isolate antibiotic-sensitive cells which we should find if the IB-less cells lost their plasmid. It is likely that these cells represent the normal Poisson distribution of cells at various stages of induction and/or product formation. We had similar experiences with other vectors derived from either pBR322 or pBR327, namely, that all of the cells obtained from the end of the fermentation contained the appropriate antibiotic markers. This is despite the fact that there may be anywhere from 5 to 10% of the cells without IBs. Plasmid instability that results from IB formation might be a greater problem in expression systems involving low copy number plasmids, but this has yet to be demonstrated. With respect to the second consequence, there is no direct evidence to imply that the cessation of product formation is related to the sequestering of RNA polymerase or ribosomes. The fact that expression levels reach 30–40% of total cell protein would argue against any seriously adverse effects of IB formation on the total level of expression of the cloned gene product.

It has been observed that *E. coli* cells containing BST inclusion bodies did not divide, even though the OD continued to increase throughout the fermentation. It is not clear why the formation of these intracellular aggregates results in a cessation of cell division. One can measure cell division by placing the cells on a glass slide in nutrient broth containing soft agar and observing the IB-containing cells over time. Cells containing IBs did grow and divide, although much more slowly than cells which lacked IBs (Kane, unpublished results; Mann, personal communication). These cell divisions may give rise to cells lacking IBs which presumably would grow more rapidly than IB-containing cells, leading eventually to an increase in the number of IB-free cells in the culture. It is also likely that under these conditions the cloned product was no longer synthesized and the degradative systems dissolved the IB and permitted cell division to occur. Apparently, IBs interfered with cell division, either by preventing migration of the chromosome or by interfering with the equatorial elongation which precedes cell division. Perhaps it is necessary for cells to begin to degrade IBs before cell division can resume.

C. Plasmid-Encoded Proteins

If the model of IB formation were true, then one would expect that proteins encoded by the plasmid would also be associated with the IB. Schoner et al. (13) reported that kanamycin phosphotransferase was the most prominent contaminating band in the washed IBs. It is interesting to note that their vector has the BST gene and the kanamycin resistance gene being transcribed in opposite directions (69). We found this to be true with some plasmid constructs containing the gene conferring resistance to ampicillin and kanamycin (Fig. 8). The most important determinant appeared to be the orientation of the gene relative to the orientation of the cloned gene, and the effectiveness of the terminator. When the genes were being transcribed in the same direction, there was a greater level of the resistance protein than when the genes were transcribed in opposite directions. In addition, the β-lactamase containing the leader peptide was present in the IB but the processed β-lactamase was not.

D. Other Contaminants

One other likely component associated with the refractile bodies is the lipopolysaccharide (LPS) or endotoxin since this is found in membrane fractions of gram-negative microorganism. Because LPS elicits an inflammatory response, it is imperative that the purification of pharmaceuticals produced by *E. coli* take this into account. Fortunately, the LPS can be removed by a detergent wash or by passage over some adsorbent material.

E. Structure of Inclusion Bodies

Previous work had indicated that there was an ordered crystalline structure in intracellular inclusion bodies containing the paracrystalline protein of *Bacillus thuringiensis* and the poly-β-hydroxybutyrate granules of *Pseudomonas*. The structure of these intracellular inclusions is in marked contrast to that of the protein aggregates in IBs. Early observations of IBs using transmission electron microscopy suggested that there was no membrane and no orderly or crystalline structure to the protein.

We examined washed IBs from several fermentations as well as from shake flasks using the technique of high-resolution ^{13}C-NMR. If there were ordered,

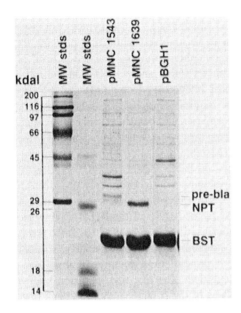

Figure 8 Antibiotic resistance proteins in BST-containing IB preparations. Lanes 1 and 2 contain molecular weight standards. IBs were isolated from *E. coli* W3110G containing pMON1543 which has the transcription of the BST gene oriented in same direction as that of β-lactamase (lane 3). Lane 4 contains IBs isolated from *E. coli* W3110G/pMON1639, which has the gene for kanamycin resistance. The directions of transcription of BST and NPT are the same. Lane 5 contains IBS from *E. coli* W3110G/pBGH1. As shown in Fig. 1, the direction of transcription for BST is clockwise whereas the direction for β-lactamase is counterclockwise. The pre-bla refers to the precursor of β-lactamase.

crystalline structures to the protein, then the bands in the NMR spectrum which are attributable to protein (C-C-N, C-C-O) should appear as sharp peaks because the environment of each would be identical in the packed crystal (70). This was not the case. No ordered structure was observed providing further support that IBs are proteinaceous aggregates.

IV. FACTORS AFFECTING IB FORMATION

A. Properties of the Protein

The conformation of the protein in the *E. coli* cytoplasm has a major effect on its fate. The protein may be stable and soluble, unstable and soluble, or form insoluble aggregates. Two examples of stable and soluble proteins are aspartase (71) and cyanase (72), both of which are *E. coli* enzymes composed of identical subunits and do not appear to be involved with other proteins for their in vivo activity. Despite the fact that these proteins reached a level of 30% of the total cell protein, both remained stable and soluble. On the other hand, penicillin acylase (8) and the host integration factor (73) are proteins composed of nonidentical subunits that require all of the subunits to be produced in order to form their proper conformation and be catalytically functional. When the individual subunits of these enzymes were cloned independently, the proteins were either rapidly degraded (unstable and soluble) or formed inclusion bodies (insoluble); these differences probably reflect the relative rates of proteolysis for each protein. When the subunits were coordinately produced in a single host, the proteins remained stable and soluble. A similar situation existed with the overproduction of the σ factor (a part of the RNA polymerase complex involved in the initiation of transcription) in *E. coli* (74). Since the σ factor is generally present in less than stoichiometric amounts relative to core polymerase (75), large amounts of free σ-protein could assume abnormal conformations, and aggregate and/or be degraded. The fact that inclusion bodies formed suggests that the rate of proteolysis of the σ factor may have been low.

The restriction enzyme *Eco*RI is another interesting example of conformation effects on protein solubility in *E. coli*. It has been reported that *Eco*RI was soluble at low levels of expression, whereas at high levels of expression the protein was found in inclusion bodies (76). This protein has been reported to form tetramers at high protein concentrations, and the properties of the tetramer are different from those of the catalytically functional dimer (77). In this case, one could hypothesize that the high concentrations of this protein led to the formation of tetrameric forms (or higher) of the protein which then formed inclusion bodies. Perhaps other properties of the protein, such as its hydrophobic properties affect the solubility in the cell. We examined this possibility by determining the relative hydrophobicity of some proteins (18). In this analysis each amino acid is assigned a hydropathy index and these indices are summed and divided by the number of amino acid residues to yield a "gravy score" (78), or the relative value for the hydrophobic residues of the protein. Although this approach discounts positional/interactive effects on non-adjacent residues, it nevertheless is some indication of the physical state of the protein. No apparent correlations were found, however, between the hydrophobicity and the solubility of the protein in *E. coli*. This may reflect our poor understanding of this area of protein chemistry. The inherent solubility of the protein at neutral pH might be critical for the formation of the inclusion bodies. Proteins that are insoluble at or near neutral pH would be expected to become insoluble more rapidly after the initiation of synthesis. Marston, however, suggested that the aggregation of proteins was not just the result of exceeding the solubility limits of the protein since strong chemical reagents were required for their solubilization (11). Although these factors have been considered independently, it is probable that interactions among these various properties determine the fate of the protein inside the cell.

B. Host Cell and Growth Conditions

An intriguing aspect to the formation of IBs is the observation that the host cell and/or growth conditions can influence the physical state of the abnormal protein. In one study Kenealy et al. (23) examined the solubility of six cloned proteins (β-lactamase, HIV-1 core antigen and envelop antigen, interferon-β, interleukin-2, polio virus 3C protease) in 11 different hosts. β-Lactamase and the core antigen, GAG-9, were the only proteins that had the largest range of solubility. The other four were essentially insoluble in all strains. The β-lactamase and GAG-9 were expressed behind the tryptophan promoter/operator (*trp* p/o). The cultures were grown in Luria broth and shifted to M9 minimal salts glucose medium containing casamino acids and some additional supplements depending on the host requirements. The cultures were grown for 24 hr in this medium at 35°C. Under these conditions one would expect physiological derepression of the *trp* p/o leading to product expression. β-Lactamase ranged from 10 to 100% soluble depending on the *E. coli* host. Four strains are of particular interest: X156 and RGC103 which are isogenic except that X156 is *proC34* and *capR+* (*cap* is the old designation for the *lon* gene) whereas RGC103 is *pro+* and *capR9*; CS122 and CAG456 are isogenic except that CAG456 contains an additional mutation in the *htpR* gene. β-Lactamase was soluble in strain X156 but insoluble in the *lon* mutant RGC103. Strains SC122 and CAG456 were grown in the same medium but at 30°C since both contained a temperature-sensitive mutation. Again, β-lactamase was insoluble in strain CAG456 which has a mutation in the *htpR* gene that controls the heat shock genes including *lon* (*cap*R). What role the heat shock genes played in the formation of IB is unclear. The La protease is significantly reduced in each host but is only part of the cellular stress or heat shock system. As mentioned above, the La protease does affect gene products from the SOS system, particularly the product of the *sulA* gene which inhibits filamentation.

These effects on solubility become somewhat more difficult to interpret since there is some evidence that the physiological state of the cell may play a role in IB formation. Cheng (38,39) reported that a cloned β-galactosidase from a spontaneously derived mutation in *lacZ* that restored the catalytic activity to the X90 protomer formed inclusion bodies in the late stationary phase but not in the exponential phase of growth. He concluded that high-level accumulation per se is not sufficient to produce these structures but some properties of the cell are critical for these aggregates to form. In support of this hypothesis is the work of Schein et al. (79), Piatek et al. (80), and O'Hare et al. (81). These investigators reported that temperature played a key role in the formation of IBs with interferon-α2 (IFN-α2) and ricin A (rRA). With IFN-α2 the effect of temperature on solubility appeared to be independent of the level of expression but, interestingly, it could be mimicked in vitro. Lysates from *E. coli* grown at 30 or 37°C would convert soluble IFN-α2 to insoluble material when incubated at 37°C but not when incubated at either 0°C or 30°C. This temperature effect was not observed in mock lysates containing bovine serum albumin as the protein, although the addition of sulfhydryl reagents to the mock lysate caused the IFN-α2 to become insoluble. These results suggest that an enzyme affects solubility in a posttranslational manner. This enzyme is present in cells grown at 30°C but is inactive, at least with respect to the activity required to precipitate the protein. With rRA, high temperatures (42°C) resulted in the formation of insoluble material but this was resolubilized by incubation at 37°C. These interconversions between soluble and insoluble forms could not be duplicated by incubating rRA in the sonicates at the appropriate temperatures. This is in contrast to the results with IFN-α2, and the authors concluded that "factors relating to the intracellular environment such as ionic conditions or local concentration or the active translation of rRA itself or another heat-induced protein might be important" (80). It does appear that these cellular effects are protein-dependent. The effect of temperature on activity of a cloned human interferon-β (IFN-β) was reported by Mizukami et al. (82). The expression level as measured by polyacrylamide gels remained relatively constant in cells grown at 20 or 33°C. The amount of active IFN-β, however, was eight to nine times higher in cells grown at 20°C than in cells grown at 33°C. The basis for this increased activity was unknown but presumably was related to the protein conformation.

At the present time it is not clear how the cells and growth conditions impact the formation of these protein aggregates. Indeed, the results suggest that empirical approaches must be tried to determine if a soluble or insoluble product forms. It appears that the heat shock system plays an important but undefined role in this process. Nevertheless, it is worth the effort to investigate the hosts and fermentation conditions to develop a protocol that will generate the product with the desired solubility characteristics.

V. CONCLUSIONS

There are major advantages to making a protein insoluble inside the cell: ease of following product formation; increased stability; ease of purification. It is not clear how the host or fermentation conditions affect refractile body formation or what property(ies) of the protein is (are) more essential for their formation. If, however, one were able to control their formation, then it would be a very useful means of increasing product expression and purity.

REFERENCES

1. Itakura K, Hiroso T, Crea R, Riggs AD, Heyneker HL, Bolivar F, Boyer HW. Science 1977; 198:1056–1063.
2. Goeddel DG, Heyneker HL, Hozumi T, Arentzen R, Itakura K, Yansura DG, Ross MJ, Miozzari G, Crea T, and Seeburg PH. Nature 1979; 281:544–548.
3. Smith J, Cook E, Fotheringham I, Pheby S, Derbyshire R, Eaton MAW, Doel M, Lilley DMJ, Pardon JF, Patel T, Lewis H, Bell LD. Nucleic Acids Res 1982; 10:4467–4482.
4. Goeddel DV, Shepard HM, Yelverton E, Leung D, Crea R, Sloma A, Pestka S. Nucleic Acids Res 1980; 8:4057–4074.
5. Taniguchi T, Grarente L, Roberts TM, Kimelman D, Douhan J III, Ptashne M. Proc Natl Acad Sci USA 1980; 77:5230–5233.
6. Shepard HM, Yelverton E, Goeddel DV. DNA 1982; 1:125–131.
7. Derynck R, Remaut E, Saman E, Stanssens P, deClerq E, Content J, Fiers W. Nature 1980; 287:193–197.
8. Schumacher G, Sizmann D, Huag H, Bock A. Nucleic Acids Res 1986; 14:5713–5727.
9. Kronheim SR, Cantrell MA, Deeley MC, March CJ, Glackin PJ, Anderson DM, Hemenway T, Merriam JE, Cosman D, Hopp TP. Biotechnology 1986; 4:1078–1082.
10. Nishimura N, Komatsubara S, Taniguchi T, Kisumi M. J Biotechnol 1987; 6:31–40.
11. Goldschmidt R. Nature 1970; 288:1151–1156.
12. Langley KE, Berg TG, Strickland TW, Fenton DM, Boone TC, Wypych J. Eur J Biochem 1987; 163:313–321.
13. Schoner RG, Ellis LF, Schoner BF. Biotechnology 1985; 3:151–154.
14. Marston FAO. Biochem J 1986; 240:1–12.
15. Lowe PA, Rhind SK, Sugrue R, Marston FAO. In Burgess R. ed. Protein purification: Micro to macro. New York: Alan R. Liss, 1987:429–442.
16. King J. Biotechnology 1986; 4:297–303.
17. Price NC. Biochem Soc. Trans 1987; 818.
18. Havel HA, Kauffman EW, Plaisted SM, Brems DN. Biochemistry 1986; 25:6533–6538.
19. Brems DN, Plaisted SM, Kauffman EW, Havel HA. Biochemistry 1986; 25:6539–6543.
20. Brems DN, Plaisted SM, Dougherty JJ Jr, Holzman TF. J Biol Chem. 1987; 262:2590–2598.
21. Fieschko J, Ritch T, Bengston D, Fenton D, Mann M. Biotechnol Progr 1985; 1:205–208.
22. Schoemaker JM, Brasnett AH, Marston FAO. EMBO J 1985; 4:775–780.
23. Kenealy WR, Gray JE, Ivanoff LA, Tribe DE, Reed DL, Korant BD, Petteway SR Jr. Dev Ind Microbiol 1987; 28:45–52.
24. Hartley DL, Kane JF. Biochem. Soc. Trans 1988; 16:101–102.
25. Kane JF. Pharmaceut Eng 1988; 8:36–41.
26. Kane JF, Hartley DL. Trends Biotechnol 1988; 6:95–101.

27. Shively JM. Ann Rev Microbiol 1974; 28:167–184.
28. Wittrup KD, Mann MB, Fenton DM, Tsai LB, Bailey JE. Biotechnology 1988; 6:423–426.
29. Seeburg PH, Sias S, Adelman J, deBoer HA, Hayflick J, Jhurani P, Goeddel DV, Heyneker HL. DNA 1983; 2:37–45.
30. Swamy KHS, Goldberg AL. Nature 1981; 292:652–654.
31. Prouty WF, Karnovsky MJ, Goldberg AL. J Biol Chem 1975; 250:1112–1122.
32. Tuggle CK, Fuchs JA. J Bacteriol 1985; 162:448–450.
33. Holmgrem A. Ann Rev Biochem 1985; 54:237–271.
34. Goldberg AL, St. John AC. Ann Rev Biochem 1976; 45:747–803.
35. Goff SA, Goldberg AL. Cell 1985; 41:587–595.
36. Rabinovitz M, Finkelman A, Reagan RL, Breitman TR. J Bacteriol 1969; 99:336–338.
37. Schachtele CA, Anderson DL, Rogers P. J Mol Biol 1968; 33:861–872.
38. Cheng Y-SE, Kwoh DY, Kwoh TJ, Soltvedt BC, Zipser D. Gene 1981; 14:121–131.
39. Cheng Y,-SE. Biochem Biophys Res Commun 1983; 111:104–111.
40. Saito Y, Yamada H, Niwa M, Ueda I. J Biochem 1987; 101:123–134.
41. Saito Y, Ishii Y, Niwa M, Ueda I. J Biochem 1987; 101:1281–1288.
42. Goddel DV, Kleid DG, Bolivar F, Heyneker HL, Yansura DG, Crea R, Hirose T, Kraszewske A, Itakura K, Riggs AD. Proc Natl Acad Sci USA 1979; 76:106–110.
43. Shen S-H. Proc Natl Acad Sci USA 1984; 81:4627–4631.
44. Masui Y, Mizuno T, Inouye M. Biotechnology 1984; 2:81–85.
45. Sassenfeld HM, Brewer SJ. Biotechnology 1984; 2:76–80.
46. Botterman J, Hofte H, Zabeau M. J Biotechnol 1987; 6:71–81.
47. Samuel KP, Flordellis CF, DuBois GC, Papas TS. Gene Anal Tech 1985; 2:60–66.
48. Weis JH, Enquist LW, Salstrom JS, Watson RJ. Nature 1983; 302:72–74.
49. Schulz M-F, Buell G, Schmid E, Movva R, Selzer G. J Bacteriol 1983; 169:5385–5392.
50. Rechsteiner M, Rogers S, Rote K. Trends Biochem Sci 1987; 12:390–394.
51. Rivett AJ, Levine RL. Biochem Soc Trans 1987; 15:816–818.
52. Herskho A, Ciechanover A. Ann Rev Biochem 1982; 51:335–364.
53. Goldberg AL, Menon AS, Goff S, Chen ST. Biochem Soc Trans 1987; 15:809–811.
54. Neidhardt FC, VanBogelin RA. In: Neidhardt FC, Ingraham JL, Low B, Magasanik B, Schaechter M and Umbarger HE. *Escherichia coli* and *Salmonella typhimurium:* Cellular and molecular biology. Volume 2. ASM Publishers. pp. 1334–1345.
55. Ananthan J, Goldberg AL, Voellmy R. Science 1986; 232:522–524.
56. Mandecki W, Powell BS, Mollison KW, Carter GW, Fox JL. Gene 1986; 43:131–138.
57. Buell GM, Schulz F, Selzer G, Challet A, Movva NR, Semon D, Escaney S, Kawashima E. Nucleic Acids Res 1985; 13:1923–1938.
58. White JH, and Richardson CC. J Biol Chem 1987; 262:8845–8850.

59. Schachtele CF, Anderson DL, Rogers P. J Mol Biol 1968; 33:861–872.
60. George J, Costellazzi M, Buttin G. Mol Gen Genet 1975; 140:309–332.
61. Gottesman S, Halpern C, Trisler P. J Bacteriol 1981; 148:265–273.
62. Simon LD, Tomczak K, St. John AC. Nature 1978; 275:424–428.
63. Simon LD, Randolph B, Irwin N, Binkowski G. Proc Natl Acad Sci USA 1983; 80:2059–2062.
64. Simon LD, Fay RB. European Patent Application 1983; 821065992.
65. Joliff G, Beguin P, Juy M, Millet J, Ryter A, Poljak R, Aubert J-P. Biotechnology 1986; 4:896–900.
66. Joliff G, Beguin P, Aubert J-P. Nucl Acids Res 1986; 14:8605–8613.
67. Paul DC, Van Frank RM, Muth WL, Ross JW, Williams DC. Eur J Cell Biol 1983; 31:171–174.
68. Mann MB, Tsai LB, Forrer PD, Curless CE, Rotgers CG, Fenton DM. Abstr. 194th Amer. Chem. Soc. Natl. Meet. 1987.
69. Schoner BE, Hsuing HM, Belagaje RM, Mayne NG, Schoner RG. Proc Natl Acad Sci USA 1984; 81:5403–5407.
70. Jacob GS, Garbow JR, Schafer J. J Biol Chem 1986; 261:16785–16787.
71. Nishimura N, Komatsubara S, Taniguchi T, Kisumi M. J Biotechnol 1987; 6:31–40.
72. Sung Y-C, Anderson PM, Fuchs JA. J Bacteriol 1987; 169:5224–5230.
73. Nash HA, Robertson CA, Flamm E, Weisberg RA, Miller HI. J Bacteriol 1987; 169:4124–4127.
74. Gribskov M, Burgess RR. Gene 1983; 26:109–118.
75. Burton ZF, Gross CA, Watanabe KK, Burgess RR; Cell 1983; 32:335–349.
76. Botterman J, Zabeau M. Gene 1985; 37:229–239.
77. Modrich P, Zabel D. J Biol Chem 1976; 251:5866–5874.
78. Kyte J, Doolittle RF. J Mol Biol 1982; 157:105–132.
79. Schein, CH, Noteborn MHM. Biotechnology 1988; 6:291–294.
80. Piatek M, Lane JA, Bjorn MJ, Wang A, Williams M. J Biol Chem 1988; 263:4837–4843.
81. O'Hare M, Roberts LM, Thorpe PE, Watson GJ, Prion B, Lord JM. FEBS Lett 1987; 216:73–78.
82. Mizukami T, Komatsu Y, Hosoi N, Itoh S, Oka T. Biotechnol Lett 1986; 8:605–610.
83. Sarmientos P, Sylvester JE, Contente S, Cashel M. Cell 1983; 32:1337–1346.

6

Methods for Removing N-Terminal Methionine from Recombinant Proteins

Arie Ben-Bassat

Cetus Corporation
Emeryville, California

I. INTRODUCTION

Production of recombinant proteins for therapeutic purposes became feasible in the late 1970s with the development of genetic engineering technology. Recombinant proteins that are already in use or being considered for therapeutic applications include lymphokines (e.g., interferons and interleukins), hormones (e.g., insulin and human growth hormone), and enzymes (e.g., tissue plasminogen activator). Today the production of these proteins has become a major thrust of modern biotechnology.

Recombinant proteins are often produced with an extra methionine at the N-terminal end of the protein (Sec. I.C). This difference between the native protein and the recombinant protein is of great concern in the production of recombinant proteins for therapeutic applications (Sec. I.D). The purpose of this chapter is to review the methods that are available for removal of methionine from the N-terminal end of proteins. This is a new technology, and the following constitutes the first review on this subject. Many of the methods cited are based on published patent applications or issued patents and therefore might be restricted for commercial use without permission.

A. Posttranslational Modification of Proteins

Protein synthesis is initiated with methionine in eukaryotes. In prokaryotes and organelles of eukaryotes, *N*-formyl methionine is used instead. The *N*-formyl moiety and the methionyl residue are cleaved off in a significant fraction of the cytosolic proteins through the action of deformylase and methionine aminopeptidase, respectively. The cleavage occurs during translation on the nascent protein (40–60 amino acids long) (1–5). The methionine can be also removed from the mature protein after the completion of translation (6). Additional processing of the N-terminal end of the proteins can proceed with or without methionine removal; the processing can include acylation of the N-terminal residue (usually with acetate) or proteolytic degradation with peptidases or proteases. The N-terminal section of secreted proteins is frequently modified by the excision of the signal peptide (generally 20–40 amino acid residues in length) with signal peptidase. In eukaryotes the secretion process may include glycosylation.

B. Specificity of Methionine Aminopeptidase

The substrate specificity and universality of methionine aminopeptidase have been inferred from sequence analysis of mature cytosolic proteins (7,8). The recent work on cloning and purification of methionine aminopeptidase from *E. coli* has facilitated in vivo and in vitro characterization of this enzyme (6). These studies confirmed earlier observations and supplied more experimental evidence for the specificity of methionine aminopeptidase. Methionine aminopeptidase is an oligopeptidase with an absolute specificity for N-terminal methionine. The enzyme activity is strongly influenced by the nature of the adjacent amino acid residue (6–8). The terminal methionine is not cleaved when

Table 1 Cleavage of N-Terminal Methionine in Extracellular Proteins

Met cleaved	Met removal varied	Met retained	
Ala	Ile	Lys	Met
Gly	Val	Arg	Try
Pro	Cys	Leu	Tyr
Ser	Thr	Asp	Glu
		Asn	Gln
		Phe	His

Note: Data show the effect of the second amino acid on methionine removal. The data are based on Refs. 6–8 and personal communication.

the side chain of the following amino acid is large or charged, and is cleaved when the side chain of the neighboring residue is small and uncharged. In vitro studies with oligopeptides (6) and circumstantial evidence from sequence analysis of different analogs of recombinant proteins suggest that the cleavage of terminal methionine is also affected by the third amino acid (6) as well as by the secondary and tertiary structure of the protein (see Table 2). The removal of the formyl group from N-terminal formyl methionine is prerequisite to the cleavage of terminal methionine (1–5). Nothing is known about the specificity of formylase. Table 1 shows how the removal of N-terminal

Table 2 Production of Intracellular Recombinant Proteins with Uncleaved Terminal Methionine

Protein	N-terminal sequence in mature protein	Methionine retained (%)	Ref.
α-Interferon	Cys-Asp-Leu-	50	10
Interleukin-2	Ala-Pro-Thr-	70	6
Interleukin-2 analog	Pro-Thr-Ser-	<5	K. Koths pers. commun.
Interleukin-2 analog	Ser-Ser-Ser-	<5	11
Ricin A chain	Ile-Phe-Pro-	40	6
Human growth hormone	Phe-Pro-Thr-	>95	12
Human growth hormone analog	Leu-Phe-Pro-	>95	13
γ-Interferon analog	Gln-Asp-Pro-	>95	13
Bovine growth hormone analog	Asp-Gln-	>95	13
Apolipoprotein E	Lys-Val-Glu	>95	13
Tumor necrosis factor	Val-Arg-Ser-	<5	L. Lin, pers. commun.
Tumor necrosis factor analog	Leu-Arg-Ser-	>95	L. Lin, pers. commun.
Tumor necrosis factor analog	Val-Glu-Ser-	50	L. Lin, pers. commun.
Tumor necrosis factor analog	Arg-Ser-Ser-	>95	L. Lin, pers. commun.
Tumor necrosis factor analog	Ser-Ser-Ser-	<5	L. Lin, pers. commun.
Tumor necrosis factor analog	Thr-Pro-Ser-	>95	L. Lin, pers. commun.

methionine is affected by the adjacent amino acid. These conclusions are based in part on qualitative sequence analysis of mature cytosolic proteins in which partial removal of the methionine has been considered a noncleavage event. Partial processing of methionine can occur, especially when the second amino acid is isoleucine, valine, cysteine, threonine, or other amino acids of intermediate size.

C. Processing of Recombinant Proteins in the Cytosolic Space

Many of the recombinant proteins that are used for therapeutic applications are produced to resemble secreted products of mammalian cells. The N-terminal signal peptides in these secretory products are cleaved during the translocation process; thus the naturally produced mature proteins do not normally contain N-terminal methionine. Processing of the N-terminal end in recombinant proteins is subject to the same specificity restrictions as those of normal cytosolic proteins. In addition, processing can be impaired in certain recombinant proteins when they are expressed at high levels (9). Incomplete processing of recombinant proteins could be due to their accumulation and precipitation as inclusion bodies, presumably making them inaccessible to methionine aminopeptidase action. Interleukin-2 expressed in *E. coli* at high levels is an example. It accumulates as inclusion bodies, with only partial removal of the terminal methionine (Sec. II.A.2). Given the known specificity of methionine aminopeptidase from *E. coli*, recombinant interleukin-2 that begins with Met-Ala-Pro- should be completely processed. Other examples of incomplete processing of N-terminal methionine in recombinant proteins are presented in Table 2.

D. Heterogeneity and Immunogenicity Concerns Associated with an Extra Methionine

The presence of nonnative N-terminal methionine in recombinant pharmaceutical proteins can lead to several problems:

1. A difference in one methionine residue at the N-terminal end is known to change the mobility of proteins in RP-HPLC and SDS-PAGE. Differences were reported for recombinant ricin A chain and interleukin-2 (B. Ferris and K. Bauer, personal communication, and Ref. 14). This type of heterogeneity among the protein molecules can cause difficulties in the purification and characterization of these proteins. In addition, heterogeneity decreases the yield of pure products.
2. Extra methionine at the N-terminal end of the protein can affect the secondary and tertiary structure of the protein. This can affect the biological activity and function of the protein, can alter its stability, and can make it

immunogenic when administered as a therapeutic product. The recent debate on whether methionine-free human growth hormone (produced by Eli Lilly and Company) is superior to methionylated human growth hormone (produced by Genentech) exemplifies the importance of this subject (15).
3. The methionine at the N-terminal end is labile to oxidation during the purification process, especially when treated at high pH and under aerobic conditions. Oxidation of the methionine residue to sulfonyl methionine can increase the immunogenicity of the protein.

Less is known about the retention of N-terminal formyl methionine in recombinant proteins. An exception is the report about the retention of formyl methionine in α and β subunits of tryptophan synthase under conditions of high expression (16). Differences in the isoelectric point between the formyl methionine and formyl methionine-free form were reported. The same concerns that exist for therapeutic proteins with an additional terminal methionine are valid for therapeutic proteins with an additional terminal formyl methionine.

II. METHODS FOR REMOVING TERMINAL METHIONINE

The removal of the extra methionine from the N-terminal end of recombinant proteins can be performed in vivo (i.e., within the host during biosynthesis) or in vitro (i.e., by treating the purified recombinant protein chemically or enzymatically). The advantage of the in vivo methods is in the ease of the purification of the recombinant proteins. The in vitro methods offer more options to remove methionine, including chemical methods or enzymatic methods with aminopeptidases that are commercially available or are easy to produce and purify. None of the existing methods is applicable to all proteins, and a choice has to be made based on the N-terminal sequence of the existing protein and comparative studies. Our discussion is focused on the processing of recombinant proteins of therapeutic value. The in vivo methods are focused on the expression in *E. coli*, a major host organism for production of recombinant proteins.

A. In Vivo Methods

1. Fusion of the Recombinant Protein with Signal Peptide

Expression of the recombinant protein fused to a signal peptide can lead to export, cleavage of the signal peptide, and accumulation of the processed native protein in the periplasmic space of gram-negative bacteria or in the extracellular space of gram-positive bacteria, yeast, fungi, or mammalian cells. Accumulation of the proteins in the extracellular space or in the periplasmic space can ease purification and, in some cases, prevent the degradation of the proteins, e.g., preproinsulin (17). The mechanism of secretion in *E. coli* and the application of this method in various microorganisms were recently reviewed (18–20). Commercial application of this method in *E. coli* and other bacteria is limited by poor yield. Applications of this method in yeast, fungi, and mammalian cell cultures have also been reported.

Several considerations restrict the use of this method in *E. coli*:

a. Many secreted mammalian proteins are not effectively exported or processed in *E. coli* and other bacteria. The reasons are probably due to differences in the mechanism of secretion between mammalian cells and prokaryotes. These differences can be reflected in different export-specific sequence information and by the presence of hydrophobic block sequences in eukaryotic proteins. Consequently, the proteins can become "stuck" in the membrane of the host bacteria and "jam" further transport or processing.
b. Many heterologous proteins that are secreted to the periplasmic space in *E. coli* are rapidly degraded by proteolytic enzymes as "foreign proteins" or because of the denatured, open structure of these proteins.
c. The specificity of the signal peptidase may pose some restrictions on the composition of the first amino acids (+ 1) at the N-terminal end of the mature protein (21,22). For example, proline at + 1 position prevents cleavage, while alanine is the preferred amino acid at + 1 position. In addition, some lack of cleavage site specificity can occur. Examples are bovine growth hormone, where 65% of the cleavage is at the alanine residue (+ 1) and 35% at the glycine residue (− 1)(23), and human interferon expressed in yeast cells where the cleavage occurs at more than one site (24).

Examples of high-level expression, efficient export to the periplasmic space, and accurate processing of recombinant eukaryotic proteins in *E. coli* are usually confined to small proteins. Fusion of the gene encoding for epidermal growth factor with the signal peptide of alkaline phosphatase allows export, processing, and accumulation of the epidermal growth factor in the periplasmic space of *E. coli*. Protein yield was 2.4 mg/liter (culture broth), and more than 95% of the protein recovered was accurately processed (25). Fusion of the gene encoding human growth hormone with the signal peptide of heat-stable enterotoxin II from *E. coli* allows export, accurate processing, and accumulation of the human growth hormone in the periplasmic space of *E. coli*. The expression level was 6–10% of the cell protein (26).

2. Expression in Host with Enhanced Methionine Aminopeptidase Activity

Hyperproduction of methionine aminopeptidase in recombinant strains of *E. coli* allows better removal of the methionine residue from the N-terminal end of recombinant proteins. Expression of interleukin-2 (Met-Ala-Pro-IL-2) and ricin A chain (Met-Ile-Phe-Ricin A) in *E. coli* results in only partial removal of the terminal methionine (30 and 60%, respectively). Expression of the same proteins in recombinant strains of *E. coli* that were hyperproducers of methionine aminopeptidase led to removal of 95% and 91% of the methionine in interleukin-2 and ricin A proteins, respectively (6).

The specificity of methionine aminopeptidase restricts the application of this method (I.B). This method can be successful for cases where sequence analysis of the recombinant protein indicates only partial removal of the methionine, in wild-type hosts, or where the N-terminal sequence of the mature, native protein predicts susceptibility to methionine aminopeptidase action (Table 1).

3. Deletion of N-Terminal Residues Which Inhibit Methionine Removal

Modification of the N-terminal sequence can improve methionine removal in vivo. Interleukin-2 as produced in *E. coli* is recovered largely as Met-Ala-Pro-Thr-Ser- at the N terminus (Section II.A.2). Expression of interleukin-2 as a Met-Pro- construct led to production of essentially homogeneous product that begins with Pro- (K. Koths, personal communication). Removal of the coding region for the first three amino acids of the natural, mature interleukin-2 resulted in a protein with N-terminal serine (11).

4. Fusion of the Recombinant Protein with Ubiquitin

The gene encoding for the recombinant protein is fused to ubiquitin and expressed in the recombinant host. The fusion is between the 5' end of the recombinant gene and the 3' end of the ubiquitin gene. A ubiquitin-specific protease in the cytosolic space cleaved the fused protein at the point of fusion and released the recombinant protein in its native form. This process was demonstrated by Bacmair in *Saccharomyces cerevisiae* using β-galactosidase as a model (39). The proteolytic cleavage has been shown to occur with the native N terminus of β-galactosidase (methionine) as well as with 18 other amino acids that substitute the methionine. These data support the generality of this method. An exception was the substitution of the methionine with proline where no proteolytic cleavage occurred. Prokaryotes are normally not suitable for this method because they lack the ubiquitin-processing system.

B. In Vitro Methods

1. Cleavage of Methionine with Cyanogen Bromide

The gene encoding a protein of interest is expressed as a fused product between a large portion of another gene, e.g., β-galactosidase, and the gene of interest. After expression the protein is cleaved from the extraneous fragment with cyanogen bromide and purified. The cyanogen bromide treatment is not specific to the methionine at the N terminus and therefore is limited to proteins that lack internal methionines. In addition, the chemical reaction itself can denature and chemically modify the protein of interest.

This method has been successfully used for the production of small proteins and peptides. Examples are somatostatin (27), human insulin (28), N-desacetylthymosin (29), leu-encephalin (30), human proinsulin (31), and α-neoendorphin (32).

2. Cleavage with Aminopeptidase from *Aeromonas proteolytica*

Removal of the N-terminal methionine is accomplished by incubating the purified recombinant protein with purified aminopeptidase. The enzyme is produced by *Aeromonas proteolytica* and is accumulated in the broth as an extracellular product (33–34). The crude enzyme has to be purified before use and heat-treated to inactivate contaminating proteolytic activities.

The enzyme is a nonspecific exopeptidase that can remove amino acids from the N-terminal end of proteins. The stop signals for the enzymes are Glu, Asp, and X-Pro, where X can be any amino acid except proline. When the stop signal is adjacent to the initiating methionine in γ-interferon (Met-Gln-Asp-) or bovine growth hormone analog (Met-Asp-Gln-), only the methionine is removed. When the stop signals are not near methionine, proteolysis by this enzyme continues until a stop signal is encountered. Examples of this kind of processing are human growth hormone analog (Met-Leu-Phe-), where methionine and leucine are sequentially removed, and apolipoprotein E (Met-Lys-Val-Glu-), where methionine, lysine, and valine are removed (13).

3. Cleavage with Methionine Aminopeptidase

The methionine is removed from the N-terminal end of recombinant proteins by incubating the purified recombinant protein with purified methionine aminopeptidase (6). The peptidase can be easily purified from a recombinant strain of *E. coli* that produced it to about 20% of the total cell protein. The specificity of the enzyme and possible applications are described in Sec. I.B and II.A.2, respectively.

Treatment of interleukin-2 with methionine aminopeptidase reduced the amount of methionine at the N-terminal end from 70% to 6%. Similar treatment of ricin A protein reduced the amount of methionine at the N-terminal end from 40% to 8%.

4. Cleavage with Aminopeptidase M

The removal of methionine is accomplished by incubating the purified recombinant protein with aminopeptidase M (EC 3.4.11.2). The enzyme is a general aminopeptidase with no hydrolytic activity against an X-Pro imide bond (14,35). This property was used for effective removal of the methionine from the N-terminal end of interleukin-2 (Met-Ala-Pro-) and human growth hormone

(Met-Phe-Pro-). This processing was accomplished by incubation with a commercial preparation of aminopeptidase M (hog kidney, microsomal fraction, Pierce) and proteolytic inhibitors, with purified aminopeptidase M and with aminopeptidase M immobilized to formyl cellulofine. The conjugation to formyl cellulofine stabilized the enzyme without affecting its activity. The immobilization of the peptidase also simplifies the purification of the recombinant product.

No data on the application of this technique to other proteins are available, and there are not enough specificity data to make definite predictions. It is likely that other proteins with proline at the position following the N terminus would be amenable to this technique.

5. Cleavage with Dipeptidyl Aminopeptidase I

The methionine-free protein is biosynthetically produced as a fused protein with a small N-terminal extension. Then the recombinant protein is purified and the external peptide is specifically cleaved with dipeptidyl aminopeptidase I (EC 3.4.14.1; cathepsin C) (9,36,37). The N-terminal extension contains an even number of amino acids. The N terminus must be different from lysine or arginine, all other amino acids at odd-number positions must be different from lysine, arginine, and proline, and all even-numbered amino acids must not be proline.

This technique has been successfully used for production of methionine-free human growth hormone (Phe-Pro-Thr-). In this protein the dipeptidyl aminopeptidase I is unable to cleave the Pro-Thr peptidyl bond. To facilitate separation of the cleaved (mature protein) from the uncleaved product, the authors included negatively charged amino acid in the N-terminal extended peptide. In addition, to avoid partial processing of the extended peptide during biosynthesis (by methionine aminopeptidase), the authors (37) recommend that the first amino acid in the extended peptide be negatively charged or have a short side-chain residue (see also Sec. I.B). An example of such extended peptides are Glu-Ala-Glu-, Ala-Glu-, Phe-Glu-Glu-, and Ala-Glu-Ala-Glu-.

The enzyme is commercially available from Boehringer. No data about the application of this technique to other proteins are available, and there are not enough specificity data to make definite predictions. It is possible that other proteins with lysine, arginine, or proline at the N terminus or proline at the second position would be amenable to this technique. Interleukin-2 is again an example.

6. Chemical Removal of Methionine as Methionyl Proline Diketopiperazine

N-teminal methionine can be removed chemically as methionyl proline diketopiperazine from proteins with Met-Pro-peptide composition, where "peptide" denotes the native structure of the mature native protein (38). The proteins can be produced by common genetic engineering techniques and purified. The diketopiperazine formation and cleavage can proceed in aprotic solvent and acidic conditions or in aqueous buffered solution, preferably at pH 8.0.

The mechanism of diketopiperazine formulation and cleavage is best explained by citations from Dimarchi and Brook (38): "The penultimate proline exerts cis-conformation on the amino terminal dipeptide and thereby fixes the terminal α-amino in a more favored location for attack at the proline carbonyl." "Nucleophilic attack at this site is favored by formation of the energetically favored cyclic amino-proline diketopiperazine with release of the desired peptide as the leaving group." Recommended conditions for processing in organic solvents are use of 1-methylpyrrolidinone and 0.33 M acetic acid as solvents and incubation for 1–24 hr at 25–40°C. Recommended conditions in aqueous environment are use of 0.8–1.0 M phosphate buffer, pH 8.0, and incubation for 2–72 hr at 35–50°C. Stearic hinderance by the third residue can interfere with the reaction. In addition, the reaction can continue past the desired cleavage site in proteins with Met-Pro-X-Pro-peptide structure, e.g., human growth hormone (Met-Phe-Pro-).

Data for protein stability, product yield, and possible chemical modifications of the proteins after these treatments are not available. Evidence for changes in the chromatographic properties of several proteins after the treatment with aprotic acidic solvent is documented (38).

7. Cleavage with Ubiquitin-Specific Protease

The recombinant protein is expressed as a fused product with ubiquitin (see Sec. II.A.4 and Ref. 39). The protein can then be purified and treated with ubiquitin specific protease. This protease cleaves the fused protein at point of fusion and releases the recombinant protein in its native form. The production of the fused protein can be conveniently done in *E. coli* or other prokaryotes. The ubiquitin gene has been cloned and methods for its purification have been published (40).

ACKNOWLEDGMENTS

I thank K. Koths and S. Chang for reviewing this article; L. Lin, K. Bauer, B. Ferris, and K. Koths for communication of unpublished results; J. Davis for editing assistance; and G. Liuzzo for typing the manuscript.

REFERENCES

1. Adams JM. On the release of the formyl group from the nascent protein. J Mol Biol 1968; 33:571–589.

2. Takeda M., Webster RE. Protein chain initiation and deformylation in *B. subtilis* homogenates. Proc Natl Acad Sci USA 1968; 60:1487–1494.
3. Pine JM. Kinetics of maturation of the amino termini of the cell proteins of *Escherichia coli*. Biochim Biophys Acta 1969; 174:359–372.
4. Housman D, Gillespie D, Loddish HF. Removal of formyl-methionine residue from nascent bacteriophage f2 protein. J Mol Biol 1972; 65:163–166.
5. Ball LA, Kaesberg P. Cleavage of the N-terminal formylmethionine residue from bacteriophage coat protein in vitro. J Mol Biol 1973; 79:531–537.
6. Ben-Bassat A, Bauer K, Chang SY, Myambo K, Boosman A, Chang S. Processing of the initiation methionine from proteins: Properties of the *Escherichia coli* methionine aminopeptidase and its gene structure. J Bacteriol 1987; 169:751–757.
7. Sherman F, Stewart JW, Tsunasawa S. Methionine or not methionine at the beginning of a protein. Bioessays 1985; 3:27–31.
8. Flinta C, Persson B, Jornvall H, von Heijne G. Sequence determinants of cytosolic N-terminal protein processing. Eur J Biochem 1986; 154:193–196.
9. Dalboge H, Dahl HHM, Pedersen J, Hansen JW, Christensen T. A novel enzymatic method for production of authentic hGH from an *Escherichia coli* produced hGH precursor. Biotechnology 1987; 5:161–164.
10. Wetzel R, Perry LJ, Esteel DA, Lin N, Levine HL, Slinker B, Fields F, Ross MJ, Shively J. Properties of human alfa-interferon purified from *E. coli*. extracts. J Interferon Res 1981; 1:381–390.
11. Liang S-M, Allet B, Rose K, Hirschi M, Kiang C-M, Thatcher DR. Characterization of human interleukin 2 derived from *Escherichia coli*. Biochem J 1985; 229:429–439.
12. Kohr WJ, Keck R, Harkins RN. Characterization and trypsin-digested biosynthetic human growth hormone by high-pressure liquid chromatography. Anal Biochem 1982; 122:348–359.
13. Blumberg S, Ben-meir D (inventors), Biotechnology general applicant. Method of removing N-terminal amino acid residues from eucaryotic polypeptide analogs and polypeptides produced thereby. European patent application, 1986, international publication number WO 86/01299.
14. Nakagawa S, Yamada T, Kato K, Nishimura O. Enzymatic cleavage of amino terminal methionine from recombinant human interleukin 2 and growth hormone by aminopeptidase M. Biotechnology 1987; 5:824–827.
15. Glasbrenner K. Technology spurt resolves growth hormone problem, ends shortage. Med News 1986; 255:581–587.
16. Tsunasawa S, Yutani K, Ogasahara K, Taketani M, Yasuoka N, Kakudo M, Sugino Y. Accumulation of amino-terminally formylated tryptophan synthase in amplifying conditions. Agr Biol Chem 1983; 47:1393–1395.
17. Talmadge K, Gilbert W. Cellular location affects protein stability in *Escherichia coli*. Proc Natl Acad Sci USA 1982; 79:1830–1833.
18. Pugsley AP, Schwartz M. Export and secretion of proteins by bacteria. FEMS Microbiol Rev 1985; 32:3–38.
19. Nicaud JM, Mackman N, Holland IB. Current status of secretion of foreign proteins by microorganisms. J Biotechnol 1986; 3:255–270.

20. Oliver D. Protein secretion in *Escherichia coli*. Ann Rev Microbiol 1985; 39:615-648.
21. Perlman D, Halvorson HO. A putative signal peptidase recognition site and sequence in eukaryotic and prokaryotic signal peptides. J Mol Biol 1983; 167:391-409.
22. von Heijne G. How signal sequences maintain cleavage specificity. J Mol Biol 1984; 173:243-251.
23. Lingappa VR, Devillers-Thiery A, Blobel G. Nascent prehormones are intermediates in the biosynthesis of authentic bovine pituitary growth hormone and prolactin. Proc Natl Acad Sci USA 1977; 74:2432-2436.
24. Hirtzman RA, Leung DW, Perry LJ, Kohr WJ, Levine HL, Goeddel DV. Secretion of human interferons by yeast. Science 1983; 219:620-625.
25. Oka T, Sakamoto S, Miyoshi K-I, Fuwa T, Yoda K, Yamasaki M, Tamura G, Miyake T. Synthesis and secretion of human epidermal growth factor by *Escherichia coli*. Proc Natl Acad Sci USA 1985; 82:7212-7216.
26. Chang CA, Rey M, Bochner B, Heyneker H, Gray G. High-level secretion of human growth hormone by *Escherichia coli*. Gene 1987; 55:189-196.
27. Itakura K, Hirose T, Crea R, Riggs AD, Heneker HL, Bolivar F, Boyer HY. Expression in *E. coli* of a chemically synthesized gene for the hormone somatostatin. Science 1977; 198:1056-1063.
28. Goeddel DV, Kleid DG, Bolivar F, Heynecker HL, Yansura DG, Crea R, Hirose T, Kraszewski A, Itakura K, Riggs AD. Expression in *E. coli* of a chemically synthesized genes for human insulin. Proc Natl Acad Sci USA 1979; 76:106-110.
29. Wetzel R. Heynecker HL, Goeddel DV, Jhurani P, Shapiro, Crea R, Low TLK, McClure JE, Thurman GB, Goldstein AL. Production of biologically active Nadesacetylthymosin in *E. coli* through expression of a chemically synthesized gene. Biochemistry 1980; 19:6096-6104.
30. Shemyakin F, Chestukhin AV, Dolganov GM, Khodkova EM, Monsatyrskaya GS, Sverdlov D. Leu-encephalin purification from *E. coli* cells carrying the plasmid with fused synthetic leu-encephalin gene. Nucleic Acids Res 1980; 8:6163-6174.
31. Wetzel R, Kleid DG, Crea R, Heyneker HL, Yansura DG, Hirose T, Kraszewski A, Riggs AD, Itakura K, Goeddel DV. Expression in *E. coli* of a chemically synthesized gene for a "mini-C" analog of human proinsulin. Gene 1981; 16:63-71.
32. Tanaka S, Oshima T, Ohsue K, Ono T, Oikawa S, Takano I, Noguchi T, Kangawa K, Minamino N, Matsuo H. Expression in *E. coli* of chemically synthesized gene for a novel opiate peptide α-neo-endorphin. Nucleic Acids Res 1982; 10:1741-1754.
33. Wilkes SH, Bayliss ME, Prescott JM. Specificity of *Aeromonas* aminopeptidase toward oligopeptides and polypeptides. Eur J Biochem 1973; 34:459-466.
34. Prescott JM, Wilkes S. *Aeromonas* aminopeptidase. Meth Enzymol 1976; 44:531-543.
35. Nakagawa S, Osamu N (inventors), assignee Takeda Chemical Industries, Ltd. Removal of the N-terminal methionine from methionylproteins. European patent application, 1986, international publication number O 204 527 A1.
36. Delange RJ Smith EL. Dipeptidyltransferase (dipeptidyl aminopeptidase I, cathepsin C), In: 'The enzymes,' Vol. 3, Boyer PD, ed. New York: Academic Press, 1971; 105-111.

37. Andersen HD, Pedersen J, Christensen T, Hansen JW, Jessen TH (inventors), Nordisk Gentofte (applicant). A process for producing human growth hormone. European patent application, 1986, international publication number WO 86/04609.
38. Dimarchi RD, Brooke GS (inventors), Eli Lilly and Company (applicant). Selective chemical removal of a protein amino-terminal residue. European patent publication 1986, international publication number 220 958 A2, 1986.
39. Bachmair A (inventor), Massachusetts Institute of Technology (applicant). Method of Regulating Metabolic Activity of Proteins. European patent application, 1988, Int. Pub. No. WO86/02406.
40. Bachmair A, Finley D, Varshavsky A. In vivo half-life of a protein is a function of its amino-terminal residue. Science 1986; 234:179–186.

III

PURIFICATION OF RECOMBINANT PROTEINS FROM *Escherichia coli*, YEAST, AND MAMMALIAN CELLS

7
Purification of Secreted Recombinant Proteins from *Escherichia coli*

Hung V. Le and Paul P. Trotta

*Schering-Plough Research
Bloomfield, New Jersey*

I. INTRODUCTION

Advances in recombinant DNA technology have provided a variety of host/vector systems for the cloning and expression of heterologous genes in microorganisms. The ideal expression system would induce the synthesis of a sufficient level of recombinant protein to allow isolation of milligram quantities of highly purified proteins. High-level expression of heterologous proteins in bacterial cells has become readily achievable with the use of strong promoters, appropriate Shine–Dalgarno sequences, and inducible multicopy plasmids. However, it is now evident that the isolation of biologically active recombinant proteins by the application of simple purification methodology is highly dependent on the nature of the host expression system. The main focus of this chapter will be a summary of the reported advantages of secretion versus cytoplasmic expression systems for the development of efficient purification protocols that are applicable on both a laboratory and an industrial scale.

Escherichia coli is the most commonly employed bacterial host for heterologous gene expression since a large body of information is available on its genetic and physiological characteristics. However, experience has shown that cytoplasmic expression in *E. coli* is not always achievable without many undesirable effects. For example, foreign proteins, especially the smaller polypeptides, are often very rapidly degraded to truncated variants by intracellular proteases (1–3). In addition, recombinant proteins can accumulate in the cytoplasm as aggregates (i.e., inclusion or refractile bodies) when the expression level exceeds approximately 10% of the cytoplasmic protein (4–14). Recovery of biologically active protein in its native conformation from these inclusion bodies could be difficult since the process involves solubilization with strong denaturants like urea or guanidine hydrochloride. The extracted protein must be renatured and refolded to the proper conformation, which is frequently difficult to achieve with high yields. However, even if the expressed protein remains soluble in the cytoplasm, isolation of this protein is a formidable problem since it must be purified from a milieu containing a variety of host cytoplasmic proteins (15). In addition, pyrogenic lipopolysaccharides derived from the bacterial cell wall represent a potential contaminant that must not be present at high levels if the recombinant protein is to be employed in clinical trials. These substances, which cause a rise in temperature when injected parenterally, are frequently extracted during disruption of the bacterial cell wall to release the cytoplasmic components. Finally, it is important to note that proteins derived from cytoplasmic expression systems often contain the initiator *N*-formyl methionine or methionine at the N terminus. This extra amino acid may affect the biological activity as well as potentially increase the immunogenicity of the molecule, which is of particular concern for the clinical application of the recombinant protein (16).

Escherichia coli as a gram-negative bacterium contains both an outer and a cytoplasmic membrane. As a consequence, a periplasmic space is created into which proteins can be secreted if they contain an N-terminal extension called a signal sequence. These proteins are generally periplasmic proteins or proteins of the outer membrane. Utilizing the signal sequence of these proteins host vector systems have been developed that result in secretion of the heterologous gene products into the cell periplasm (17–20). In some cases there is a subsequent release into the culture medium (21,22). It is notable that secretion of heterologous proteins is not restricted to prokaryotic cells since it has been observed that eukaryotic hosts transfected with plasmids containing gene sequences for preproteins will also secrete mature proteins into the culture medium (23–25). However, the relative simplicity and economic advantages of fermenting bacteria on a large scale compared to mammalian cell culture provides a distinct advantage for a bacterial expression system. A number of excellent reviews on bacterial secretion systems are available (26,27).

The secretion of heterologous proteins into the periplasmic space provides clear advantages for the isolation and purification of the heterologous protein in high yield. Most of the problems outlined above for the cytoplasmic expression systems (i.e., use of denaturants, the presence of N-terminal methionine, and susceptibility to proteases) are avoided in the secretion systems. We review in this chapter recent data on the purification methodology that has been applied to the isolation of proteins expressed in *E. coli* with secretion plasmids. Secretion will refer not only to the phenomenon of transport into the periplasmic space but also to subsequent leakage through the outer membrane into the culture medium.

II. PURIFICATION METHODOLOGY

The first step in the isolation of a secreted heterologous protein is release from the periplasm. The most commonly employed procedure has been osmotic shock (28). This process is difficult to apply on a large scale and hence poses special difficulties for industrial manufacture. Alternatively, "leaky" mutants have been developed that result in the release of part or all of the periplasmic protein into the extracellular space (29). The released protein may be less contaminated with host proteins compared to the starting material generated by osmotic shock if the leakage is selective for only a portion of the periplasmic fraction. However, since leaky strains have frequently been demonstrated to be difficult to grow, they have not represented a common source for secreted proteins. Interestingly, a portable "secretion" gene has also been cloned into *E. coli* that apparently renders the outer membrane permeable (30). The utility of the latter system for different heterologous proteins remains to be demonstrated.

Published data on purification and characterization of secreted heterologous proteins are available for a relatively few systems, which are summarized in Table 1. Although genes for other proteins have been expressed in *E. coli*,

Table 1 Proteins Expressed in *E. coli* with Secretion Vectors and Purified to Homogeneity

Protein	Signal peptide	Ref.
hGH	Alkaline phosphatase	31
	ompA	32, 33
muIL-2	ompA	35
Human GM-CSF	ompA	44
IGF-1	LamB	45
	Staphylococcal protein A	21, 47
Hirudin	Alkaline phosphatase	51
hEGF	Alkaline phosphatase	54
Human β-endorphin	ompF	22

detailed information on purification protocols and recovery is not available. The data do demonstrate that relatively simple purification methodology can result in a high degree of purification without the use of strong denaturants for extraction. Appropriate characteristics of the purified protein will be cited that provide evidence for the presence of a native, biologically active conformation.

A. Human Growth Hormone (hGH)

Studies have been performed in the secretion and processing of hGH expressed in *E. coli* with secretion vectors. Gray and colleagues (31) constructed expression vectors encoding either (a) the naturally occurring hGH precursor under the control of the *E. coli trp* promoter or (b) the mature hGH coding sequence fused to the *E. coli* alkaline phosphatase promoter and alkaline phosphatase signal sequence codons. In both cases, osmotic shock followed by assay of the protein released for hGH content by radioimmunoassay indicated an export of ~80% of the hGH to the periplasm. These data demonstrate the effectiveness of the bacterial leader sequence in mediating periplasmic transport. Immunoaffinity chromatography was employed as a one-step procedure to purify each of the hGH proteins from the osmotic shock fluids. Migration of the purified protein on SDS-PAGE confirmed that the molecular weights were similar to that of authentic methionine-hGH. However, definitive evidence for correct processing was obtained by amino acid sequencing, which indicated the Phe-Pro-Thr-Ile sequence at the N terminus. This sequence is exactly as predicted for the mature, correctly processed hGH.

Another issue that was also addressed by Gray et al. (31) is whether the secreted hGH proteins have correctly paired disulfide bonds, which are expected to occur at Cys^{53}–Cys^{165} and Cys^{182}–Cys^{189}. Although these authors did not perform an analysis of disulfide pairing, it was demonstrated that the migration of the periplasmic hGH was similar to authentic hGH on SDS-PAGE in both reduced and nonreduced states. Similarly, the hGH was homogeneous and monomeric under both conditions. In contrast, cytoplasmic methionine hGH displayed a variety of electrophoretic forms, including oligomers, on SDS-PAGE of the nonreduced protein.

An efficient, high-level secretion of hGH into the periplasm of *E. coli* was also achieved with a secretion-cloning vector containing a gene coding for the *ompA* signal (32). This vector was constructed from the plasmid pIN-III (33). hGH was released from the periplasm by osmotic shock and purified by sequential anion exchange and gel filtration chromatographies. The results of this purification procedure are presented in Table 2. Densitometric scanning of the periplasmic fraction indicated a percentage purity of ~30%, supporting a significant enrichment of hGH simply by application of osmotic shock. Anion exchange chromatography resulted in a preparation of ~90% purity, which could be further improved to greater than 90% purity by gel filtration.

Table 2 Purification of hGH from the Periplasmic Fraction of *E. coli* RV308/pOmpA-hGH2[a]

Step	Protein (mg)	hGH (mg)	Specific activity (%)	Purification (fold)	Yield (%)
Total cell lysate	ND	ND	6[b]	1	—
Periplasmic fraction	93.4	28.4	30.4	5	100
Q-Sepharose Fast Flow	24.6	22.3	90.6	15	78
Sephacryl S-200	N.D.	20.2	>90[b]	>15	71

[a]Human growth hormone was purified from the periplasmic fraction derived derived from 1.0 liters of culture grown to a cell density of $A_{600} = 1.6$. N.D., not determined.
[b]The specific activities of the total cell lysate and of the Sephacryl S-200 pool were estimated by densitometric scanning of an SDS-PAGE.
Source: Ref. 32.

Tryptic mapping of the secreted hGH demonstrated that the two expected disulfide bonds were present in the secreted hGH. The tryptic map was indistinguishable from that produced by authentic hGH. Further, as described above, N-terminal sequencing demonstrated that *E. coli* correctly processed the precursor, since the expected sequence (Phe-Pro-Thr-Ile) was observed at the N terminus. It was consistent with this observation that migration on SDS-PAGE was identical for the periplasmically derived hGH versus an authentic hGH standard.

B. Murine Interleukin-2 (muIL-2)

Based on the sequence of the cloned cDNA (34), the mature protein sequence of muIL-2 is predicted to consist of 149 amino acids, including three cysteines. muIL-2 has been purified from an *E. coli* strain engineered with a secretion vector containing an ompA signal peptide (35). In these studies the *E. coli* secretion strain 294 pOmpA mIL-2 was subjected to osmotic shock according to a procedure described by Neu and Heppel (28). Characterization of the extracted periplasmic fraction by SDS-PAGE showed the presence of muIL-2 at a level of ~10% of total extracted protein. As summarized in Table 3 purification to a high degree of homogeneity was achieved by DEAE-Sepharose and Sephadex G-100 chromatography. The specific activity of the purified muIL-2 (i.e., 3.5×10^7 U/mg in a T-cell proliferation assay) was comparable to that reported for naturally occurring muIL-2 isolated from EL-4 thymoma cells (36). N-terminal sequencing of the purified protein demonstrated that the ompA signal peptide had been correctly processed, as evidenced by the fact that the first five N-terminal amino acids were exactly as predicted for the mature muIL-2.

Table 3 Purification of Recombinant muIL-2 Derived from *E. coli* Strain 294/pOmpA-mIL-2[a]

Step	Protein (mg)	Units × 10^6 Per mg	Units × 10^6 Total	Recovery (%)
Osmotic lysate	665	4.2	280	(100)
DEAE-Sepharose	31	50	160	57
Sephadex G-100	7.5	35	26	9.7

[a]All steps were performed at 4°C.
Source: Ref. 35.

A homogeneous preparation of muIL-2 from the secretion strain was analyzed for the presence of disulfide bonds and free cysteines. It was observed that 0.7 mol of disulfides was present per mol of protein when titration was performed with 2-nitro-5-thiosulfobenzoic acid (37). Titration with 4,4'-dithiodipyridine (38) indicated the presence of 1.8 mol of free sulfhydryls per mol of protein. Summation of these values resulted in a total content of cysteine (i.e., 3.2 mol) in good agreement with the cDNA prediction of 3.0 (34). The fact that only 0.7 mol of disulfide was titratable per mol of muIL-2 suggests the existence of a small subpopulation in which all cysteines are reduced. It has not yet been established whether the fully reduced muIL-2 is biologically active.

In the same study (35), muIL-2 was also extracted from *E. coli* engineered with a cytoplasmic vector and the properties of muIL-2 from the two strains were compared. It was observed that inclusion of 1% deoxycholate in the extraction buffer during Manton–Gaulin homogenization resulted in an approximately fivefold increase in the amount of muIL-2 extracted. These data suggested that at least a portion of the muIL-2 in this strain was in an aggregated form. Further evidence for the aggregated state of muIL-2 was provided by the fact that gel filtration of a partially purified preparation indicated a molecular weight in excess of 250,000. In contrast, muIL-2 purified from the secretory strain exhibited an apparent molecular weight of 30,000 under the same conditions of gel filtration. Based on the molecular weight of the polypeptide chain predicted from the cDNA (34), the value of 30,000 suggested the presence of a dimer.

Experiments employing SDS-PAGE supported the idea that disulfide bonds contributed to the aggregation of muIL-2 derived from the cytoplasmic strain. In the absence of prior treatment with a reducing agent muIL-2 derived from the cytoplasmic strain failed to demonstrate significant penetration into a 15% polyacrylamide gel (Fig. 1, lanes E and F). However, reduction of the preparation with 2-mercaptoethanol resulted in a band on SDS-PAGE with an apparent molecular weight of 19,200 (Fig. 1, lanes B and C). In distinction, muIL-2 derived from the secretory strain migrated on SDS-PAGE with the same apparent molecular weight with or without prior reduction (Fig. 1, lanes D and G).

The specific biological activity of muIL-2 in the deoxycholate extract of the cytoplasmic strain was 10-fold lower than observed for the specific activity of muIL-2 derived from the secretion strain (Table 3). In contrast, the percentage of muIL-2 in the extracts of the cytoplasmic and secretion strains was determined from densitometric scanning of the SDS-PAGE to be similar (i.e., ~10% of the total protein in the extract). This result suggests an inherent difference in the biological activity of the muIL-2 from the two strains. It is tempting to speculate that the lower specific activity is related to the high degree of aggregation. However, it was also determined that the N-terminal amino acid of muIL-2 from the cytoplasmic strain was methionine, whereas muIL-2 from the secretory strain lacked this residue (35). The small difference in migration of the two muIL-2 preparations on SDS-PAGE of reduced preparations (Fig. 1, lanes B and D) may reflect the presence of this extra amino acid. Thus, an additional hypothesis to explain the data is that the presence of an N-terminal methionine may suppress biological activity.

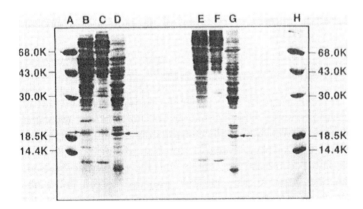

Figure 1 SDS-PAGE of *E. coli* extracts of muIL-2 with and without prior reduction with 2-mercaptoethanol. Samples in lanes A, B, C, D, and H were boiled for 2 min with 1% 2-mercaptoethanol compared to samples in lanes E, F, and G, which were not treated with 2-mercaptoethanol. (A, H) molecular weight standards; (B, E) phosphate-buffered saline extract of *E. coli* 294/exTAC-mIL-2; (C,F) phosphate-buffered saline extract of *E. coli* 294/exTAC-mIL-2 partially purified; (G,D) osmotic lysate of *E. coli* 294/pOmpA-mIL-2. The gel was stained with Coomassie Brilliant Blue R-250. MuIL-2 is identified with an arrow.

In summary, the available evidence indicates that muIL-2 was efficiently exported into the periplasmic space following cleavage of the ompA signal peptide. The environment of the periplasmic space was conducive to refolding of the polypeptide into a native form. These results suggest a clear advantage for the secretion expression system for the isolation of muIL-2.

C. Granulocyte-Macrophage Colony-Stimulating Factor (GM-CSF)

GM-CSF is one of a group of glycoproteins that can stimulate the proliferation and differentiation of granulocyte and macrophage progenitor cells. GM-CSF has been expressed in a number of host/vector systems with only limited success. For example, transfection of COS-7 cells with the pcD plasmid containing the gene for human GM-CSF resulted in a low, transient level of expression of 1 μg/ml (39). Expression of the GM-CSF gene in yeast (40) resulted in secretion of GM-CSF into the medium at the higher level of 10 μg/ml. However, expression in yeast cells may result in a higher degree of glycosylation than observed in the naturally occurring protein, which could impair the full expression of biological activity of the polypeptide (41). Human and murine GM-CSF have both been expressed cytoplasmically in *E. coli* at a high level although in an insoluble form (42,43). Purification of these proteins required extraction of the insoluble aggregates with a strong denaturant (e.g., guanidine-HCl) followed by refolding of the protein in a medium containing urea.

Human GM-CSF has been expressed from a high-level secretion vector (pIN-III-ompA$_3$) in which a cDNA fragment encoding mature GM-CSF was fused either directly or through a synthetic oligonucleotide encoding four extra amino acids (44). The constructions were under the transcriptional control of the tandem lipoprotein promoter/lactose promoter-operator and was regulated by the lactose repressor. In principle, both constructions should result in the export of GM-CSF into the periplasmic space. However, treatment of the cells with osmotic shock failed to release the protein into the cold-water fraction. Consequently, in an attempt to release insoluble protein, the cell pellet obtained after osmotic shock was treated with high salt, but this treatment did not result in solubilization. This result provided evidence for the high degree of insolubility of human GM-CSF, which should have accumulated in the periplasmic space. Subsequent cellullar fractionation followed by studies of the distribution of GM-CSF by SDS-PAGE and Western blots showed that GM-CSF was associated with a membrane fraction. Western blot analysis with antibodies to GM-CSF revealed that a low level of GM-CSF containing the uncleaved ompA signal peptide also was present in the membrane fraction. Identification of which membrane was involved was not obtained in this study.

However, Greenberg et al. (45), who recently expressed human GM-CSF in *E. coli* utilizing a similar secretion vector, demonstrated that insoluble GM-CSF was located in the outer membrane fraction. Thus, it appears that the ompA signal peptide directed GM-CSF out of the cytoplasm and into the outer membranes. The level of expression with the ompA secretion vector was estimated at 20 mg/liter of culture.

Extraction of GM-CSF from the membrane fraction was accomplished with 8 M urea. However, the extraction of GM-CSF from the membranes did not require the addition of reducing agent to the extraction buffer. Membrane-associated GM-CSF is therefore apparently not aggregated through intermolecular disulfide bonds. It is plausible that formation of the two intramolecular disulfide bonds (Cys^{54}–Cys^{96} and Cys^{88}–Cys^{121}, respectively) had already occurred during or after translocation of the polypeptide from the cytoplasm through the inner membrane. The solubilized protein was purified by conventional chromatography on DEAE-Sephacel and phenyl-Sepharose CL-4B. Two additional reversed-phase high-performance liquid chromatography steps using a Vydac C4 resin were employed to achieve homogeneity.

D. Insulinlike Growth Factor-I (IGF-1)

IGF-1 (somatomedin C) shares many common biological effects with hGH. The production of this 70-amino-acid polypeptide by recombinant DNA technology has been problematic. Unsuccessful attempts to express the gene for IGF-1 in *E. coli* using cytoplasmic expression vectors have been attributed to the susceptibility of the polypeptide to intracellular proteases. Construction of a secretion vector system allowing export of IGF-1 to the outer membrane of *E. coli* has provided a solution to this problem. Thus, a synthetic human IGF-1 gene sequence was inserted immediately downstream from the signal peptide sequence of the outer membrane protein LamB. The hybrid gene was introduced into a pBR327 plasmid under the control of the RecA promoter with the bacteriophage T7 gene 10 leader region located 5' to the coding sequence (46). After induction by naldixic acid the level of expression of IGF-1 in the periplasm was 16–20 μg/OD/ml, which was determined by densitometric scanning of Coomassie Blue-stained SDS-PAGE. However, 80% of the processed IGF-1 was insoluble in the form of refractile bodies and 20% was a soluble form that could be released by osmotic shock. A small percentage (3–5%) of the processed form of IGF-1 appeared to leak through the outer membrane and could be detected in the culture medium by Western blot analysis. The soluble form of IGF-1 constituted the starting material for purification from this expression system. Sequencing of the purified IGF-1 indicated an N-terminal sequence identical to authentic IGF-1.

A novel expression system that facilitates large-scale production of IGF-1 by allowing efficient excretion into the culture medium of *E. coli* has been reported

(47). The secretion plasmid contained a "heat shock"-induced staphylococcal protein A promoter/signal sequence and two repeats of the IgG-binding domain linked to a synthetic IGF-1 gene (21). The peptide bond linking the IgG-binding domain to IGF-1 was constructed as the sequence Asn-Gly by oligonucleotide-directed in vitro mutagenesis, thereby allowing for cleavage of the fusion product by hydroxylamine. For reasons yet to be elucidated, the plasmid caused the secretion of the majority (i.e., 85%) of the fusion protein into the culture medium and the remainder was sequestered in the periplasm.

In a large-scale fermentation (1000-liter) the IGF-1 protein was secreted into the culture medium at a level of 75 mg/liter of fermentation following a heat shock treatment that consisted of shifting the incubation temperature from 37°C to 44°C during stationary phase. Cells were separated from the culture medium by a combination of centrifugation and cross-flow ultrafiltration through a 0.2-μm membrane. The sterile culture medium supernatant was applied to an IgG-Sepharose Fast Flow column from which the fusion protein was eluted with 0.5 M ammonium acetate, pH 3, and the process of adsorption and elution was repeated for 21 cycles. However, based on the reported data, processing of the culture medium from a 1000-liter fermentation can be calculated to require an average time of 3-½ months. Thus, the efficiency of this affinity chromatography step clearly requires further optimization. The pooled eluate was lyophilized, redissolved in water, and treated with hydroxylamine, which resulted in 80% cleavage at the Asn-Gly peptide bond. The cleavage mixture was desalted and passed through the IgG affinity resin to remove uncleaved starting material. The protein fraction that failed to adsorb consisted of pure IGF-1, as demonstrated by SDS-PAGE. The purified recombinant IGF-1 exhibited a specific activity in a receptor-binding assay similar to that of the native hormone purified from human serum (48).

E. Hirudin

Hirudin is a polypeptide containing 65 amino acids that has been obtained from the salivary gland of the leech *Hirudo medicinalis*. It has generated considerable interest as an antithrombolytic agent since it is a potent and specific inhibitor of thrombin. Several laboratories have attempted to clone and express the gene for hirudin in *E. coli* using cytoplasmic expression vectors (49,50). However, these attempts uniformly resulted in low level of expression ranging from 15 to 28 ng/liter/OD_{578} unit of cell. Although the cause for low expression was not investigated, the possibility exists that newly synthesized hirudin was rapidly degraded by the intracellular proteases of the host *E. coli* strain.

Dodt et al. (51) introduced a synthetic gene for hirudin into a secretion plasmid using the alkaline phosphatase (phoA) signal peptide under the control of the *tac* promoter. Hirudin was predominantly secreted into the periplasmic space of *E. coli* at a level of ~4 mg/liter of fermentation broth although a significant quantity was also found in the culture medium. This level of expression constituted an improvement of 3000- to 4000-fold over the cytoplasmic expression systems, in which hirudin was expressed as the mature protein or as a fusion protein with β-galactosidase (49). Hirudin from the periplasmic fraction was readily purified to homogeneity by anion exchange chromatography on DEAE-Sephadex followed by reversed phase HPLC. Sequencing studies confirmed that the phoA signal peptide was correctly processed during translocation of the preprotein into the periplasmic space. The secreted hirudin did not contain the sulfated tyrosine at position 63 that is characteristic of the naturally occurring polypeptide isolated from leeches. However, the specific activity of the recombinant product with respect to thrombin inhibitory activity was comparable to that obtained by desulfatation of naturally occurring hirudin (51). Another form of hirudin that had three additional amino acids at the N terminus, which were derived from the hirudin gene, was also isolated, but it displayed very low biological activity.

It is notable that approximately 30% of the secreted hirudin was detected in the culture medium. Thus, the culture medium itself could have provided an excellent starting material for purification. The mechanism by which hirudin is secreted into the culture medium remains unknown. It can be speculated that the low molecular weight and hydrophilicity of the molecule facilitated its passage through the outer membrane.

F. Human Epidermal Growth Factor (hEGF)

hEGF is a polypeptide hormone containing 52 amino acids that stimulates the proliferation of epithelial cells and displays a potent inhibitory effect on gastric secretion (52). Its primary structure is identical to that reported for urogastrone (53). A synthetic gene of hEGF, linked to the alkaline phosphatase (phoA) signal sequence and promoter, was expressed in several *E. coli* hosts by Oka et al. (54). Virtually all of the processed hEGF in the periplasm could be recovered in the cold-water wash of the cells that followed the osmotic shock procedure. However, the yield of hEGF in the periplasm was shown to be host-dependent. After osmotic shock treatment of the cells, the cold-water wash of strains C-600 and YK537 contained 0.2 and 1.0 mg/liter of culture, respectively. Further improvement in the yield of hEGF from strain YK537 to 2.4 mg/liter of culture was achieved by increasing the incubation period at 30°C from 6 to 9 hr.

Purification of hEGF from the periplasmic fraction was accomplished by a two-step procedure consisting of gel filtration chromatography followed by reversed phase HPLC using a C18 column. The final product was shown to consist of a single component on SDS-PAGE. Amino acid analysis and N-terminal sequencing indicated that the recombinant hEGF was identical to the authentic material. The high biological activity, which was comparable to that of human urogastrone, suggested that the expected three disulfide bonds were correctly paired, although this pairing was not directly demonstrated.

It was also shown in these studies (54) that when hEGF was expressed with a cytoplasmic expression vector in which the hEGF gene was under the control of the same promoter as the secretion vector but did not contain the phoA signal sequence, only very low levels of hEGF were detected in either the cytoplasm or the periplasm of the *E. coli* host. For example, the total cell lysate of *E. coli* strain YK537 transfected with the cytoplasmic vector produced hEGF at 0.02 mg/liter of culture. This low expression level was probably not due to the inability of the plasmid to support efficient expression of hEGF since other heterologous proteins have been expressed under the control of the phoA promoter (55), but may have resulted from proteolytic degradation of hEGF in the cytoplasm.

G. OmpF-Human β-Endorphin Fused Polypeptide

β-Endorphin is a neuropeptide containing 63 amino acids that displays a high affinity for the opioid receptor and is derived from the corticotrophin-β-lipotropin precursor. Since it is a relatively small polypeptide, β-endorphin has been readily obtained directly by solid phase peptide synthesis (56). Its production by recombinant DNA technology was recently achieved by Nagahari et al. (22), who reported the secretion into the culture medium of *E. coli* of a fused polypeptide consisting of a terminal fragment of the ompF protein, which is a major protein of the outer membrane of *E. coli*, followed by β-endorphin. A plasmid was constructed in which the coding sequence for human β-endorphin was preceded by the upstream region of the ompF gene. This region contained the coding sequences for the ompF signal peptide and eight amino acid residues from the N terminus of the ompF protein. The purpose of this construct was to use the ompF signal peptide for the passage of the product through the cytoplasmic membrane and to use the N terminus of the ompF protein for access to the outer membrane. An ompF/β-endorphin fused polypeptide was synthesized in two strains containing this plasmid and was secreted into the culture medium by passage through both the cytoplasmic and outer membranes. The fused polypeptide could not be detected in the periplasm.

Under optimal conditions the concentration of the fused polypeptide was 1.1 mg/liter of culture medium as determined by radioimmunoassay. Purification from 1 liter of culture medium was accomplished by gel filtration chromatography followed by a reversed phase HPLC. Four distinct immunoreactive polypeptides were resolved by reversed phase HPLC. The two most abundant polypeptides were shown by amino acid analysis and N-terminal sequencing to correspond to the mature fused peptide and a variant truncated at the C terminus between Lys^{31} and Asn^{32}. Two other minor components consisted of a variant terminating at Lys^{36} and a C-terminal fragment corresponding to residues 32–43. The absence of the ompF signal peptide indicated that processing had occurred as expected during translocation through the cytoplasmic membrane.

The mechanism by which the β-endorphin-fused polypeptide is secreted into the culture medium remains unexplained. However, it is known that the process is highly selective since substantial amounts of marker proteins of the periplasm (i.e., β-lactamase and alkaline phosphatase) did not leak into the culture medium. The selective secretion through the outer membrane can be attributed to the small size and hydrophilic nature of the fused polypeptide. Alternatively, the fused polypeptide could have been secreted by the mechanism normally controlling secretion of *E. coli* proteins like hemolysin.

III. SUMMARY AND CONCLUSIONS

Secretion systems engineered for the expression of heterologous protein in *E. coli* provide several advantages for subsequent isolation of purified product. Proteins released from the periplasmic space, which represent a small fraction (i.e., 4–10%) of total cell protein, can readily be separated from other cellular proteins by centrifugation of the remaining cellular debris or cross-flow ultrafiltration. The starting material derived from secretion systems is generally of higher purity than comparable material produced from strains expressing cytoplasmically for systems exhibiting similar expression levels. The available evidence suggests that recombinant proteins derived from the periplasm are generally, but not always (44–46), soluble in a nonaggregated form. Consequently, simple purification protocols can be effectively employed for producing homogeneous product with a high yield. The majority of the secreted recombinant proteins reviewed in this chapter were purified by simple one- or two-step chromatography procedures. High-resolution techniques such as reversed phase HPLC were found necessary only in cases where the secreted polypeptides were contaminated with proteolytic degradation variants, e.g., hirudin (51) and β-endorphin (22).

The fact that a high level of biological activity has been shown to be characteristic of purified recombinant proteins secreted into the periplasmic space suggests the presence of a native conformation stabilized by the expected disulfide linkages. Intramolecular disulfide bonds most probably form either as the polypeptide is translocated through the cytoplasmic membrane into the periplasm or within the periplasmic compartment, which has a higher oxidation potential than that found in the cytoplasm (57). Studies performed with hGH (31) and muIL-2 (35) provide excellent examples of differences observed in protein folding and disulfide bond formation between heterologous proteins expressed in the cytoplasmic and periplasmic compartments. Thus, hGH and muIL-2 extracted from the cytoplasm of *E. coli* have been characterized as high molecular weight disulfide-bonded oligomers. It is likely that oligomerization occurs as the polypeptides are released from the reducing environment of the cytoplasm. In contrast, secreted hGH and muIL-2 extracted from the periplasm of *E. coli* by osmotic shock displayed the properties of a properly folded native protein with correct disulfide pairing. In the case of muIL-2 only a small residual fraction ($\sim 15\%$) of the purified secreted protein exhibited incomplete oxidation of cysteine (35).

Secretion of heterologous proteins into the periplasm prevents their exposure to the action of proteases located in the cytoplasm of *E. coli* (58). The smaller polypeptides such as somatostatin (59), IGF-1 (46), and hEGF (54) are known to be particularly susceptible to intracellular degradation. Interestingly, Goff et al. (2) reported that cytoplasmic expression of tissue plasminogen activator increased the level of ATP-dependent proteases in *E. coli*. Despite the high level of general proteolytic activity in the cytoplasm, heterologous proteins expressed in this compartment have often been observed to retain the N-terminal methionine on at least a portion of the molecules as a result of a lack of proper processing by the appropriate N-terminal methionine peptidases. The presence of the extra methionine could be problematic for recombinant proteins targeted for therapeutic applications since this residue represents a nonnative addition to the polypeptide chain and hence may be a potential site for antibody formation. The secretion plasmids offer a clear solution to this dilemma since transport of newly synthesized preproteins into the periplasm must be accompanied by cleavage of signal peptides, thereby completely eliminating the possibility of residual N-terminal methionine.

Although secretion systems are generally superior to cytoplasmic expression systems as outlined above, various shortcomings have been noted. First, tagging a heterologous protein with a signal peptide does not necessarily guarantee its secretion into the periplasm. Studies with the LamB-LacZ fusion proteins demonstrated that the amino acid sequence located at the N terminus of the mature sequence is an important factor determining whether a heterologous protein will be efficiently secreted (60,61). It is notable that two proteins (62,63) have been shown to fail to be secreted when they were fused with bacterial signal peptides. When the gene coding for triosephosphate isomerase was joined to the signal sequence for β-lactamase, the resulting hybrid protein accumulated in the cytoplasm in an unprocessed form (62). In addition, a fusion protein between prealkaline phosphatase and human α-neoendorphin remained in the cytoplasm in spite of the proper cleavage of its signal peptide (63). However, even under conditions where the signal peptide has been processed correctly, the mature protein may remain in either the inner or outer membranes of the bacterial envelope, as has been observed for GM-CSF (44) and IGF-1 (R. Seetharam, personal communication). In such cases a high concentration of a chaotropic agent may be required for extraction, potentially resulting in irreversible denaturation.

A potential solution to preventing the entrapment of secreted proteins in the membrane fraction is the use of periplasmic-leaky mutants as hosts (29,64). These mutants, which may be induced by ultraviolet irradiation or nitrosoguanidine treatment, release periplasmic enzymes directly into the extracellular milieu. This phenotype has been described in strains of *E. coli* containing reduced amounts of the outer membrane protein ompF (29) as well as those deficient in lipoprotein and lipopolysaccharide (65,66). Periplasmic-leaky strains have been utilized for the production of alkaline phosphatase, an endogenous protein in the *E. coli* periplasm, and cellulase cloned from *Cellulomonas fimi* (27). In principle, continuous release of heterologous proteins into the medium during fermentation offers distinct advantages over osmotic lysis of harvested cells for the preparation of purified proteins. However, the general utility of such a host for the synthesis and secretion of a variety of heterologous proteins has yet to be established.

REFERENCES

1. Prouty WF, Karnovsky M, Goldberg AL. Degradation of abnormal proteins in *Escherichia coli*. J. Biol. Chem. 1975; 250:1112–1123.
2. Goff SA, Goldberg AL. Production of abnormal proteins in *E. coli* stimulates transcription of Lon and other heat shock genes. Cell 1985; 41:587–595.
3. Goff SA, Goldberg AL. An increase content of protease La, the Lon gene product, increases protein degradation and block growth in *E. coli*. J Biol Chem 1987; 262:4508–4515.
4. Williams DC, Van Frank RM, Muth WL, Burnett JP. Cytoplasmic inclusion bodies in *Escherichia coli* producing biosynthetic human insulin proteins. Science 1982; 215:687–689.
5. Schoner RG, Ellis LF, Schoner BE. Isolation and purification of protein granules from *Escherichia coli* cells overproducing bovine growth hormone. Biotechnology 1985; 3:151–154.
6. Gribskov M, Burgess RR. Overexpression and purification of the sigma subunit of *Escherichia coli* RNA polymerase. Gene 1983; 26:109–119.
7. Ho YS, Lewis M, Rosenberg M. Purification and properties of a transcriptional activator. J Biol Chem 1982; 257:9128–9134.
8. Simons G, Remaut E, Allet B, Devos R, Fiers W. High level expression of human interferon gamma in *Escherichia coli* under control of the PL promoter of bacteriophage lambda. Gene 1984; 28:55–64.
9. Devos R, Plaetinck G, Cheroutre H, Simons G, Degrave W, Tavernier J, Remaut E, Fier W. Molecular cloning of human interleukin 2 cDNA and its expression in *Escherichia coli*. Nucleic Acids Res 1983; 11:4307–4323.

10. Cheng YSE, Kwoh DY, Kwoh TJ, Soltvedt BC, Zipser D. Stabilization of a degradable protein by its overexpression in *Escherichia Coli*. Gene 1981; 14:121–130.
11. Itakura K, Hirose T, Crea R, Riggs AD, Heynecker HL, Bolivar F, Boyer. Expression in *Escherichia coli* of a chemically synthesized gene for the hormone somatostatin. Science 1977; 198:1056–1063.
12. Kleid DG, Yansura D, Small B, Dowbenko D, Moore DM, Grubman MJ, McKercher PD, Morgan DO, Robertson BH, Bachrach HL. Cloned viral protein vaccine for foot-and-mouth disease: Responses in cattle and swine. Science 1981; 214:1125–1129.
13. Paul DC, Van Frank RM, Muth WL, Ross JW, Williams DC. Immunocytochemical demonstration of human proinsulin chimeric polypeptide within cytoplasmic inclusion bodies of *Escherichia coli*. Eur J Cell Biol 1983; 31:171–174.
14. Masui Y, Mizuno T, Inouye M. Novel high-level expression cloning vehicles: 104-fold amplification of *Escherichia coli* minor protein. Biotechnology 1984; 2:81–85.
15. Trotta PP, Le HV, Sharma B, Nagabhushan TL. Isolation and purification of human alpha interferon, A recombinant DNA protein. In: Development in industrial microbiology, Vol. 27, Pierce, G. ed., Amsterdam: Elsevier, 1987:53–64.
16. Glasbrenner K. Technology spurt resolves growth hormone problem, ends shortage. JAMA 1986; 255:581–587.
17. Talmadge K, Stahl S, Gilbert W. Eukaryotic signal sequence transports insulin antigen in *Escherichia coli*. Proc Natl Acad Sci USA 1980; 77:3369–3373.
18. Talmadge K, Kaufman J, Gilbert W. Bacterial mature preproinsulin to proinsulin. Proc Natl Acad Sci USA 1980; 77:3988–3992.
19. Ghrayeb J, Kimma H, Takahara M, Hsiang H, Masui Y, Inouye M. Secretion cloning vectors in *Escherichia coli*. EMBO J 1984; 2437–2442.
20. Takahari M, Hibler DW, Barr PJ, Gerlt JA, Inouye M. The OmpA signal peptide directed secretion of staphylococcal A by *Escherichia coli*. J Biol Chem 1985; 260:2670–2674.
21. Abrahmsen L, Moks T, Nilsson B, Uhlen M. Secretion of heterologous gene products to the culture medium of *Escherichia coli*. Nucleic Acids Res 1986; 14:7487–7500.
22. Nagahari K, Kanaya S, Munakata K, Aoyagi Y, Mizushima S. Secretion into the culture medium of a foreign gene product from *Escherichia coli*: Use of the OmpF Gene for secretion of human β-endorphin. EMBO J 1985; 4:3589–3592.
23. Kingsman SM, Kingsman AJ, Dobson MJ, Mellor J, Roberts NA. Heterologous gene expression in *Saccharomyces cerevisiae*. Biotechnol Genet Eng Rev 1985; 3:377–416.
24. Cullen D, Gray GL, Wilson LJ, Hayenga KJ, Lamsa MH, Rey MW, Norton S, Berka RM. Controlled expression and secretion of bovine chymosin in *Aspergillus nidulans*. Biotechnology 1987; 5:369–376.
25. Umeda M, Koyama H, Minowada J, Oishi M, eds. Biotechnology of mammalian cells. New York: Springer-Verlag, 1987.
26. Barry Holland I, Mackman N, Nicaud J-M. Secretion of proteins from bacteria. Biotechnology 1986; 4:427–431.
27. Nicaud J-M, Mackman N, Holland IB. Current status of secretion of foreign proteins by microorganisms. J Biotechnol 1986; 3:255–270.

28. Neu HC, Heppel LA. The release of enzymes from *Escherichia coli* by osmotic shock and during the formation of spheroplasts. J Biol Chem 1965; 240:3685–3692.
29. Lazzaroni J-C, Portalier RC. Genetic and biochemical characterization of periplasmic-leaky mutants of *Escherichia coli* K-12. J Bacteriol 1981; 145:1351–1358.
30. Kudo T, Kato C, Horikoshi K. Excretion of the penicillinase of an alkalophilic *Bacillus* sp. through the *Escherichia coli* outer membrane, J Bacteriol 1983; 156:949–951.
31. Gray GL, Balridge JS, McKeown KS, Heynecker HL, Chang CN. Periplasmic production of correctly processed human growth hormone in *Escherichia coli*: Natural and bacterial signal sequences interchangeable. Gene 1985; 39:247–254.
32. Becker GW, Hsiung HM. Expression, secretion and folding of human growth hormone in *Escherichia coli*. FEBS Lett 1986; 204:145–150.
33. Hsiung HM, Mayne NG, Becker GW. High level expression, efficient secretion and folding of human growth hormone in *Escherichia coli*. Biotechnology 1986; 4:991–995.
34. Yokota T, Arai N, Lee F, Rennick D, Mossman N, Arai K-I. Use of a cDNA expression vector for the isolation of mouse interleukin-2 cDNA clones: Expression of T-cell growth-factor activity after transfection of monkey cells. Proc Natl Acad Sci USA 1985; 82:68–72.
35. Le HV, Syto R, Mays C, Reichert P, Narula S, Gwain K, Greenberg R, Kastelein R, Van Kimmenade A, Nagabhushan TL, Trotta P. Isolation of *E. coli*-derived murine interleukin-2 from intracellular and secretory expression systems. In: Burgess R., ed., Protein purification: Micro to macro. New York: Alan R. Liss, 1987:383–391.
36. Riendau D, Harnish DG, Bleackley RC, Paetkau V. Purification of mouse interleukin 2 to apparent homogeneity, Nature 1983; 258:12114–12117.
37. Thannhauser TW, Konishi Y, Scheraga H. Sensitive quantitative analysis of disulfide bonds in polypeptides and proteins, Anal Biochem 1984; 138:181–188.
38. Grassetti DR, Murray JF Jr. Determination of sulfhydryl groups with 2,2′- or 2,4′-dithiodipyridine, Arch Biochem Biophys 1967; 119:41–49.
39. Lee F, Yokota T, Otsuka T, Gemmel L, Larson N, Luh J, Arai K-I, Rennick D. Isolation of cDNA for a human granulocyte-macrophage colony-stimulating factor by functional expression in mammalian cells. Proc Natl Acad Sci USA 1985; 4360–4364.
40. Cantrell MA, Anderson D, Cerretti DP, Price V, McKerreghan K, Tushinski RJ, Mochizuki DY, Larsen A, Grabstein K, Gillis S, Cosman D. Cloning, sequence and expression of a human granulocyte/macrophage colony-stimulating factor. Proc Natl Acad Sci USA 1985; 82:6250–6254.
41. O'Hara PJ, Hart CE, Forstrom JW, Hagen FS, Role of carbohydrate in the function of human granulocyte-macrophge colony-stimulating factor. Biochemistry 1987; 26:4861–4867.
42. Delamarter JF, Mermod J-J, Liang C-M, Eliason JF, Thatcher DR. Recombinant murine GM-CSF from *E. coli* has biological activity and is neutralized by a specific antiserum. EMBO J 1985; 4:2575–2581.

43. Schrimsher JL, Rose K, Simona MG, Wingfield P. Characterization of human and mouse granulocyte-macrophage colony-stimulating factors derived from *Escherichia coli*. Biochem J 1987; 247:195–199.
44. Libby RT, Braedt G, Kronheim SR, March CJ, Urdal DL, Chiaverotti TA, Tushinski RJ, Mochizuchi DY, Hopp TP, Cosman D. Expression and purification of native human granulocyte-macrophage colony-stimulating factor from an *Escherichia coli* secretion vector. DNA 1987; 6:221–229.
45. Greenberg R, Lundell D, Alroy Y, Bonitz S, Condon R, Fossetta J, Frommer B, Gewain K, Katz M, Leibowitz PJ, Narula SK, Kastelein R, Van Kimmeuade A. Expression of biologically active, mature human granulocyte-macrophage colony stimulating factor using an *E. coli* secretory expression system. Curr Microbiol 1988; 17:321–332.
46. Wong EY, Bradford SB, Klein BK, Heeren RA, Seetharam R, Siegel NR, Tacon WC. Secretion of IGF-1 in *E. coli*. In: Brew K, et al. eds. Advances in gene technology: Protein engineering and production. Washington, D.C.: IRL Press, 1988:104.
47. Moks T, Abrahmsen L, Ossterlof B, Josephson S, Ostling M, Enfors SO, Persson I, Nilsson B, Uhlen M. Large scale affinity purification of human insulin-like growth factor 1 from culture medium of *Escherichia coli*. Biotechnology 1987; 5:379–382.
48. Humbel RE. Insulin-like growth factors, somatomedins, and multiplication stimulating activity chemistry. In: Li CH, ed. Hormonal proteins and peptides. Volume 12. New York: Academic Press, 1984:57–59.
49. Bergmann C, Dodt J, Kohler S, Fink E, Gassen HG. Chemical synthesis and expression of a gene coding for hirudin, the thrombin specific inhibitor from the leech hirudo medicinalis. Biol Chem Hoppe-Seyler 1986; 367:731–740.
50. Harvey RP, Degryse E, Stefani L, Schamber F, Cazenave JP, Courtney M, Tolstoshev P, Lecocq JP. Cloning and expression of a cDNA coding for the anticoagulant hirudin from the blood sucking leech, *Hirudo medicinalis*. Proc Natl Acad Sci USA 1986; 83:1084–1088.
51. Dodt JC, Schmitz T, Schafer T, Bergmann C. Expression, secretion and processing of hirudin in *E. coli* using the alkaline phosphatase signal sequence. FEBS Lett 1986; 202:373–377.
52. Elder JB, Ganguli PC, Gillespie IE, Delamore I, Gregory H. Effect of urogastrone in the Zollinger–Ellison syndrome. Lancet 1975; 2:424–427.
53. Gregory H. Isolation and structure of urogastrone and its relationship to epidermal growth factor. Nature 1975; 257:325–327.
54. Oka T, Bakamoto S, Miyashi K-I, Fuwa T, Yoda K, Yamasaki M, Tamura G, Miyake T. Synthesis and secretion of human epidermal growth factor by *Escherichia coli*. Proc Natl Acad Sci USA 1985; 82:7212–7216.
55. Miyake T, Oka T, Nishizawa T, Misoka F, Fuwa T, Yoda K, Yamasaki M, Tamura G. Secretion of human interferon-α induced by using secretion vectors containing a promoter and signal sequence of alkaline phosphatase gene of *Escherichia coli*. J Biochem 1985: 97:1429–1436.

56. Yamashiro D, Li CH. β-Endorphin structure and activity. In: Udenfriend S, Meienhofer, J. eds. The peptides. Volume 6. New York: Academic Press, 1984:191–217.
57. Pollitt S, Zalkin H. Role of primary structure and disulfide bond formation in β-lactamase secretion. J Bacteriol 1983; 153:27–32.
58. Talmadge K, Gilbert W. Cellular location affects protein stability in *Escherichia coli*. Proc Natl Acad Sci USA 1982; 79:1830–1833.
59. Itakura K, Hirose T, Crea R, Riggs AD, Heynecker HL, Bolivar F, Boyer HW. Expression in *Escherichia coli* of a chemically synthesized gene for the hormone somatostatin. Science 1977; 198:1054–1063.
60. Moreno F, Fowler AV, ZHall M, Silhavy TJ, Zabin I, Schwartz M. A signal sequence is not sufficient to lead β-galactosidase out of the cytoplasm. Nature 1980; 286:356–359.
61. Benson SA, Silhavy TJ. Information within the mature lamb protein necessary for localization to the outer membrane of *E. coli* K12. Cell 1983; 32:1325–1335.
62. Kadonaga JT, Gautier AE, Straus DR, Charles AD, Edge MD, Knowles JR. The role of the beta-lactamase single sequence in the secretion of proteins by *Escherichia coli*. J Biol Chem 1984; 259:2149–2154.
63. Ohsuye K, Nomura M, Tanaka S, Kubota I, Nakazato H, Shinagawa H, Nakata A, Noguchi T. Expression of chemically synthesized alpha-neo-endorphin gene fused to *E. coli* alkaline phosphatase. Nucleic Acids Res 1982; 10:1741–1754.
64. Gilkes NR, Kilburn DG, Miller RC. Jr, Warren RAJ. A mutant of *Escherichia coli* that leaks cellulase activity encoded by cloned cellulase genes from *Cellulomonas fimi*. Biotechnology 1984; 2:259–263.
65. Havekes LM, Lugtenberg JJ, Hoekstra WPM. Conjugation deficient *Escherichia coli* F mutants with heptose-less lipopolysaccharide. Mol Gen Genet 1976; 146:43–50.
66. Hirota Y, Suzuki H, Nishimura Y, Yasuda S. On the process of cell division in *Escherichia coli*: A mutant of *Escherichia coli* lacking a murein lipoprotein. Proc Natl Acad Sci USA 1977; 74:1417–1420.

8

Purification of Recombinant Proteins from Yeast

Roger G. Harrison, Jr.

University of Oklahoma
Norman, Oklahoma

I. INTRODUCTION

Since 1979, an effort has been underway to express recombinant proteins in yeast. This work has been motivated largely by a desire to overcome the shortcomings of the widely used *Escherichia coli* expression systems. Almost all of the work has been done with *Saccharomyces cerevisiae*. In contrast to *E. coli*, *S. cerevisiae* lacks endotoxins and lytic viruses, has no known pathogenic relationship with man, and is a GRAS (generally recognized as safe) organism. Yeast is attractive for large-scale work because of its long and successful history of use in the baking and brewing industries.

A major impetus in the work with yeast has been to develop efficient secretion systems. Because the yeast medium is relatively protein-free [normal yeast proteins represent only 0.5% of the total cellular proteins for *S. cerevisiae* (1)], yeast is an ideal system in which to develop a secretion system. Another feature that has been sought by using yeast for expression is the ability of the organism to carry out desired posttranslational modifications, such as correct disulfide bond formation and proteolysis of signal peptides. These posttranslational modifications, which are common in eukaryotes, do not take place in *E. coli*.

The expression system often has a large and critical impact on the ease of purification of the recombinant protein being expressed. Therefore, the purpose of this chapter is to do an in-depth review of the purification methods that have been used to purify recombinant proteins expressed in yeast. The success of yeast in performing posttranslational modifications will be evaluated. This chapter will cover work where the protein of interest has been purified essentially to homogeneity (at least 95% pure). Future directions that seem appropriate for work that will lead to more efficient purification of recombinant proteins in yeast will also be discussed.

All of the work to be discussed employed *S. cerevisiae* as the host organism. Besides the details of purification, the following information will be reported for the recombinant proteins if available: gene regulation system, expression level in fermentation broth (absolute concentration and percentage of soluble, secreted, or total protein), molecular weight, final purity, method for determination of final purity, and percentage recovery.

II. PROTEINS EXPRESSED INSIDE CELL

It was anticipated that there might be differences in the purification work on proteins expressed intracellularly with work on secreted proteins. To facilitate comparison, the purification of proteins expressed inside the cell is discussed as a group in this section, and secreted proteins are discussed as a group in the following section. For the sake of comparison and contrast, both sections, have been subdivided into work where only classical protein purification methods were used (e.g., ammonium sulfate precipitation and conventional chromatography) and into work where high-resolution methods were used (e.g., affinity chromatography and HPLC).

A. Purifications Using Classical Methods Exclusively

A rather simple procedure was used by Barr et al. (2) to purify two polypeptides representing different domains of the envelope gene product of human immunodeficiency virus (HIV). The glyceraldehyde-3-phosphate dehydrogenase (GAPDH) gene promoter and terminator were used on a yeast plasmid. One of the polypeptides, designated env-2, has a molecular weight of approximately 53,000. The other polypeptide, env-5, has a molecular weight of approximately 15,000 and was expressed as a fusion protein with superoxide dismutase (fusion product designated SOD env-5). To begin purification, the cells were disrupted with glass beads and the cell lysate was centrifuged at $39,000g$. The pellet was resuspended in buffer containing 0.1% sodium dodecyl sulfate (SDS) and recentrifuged. The pellet was solubilized by boiling for 10 min in buffer containing 2.3% SDS and 5% β-mercaptoethanol. The solubilized proteins were chromatographed on an ACA-34 gel filtration column equilibrated with buffer containing 0.1% SDS. Fractions containing immunoreactive recombinant material were concentrated by ultrafiltration. The purity for both env-2 and SOD env-5 was judged to be 95% by SDS-PAGE with Coomassie Blue staining.

Two different types of chromatography, anion exchange and gel filtration, were used in a study by Janoff et al. (3) to purify α_1-proteinase inhibitor (α1PI; molecular weight 44,000). The GAPDH gene promoter and terminator were used on a yeast plasmid. The yeast cells were lysed with glass beads and then centrifuged. The supernatant was loaded on a DEAE-Sephacel column, and α1PI was eluted with a 0–0.3 M NaCl gradient. The fractions containing antielastase activity were pooled, concentrated, and applied to a Sephadex G-75 gel filtration column. The α1PI fractions were again pooled and concentrated. The purified material was estimated to be greater than 95% pure based on its antielastase activity.

B. Purifications that Included High-Resolution Methods

Affinity chromatography was used as the last step by Barr et al. (4) for the purification of human immunodeficiency virus reverse transcriptase (HIV RT; molecular weight 66,000) from yeast. For expression a plasmid was used that contained the HIV RT gene flanked by promoter and terminator sequences of the GAPDH gene. The HIV RT in crude lysate amounted to only 0.05% of total soluble protein and had a concentration of 0.8 mg/liter. The cell lysis procedure was more sophisticated than those used in the other studies reviewed here. The cell pellet was resuspended in a buffer that included 1.2 M sorbitol and 200 mg/liter zymolyase. Spheroplast formation was allowed to proceed for 90 min at 30°C. The spheroplasts were then centrifuged and lysed in a buffer containing 0.1% Triton X-100 and 1 mM dithiothreitol at room temperature. The lysate was clarified by centrifugation and the supernatant was fractionated by ammonium sulfate precipitation. The 0–30% ammonium sulfate-insoluble fraction, which contained greater than 90% of the RT, was resuspended in a buffer that included 20% (v/v) glycerol and 50 mM KCl. After desalting using an ultrafilter, cation exchange chromatography was carried out on a column of cellulose phosphate using a gradient of 50–800 mM KCl for elution. The peak RT fractions were desalted on an ultrafilter and applied to a single-stranded DNA cellulose column. Elution was done with a gradient of 50–800 mM KCl. The peak RT fractions were pooled. The RT was purified to homogeneity as judged by SDS-PAGE. The percentage recovery of RT was 43% from the crude lysate. There was good evidence that during purification, processing of the 66,000 molecular weight RT protein occurred giving a second major species of molecular weight 51,000. It appeared that part of the C-terminal end of RT was clipped off.

HPLC was used to purify human Cu,Zn superoxide dismutase (HSOD; molecular weight 32,000) from yeast by Hallewell et al. (5). Again, the GAPDH promoter was used in expressing the HSOD gene on a yeast plasmid. The level of HSOD varied between 30 and 70% of total cell protein in stationary phase yeast. The cells were lysed with glass beads and the cell debris was removed by centrifugation. The supernatant, which contained HSOD, was purified using a Vydac C4 reverse phase column on a Waters HPLC system. A gradient of 20–50% acetonitrile in 0.05% trifluoroacetic acid was used. The peak containing HSOD was concentrated on a vacuum evaporator. Analysis of the concentrate by SDS-PAGE with Coomassie Blue revealed no contaminating proteins. A very interesting finding in this study was that the N-terminal amino acid of the purified HSOD from yeast was acetylated, which is also the case for HSOD from human erythrocytes. Thus, for clinical purposes it may be preferable to use HSOD from yeast rather than HSOD from *E. coli*, which is not acetylated.

Three different affinity chromatography columns were used in a well-documented study by Hoylaerts et al. (6) of the purification of α1PI from yeast. The α1PI gene was under the control of yeast ARG3 regulatory signals in a yeast recombinant plasmid. The α1PI represented 1% of the total soluble protein. After disruption of the cells in a bead mill, PEG 1000 was added to the 7% level to precipitate impurities. The precipitate was resuspended in buffer containing 7% PEG 1000. The centrifuge supernatants from both precipitations were pooled and chromatographed on a DEAE-Sepharose Fast Flow column using steps of 150 mM NaCl for elution. Eluate containing α1PI was passed through a column of human immunological light chains, idiotype κ, insolubilized on Sepharose 4B. The α1PI was eluted with TNB [1 mg/ml bis(4-nitrophenyl)disulfide-3,3' + 0.25 mg/ml dithiothreitol in buffer]. Pooled fractions containing α1PI were desalted by chromatography on Sephadex G-25 and then applied to a heparin-agarose column. The flow-through from this column was connected to the inlet of a column of chelating Sepharose charged with Zn^{2+} ions. A gradient elution with histidine (0–25 mM) was used to remove the α1PI from the column. Finally, the fractions containing α1PI were dialyzed against buffer containing 25 mM NaCl and chromatographed on an aminohexyl agarose ion exchange column using a gradient of NaCl (25–300 mM) for elution. The final purity of α1PI was greater than 95% based on an enzyme assay, and the overall recovery was 25%.

One conventional chromatography and one affinity chromatography step were all that was needed to purify two forms of α1PI from yeast by Travis et al. (7). The GAPDH promoter and transcription terminator were used in a yeast expression plasmid. One form of α1PI was analogous to the human plasma protein, and the other form had the methionine at position 358 replaced by valine. The α1PI variants represented an average of 10% of the soluble cell protein. The cells, contained in an extraction buffer that included 0.05 M NaCl, were broken by vortexing with glass beads, and the cell debris was removed by centrifugation at 24,000g. The supernatant was applied to a column of Cibracon Blue Sepharose, and the α1PI was eluted by further washing of the column with extraction buffer. Fractions containing α1PI were chromatographed on a column of DEAE-cellulose, equilibrated with extraction buffer, and elution was carried out using a 0–0.2 M NaCl gradient. Both the methionine and the valine α1PI variants were purified to homogeneity, as judged by agarose gel electrophoresis and by SDS-PAGE using Coomassie Blue staining. One interesting finding was that the valine variant was stable to oxidation, while the methionine variant was readily inactivated.

III. PROTEINS SECRETED FROM CELL

A. Purification Using Classical Methods Exclusively

Three different chromatography steps were used by Van Den Bergh et al. (8) to purify porcine pancreatic prophospholipase A_2 (pro-PLA; molecular weight 19,000) secreted from yeast. To achieve secretion, the promoter and prepro genes for α-mating factor were fused upstream of the the pro-PLA gene on a plasmid and the GAPDH terminator sequence was used downstream. The concentration of pro-PLA in the yeast medium was approximately 0.6 mg/liter. The cells were removed by centrifugation and the supernatant was acidified to pH 3.5. SP-Sephadex C-25 cation exchanger was added, and mixing was done until the pro-PLA was completely absorbed. The SP-Sephadex beads were placed on top of an SP-Sephadex column, which was eluted with a 0–0.5 M NaCl gradient. The pro-PLA-containing fractions were dialyzed, lyophilized, and further purified on a CM-cellulose cation exchange column at pH 6.0. A 0–0.4 M NaCl gradient was used for elution. The fractions with pro-PLA were desalted and freed from traces of colored material using a Sephadex G-50 gel filtration column. The purified material was judged to be greater than 95% pure by SDS-PAGE and by analysis on an FPLC system with Mono Q and Mono S columns. The N-terminal sequence of the pure protein was in accordance with that predicted and also proved that the KEX2 protease cleaved off the prepro α-mating factor peptide. In addition, the pure protein showed the same specific activity as native pro-PLA, indicating that the enzyme was correctly folded and that all seven disulfide bridges were correctly formed. The overall recovery of pro-PLA from the culture medium was 65%.

B. Purifications that Included High-Resolution Methods

A combination of conventional chromatography and HPLC were used by Craig and Wondrack (9) to purify growth hormone-releasing factor (GRF; molecular weight 4500) secreted from yeast. The α-mating factor promoter and prepro sequence were used on a plasmid. Levels of GRF in the yeast broth were in the range 5–30 mg/liter as monitored by a radioimmune assay. The cells were removed by tangential microfiltration using Millipore membranes with nominal 0.45-μm pore size. To concentrate GRF, the cell-free media at pH 2.0 was passed through a Vydac C18 column. After a wash with buffer at pH 3.0, GRF was eluted with a 50:50 (v/v) mixture of buffer and isopropanol. The volume of eluate was reduced to near dryness with rotary evaporation. This concentrate was chromatographed on a size exclusion column containing Sephadex G-50 (bottom 50%) and Sephadex G-25 (top 50%). Fractions containing 90% of the GRF activity were pooled and applied to a preparative scale C8 HPLC column previously equilibrated in a 75:25 (v/v) mixture of buffer at pH 3.0 and acetonitrile. A 25–40% acetonitrile linear gradient was used to elute GRF. Rotary evaporation was used to remove the organic reagent from fractions containing GRF. A final HPLC was carried out on a semipreparative C18 column previously equilibrated with a 75:25 (v/v) mixture of 0.05% aqueous trifluoroacetic acid (TFA) and 0.05% TFA in acetonitrile. GRF was eluted with a 25–50% acetonitrile linear gradient. Purity was judged to be greater than 95% using SDS-PAGE. Tyrosine was shown to be the N-terminal amino acid, which is correct for GRF.

Only HPLC was required to purify granulocyte-macrophage colony-stimulating factor (GM-CSF; molecular weight 21,000) from yeast media by Mochizuki et al. (10). The α-mating factor leader sequence under control of the alcohol dehydrogenase II promoter was used. The media was applied to a Vydac C4 column, and the column was washed with 0.1% TFA. Elution was done with a gradient of acetonitrile in 0.1% TFA. The fractions containing GM-CSF were pooled, diluted with two volumes of 0.1% TFA, and applied to a Vydac C18 column. Again elution was done with a gradient of acetonitrile in 0.1% TFA. Based on SDS-PAGE, the GM-CSF was purified to homogeneity.

A somewhat similar strategy was used by Grabstein et al. (11) to purify GM-CSF from yeast media. The α-mating factor promoter and leader sequences directed synthesis and secretion. The media was made 0.1% in TFA and applied to a Waters C18 column previously equilibrated in 0.1% TFA in water. Proteins were eluted by a linear gradient of acetonitrile in 0.1% TFA. The GM-CSF peak fraction was rechromatographed on the same C18 column equilibrated with 0.9 M acetic acid and 0.2 M pyridine. Elution was by a gradient of N-propanol. The GM-CSF was purified to homogeneity as judged by SDS-PAGE. There were two bands of GM-CSF, representing glycosylated and unglycosylated forms.

Chromatofocusing was the last of a series of steps used by Zsebo et al. (12) to purify consensus interferon (IFN-Con; molecular weight range 18,000–22,000 for α-interferons) secreted from yeast. Secretion was directed by the prepro gene for α-mating factor using a yeast plasmid. The IFN-Con in the yeast media was estimated to represent 15–20% of the total secreted proteins, which translates to an IFN-Con concentration of 15–20 mg/liter. The cells in 29 liters of fermentation broth were removed by tangential microfiltration using Millipore membranes with nominal 0.45-μm pore size. The cell-free broth was concentrated by ultrafiltration, and the concentrate was chromatographed on a DEAE-Trisacryl column using a 0–0.3 M NaCl gradient for elution. The factions containing IFN-Con were loaded onto a Sephadex G-75 gel filtration column, which was eluted with 50 mM imidazole at pH 6.8. The IFN-Con-containing fractions were diluted 1:1 with distilled water and chromatographed on a chromatofocusing column (PBE 94, Pharmacia). The elution buffer was 5 mM Buffalyte 4–8/Cl (Pierce) at pH 3.8. The IFN-Con was purified to homogeneity as indicated by Coomassie Blue staining. For the purified protein the N-terminal sequence was verified to be correct for the first five amino acids using automated sequence analysis, and the disulfide structure was found to be identical with that reported for the native human α-interferon. The overall recovery of IFN-Con from the cell-free broth was 13%.

Affinity chromatography was the only type of chromatography used by Gardell et al. (13) to purify carboxypeptidase A (CPA; molecular weight 34,000) and a variant carboxypeptidase A that had phenylalanine rather than tyrosine at position 248 (designated CPA-Phe-248). The α-mating factor promoter, leader sequence, and terminater were used on a plasmid. The culture media contained approximately 0.5 mg/liter of CPA or CPA-Phe-248. Based on SDS-PAGE with Coomassie Blue staining, it appeared that CPA or CPA-Phe-248 constituted 1–5% of the total protein in the media. The culture media was concentrated by ammonium sulfate precipitation (80% saturated) in the presence of bovine serum albumin (0.5 g/liter). The precipitate was dissolved in buffer and incubated with trypsin at 37°C to effect proenzyme maturation. After dialysis to remove salt, affinity chromatography was performed on a glycyl-L-tyrosylazobenzyl succinate Sepharose column, which was eluted with 0.5 M NaCl in buffer. A second affinity chromatography was done using a Sepharose column to which the potato carboxypeptidase inhibitor was immobilized. For elution 0.1 M Na_2CO_3 plus 0.5 M NaCl were used. Both CPA and CPA-Phe-248 were purified to homogeneity as judged by SDS-PAGE with Coomassie Blue staining.

IV. CONCLUSIONS

Some of the work discussed in this chapter demonstrated that yeast can accomplish certain tasks that *E. coli* cannot. Hallewell et al. (5) showed that human superoxide dismutase (HSOD) produced by yeast was acetylated at the N terminus, as is also the case for HSOD from human erythrocytes. In comparison, HSOD made in *E. coli* is not acetylated. The disulfide structure was found to be correct for two proteins secreted from yeast, porcine pancreatic prophospholipase A_2 [Van Den Bergh (8)] and consensus α-interferon [Zsebo (12)]. For four of the works discussed (5,8,9,12) it was shown that the N-terminal amino acid was correct and thus an extra methionine had not been added. An extra methionine is often added at the N terminus in *E. coli*. The recombinant protein was secreted in three of the works cited where the N-terminal amino was shown to be correct (8,9,12), indicating correct processing of the α-mating factor signal.

Most of the work discussed involved the use of high-resolution methods, such as affinity chromatography and HPLC. Often it was necessary to resort to more than one high-resolution step for purifying a given protein (6,9–11,13).

Comparing the work on intracellular expression to that on secretion from the cell, several conclusions can be drawn. It can be readily seen that the work is about evenly split between these two types of expression and that the work was all done over about the same period of time to the present. Another observation is that neither type of expression seems to give a significant advantage in purification (e.g., fewer steps or easier steps). This is undoubtedly because neither mode of expression gave a consistently higher expression level based on percentage of soluble protein, which should in general be a good indicator of the ease (or cost) of purification (i.e., as expression level increases, ease of purification increases). The expression level ranged from 0.05 to 70% of soluble protein for proteins expressed inside the cell and from 1 to 20% for secreted proteins (note that expression level was not reported for some of the cases).

There was only one case [Barr et al. (2)] where the recombinant proteins were insoluble within the cell. This occurred for two polypeptides representing different domains of the envelope gene product of human immunodeficiency virus (one was expressed as a fusion protein with superoxide dismutase). However, purification was relatively simple with only one chromatography step required. Thus, the proteins being insoluble appeared to facilitate purification in this instance.

V. FUTURE DIRECTIONS

An obvious direction for the future is to find ways to increase expression levels relative to total soluble protein for both intracellular expression and secretion. If this can be accomplished, it will make the task of purification easier. Also, systems need to be developed that will achieve more predictable levels of expression for a wide variety of proteins.

Another need that this chapter has brought out is that less expensive protein purification procedures must be developed. There was a heavy reliance on costly affinity chromatography and HPLC steps in the work discussed in this chapter. The use of these steps at large scale would clearly not be economic for all cases.

REFERENCES

1. Sharma SK. On the recovery of genetically engineered proteins from *Escherichia coli*. Sep Sci Technol 1986; 21:701-726.
2. Barr PJ., Steimer KS, Sabin EA, Parkes D, George-Nascimento C, Stephans JC, Powers MA, Gyenes A, Van Nest GA, Miller ET, Higgins KW, Luciw PA. Antigenicity and immunogenicity of domains of human immunodeficiency virus (HIV) envelope polypeptide expressed in the yeast *Saccharomyces cerevisiae*. Vaccine 1987; 5:90-101.
3. Janoff A, George-Nasimento C, Rosenberg S. A genetically engineered, mutant human alpha-1-proteinase inhibitor is more resistant than the normal inhibitor to oxidative inactivation by chemicals, enzymes, cells and cigarette smoke. Am Rev Respir Dis 1986; 133:353-356.
4. Barr, PJ, Power MD, Lee-Ng CT, Gibson HL, Lucin PA. Expression of active human immunodeficiency virus reverse transcriptase in *Saccharomyces cerevisiae*. Biotechnology 1987; 5:486-489.
5. Hallewell RA, Mills R, Tekamp-Olson P, Blacher R, Rosenberg S, Otting F, Masiarz FR, Scandella CJ. Amino terminal acetylation of authentic Cu,Zn superoxide dismutase produced in yeast. Biotechnology 1987; 5:363-366.
6. Hoylaerts M, Chuchana P, Verdonck P, Roelants P, Weyens A, Loriau R, De Wilde M, Bollen A. Large scale purification and molecular characterization of human recombinant α_1-proteinase inhibitor produced in yeasts. J Biotechnol 1987; 5:181-197.
7. Travis J, Owen M, George P, Carrell R, Rosenberg S, Hallewell RA, Barr PJ. Isolation and properties of recombinant DNA produced variants of human α_1-proteinase inhibitor. J Biol Chem 1985; 260:4384-4389.
8. Van Den Bergh CJ, Bekkers AC, De Geus P, Verheij HM, De Haas GH. Secretion of biologically active prophospholipase A_2 by *Saccharomyces cerevisiae*. Eur J Biochem 1987; 170:241-246.
9. Craig WS, Wondrack LM. Purification of growth hormone releasing factor from yeast strains producing same. European patent application, 1987; 0233645.
10. Mochizuki DY, Eisenman JR, Conlon PJ, Park LS, Urdal DL. Development and characterization of antiserum to murine granulocyte-macrophage colony-stimulating factor. J Immunol 1986; 136:3706-3709.
11. Grabstein KH, Urdal DL, Tushinski RJ, Mochizuki DY, Price VL, Cantrell MA, Gillis S. Conlon PJ. Induction of macrophage tumoricidol activity by granulocyte-macrophage colony-simulating factor. Science 1986; 232:506-508.
12. Zsebo KM, Lu HS, Fieschko JC, Goldstein L, Davis J, Duker K, Suggs SV, Lai PH, Bitter GA. Protein secretion from *Saccharomyces cerevisiae* directed by the prepro-α-factor leader region. J Biol Chem 1986; 261:5858-5865.
13. Gardell SJ, Craik CS, Hilvert D, Urdea MS, Rutter WJ. Site directed mutagenesis shows that tyrosine 248 of carboxypeptidase A does not play a crucial role in catalysis. Nature 1985; 317:551-555.

9

Production of Recombinant Proteins in the Methylotrophic Yeast *Pichia pastoris*

M. J. Skogen Hagenson

Phillips Petroleum Company
Bartlesville, Oklahoma

I. INTRODUCTION

In the production of recombinant proteins, yeasts offer advantages of both bacterial and mammalian systems. Yeasts are simple to manipulate, have a relatively rapid growth rate, and can be grown to higher cell densities than bacteria. Being eukaryotic, they are capable of higher forms of posttranslational processing of proteins and glycosylation of secreted proteins at a fraction of the cost of mammalian expression systems.

Saccharomyces cerevisiae, the best characterized yeast, was the first to be used for the production recombinant proteins, including interferons (1, 2), hepatitis surface antigen (3), chymosin (4), and epidermal growth factor (5). More recently, means of gene transfer and expression vectors have been developed for other yeasts, including *K. lactis* (6) and the methylotrophic yeast *H. polymorpha* (7). Differences in processing, secretion, and glycosylation of foreign proteins in various yeast hosts have been observed. For production of pharmaceutical proteins, it will be critical to choose the yeast host which yields the most "mammalian-like" form of the protein.

This chapter describes the development of a recombinant expression system using the industrial yeast, *Pichia pastoris*. Stable, high-cell-density fermentation technology on simple, defined media had been established before embarking on the development of this yeast into a recombinant host. A brief overview of methylotrophic yeast and the development of a single-cell-protein process is given in Sec. II. Section III describes the various components of the expression system, including host strains, methanol-inducible promoters, selectable markers, and expression vectors. The production of various intracellular and secreted proteins is summarized in Sec. IV and V, respectively, and contrasted with production of these proteins in other yeast hosts when possible. A section is devoted to *P. pastoris* glycosylation as this yeast has been found to glycosylate in a more mammal-like manner than *S. cerevisiae*.

II. BACKGROUND

Methylotrophic bacteria have been known since the beginning of this century, but it wasn't until 1969 that methanol utilization by yeast was first reported by Ogata (8). Since that time, it has been found that only several of the 39 genera of yeast can grow on methanol, with most methylotrophic yeast being species of *Pichia, Hansenula, Candida*, and *Torulopsis*.

Most aspects of methanol metabolism and the physiology of methylotrophic yeast have been the subject of excellent reviews to which the reader is referred (9-11). In methylotrophic yeast, the initial steps in methanol oxidation take place in membrane-enclosed organelles, called peroxisomes, where the key enzymes, alcohol oxidase, catalase, and dihydroxyacetone synthase, are sequestered. Targeting of these proteins to peroxisomes has been intensively studied, but as yet a specific targeting "signal" has not been identified. Peroxisomes proliferate during growth on methanol and can constitute 80% of the cytoplasmic volume of the cell.

The first industrial application of methylotrophic yeast was seen in the early 1970s with the production of single-cell protein (SCP). Early development of an SCP process at Phillips Petroleum included screening of available yeast strains for growth on methanol as the sole source of carbon and energy. From the many yeast screened, a strain of *P. pastoris* was selected on the basis of higher cell mass yield from methanol, higher protein content, and stable fermentation characteristics (11). Assays of the protein content of *Pichia* revealed that nearly 60% of the cell mass was crude protein. Continued development of fermentation technology utilizing high oxygen and heat transfer has further improved the cell yield, and densities are routinely obtained which are in excess of 125 g dry wt/liter of broth.

During growth of *P. pastoris* on methanol, alcohol oxidase (AO) is the most abundant protein in the cell and may compose up to 30% of the total cellular protein (12). It is the first enzyme in the methanol utilization pathway and converts methanol to formaldehyde in the presence of molecular oxygen. Expression of AO is tightly controlled by the carbon source, with no detectable expression of the enzyme during growth on excess glucose or glycerol. Growth on methanol results in rapid induction of AO and other enzymes in the methanol utilization pathway accompanied by proliferation of peroxisomes by a budding process (13). The transfer of methanol-grown cells to media containing a repression carbon source, such as glucose or glycerol, results in peroxisomal degradation and inactivation of the enzymes contained therein.

III. DEVELOPMENT OF THE *Pichia pastoris* EXPRESSION SYSTEM

A. Methanol-Inducible Regulatory Sequences

Use of *P. pastoris* for the expression of heterologous proteins has centered around the use of the methanol-inducible promoter for alcohol oxidase which was isolated by Ellis et al. (13). The authors also described the isolation of two other methanol-regulated genes, P40 and P76, which encoded for proteins of 40,000 and 76,000 Da, respectively. P76 was subsequently identified as dihydroxyacetone synthase (DAS) (G. Thill, personal communication). Ellis and coworkers found that expression of these genes was controlled at the level of transcription. Sequencing of the upstream region of the alcohol oxidase gene revealed DNA regions which resemble yeast consensus transcriptional and translational promoter sequences.

Methanol-responsive regulatory sequences from the alcohol oxidase (*AOX*1) and *DAS* genes have been incorporated in expression vectors for production of heterologous proteins in *P. pastoris*. The *AOX*1 promoter was isolated from the *P. pastoris* genome as a 1000-bp fragment immediately preceding the alcohol oxidase structural gene.

B. *Pichia* Host Strains

Pichia pastoris histidinol dehydrogenase mutant, GS115 (also referred to as GTS115), has served as the host strain in most gene expression studies (14). This mutant was generated by treatment of wild-type *P. pastoris* with nitrosoquanidine.

Cregg reports development of mutants of *P. pastoris* which are defective in methanol metabolism due to gene disruptions (15). Once such strain, KM71, in which *AOX*1 has been disrupted, grows slowly on methanol. Using a similar approach, a strain designated KM7121 which is defective in both *AOX*1 and *AOX*2 was generated. This double mutant contains no alcohol oxidase and cannot grow on methanol.

Generally higher expression levels of heterologous genes are obtained in KM71 as compared to GS115. Increases in the range of two- to sixfold have been reported for specific proteins (see Secs. IV and V), but the reason for this is not understood at present.

C. Selectable Markers

The *P. pastoris HIS4* gene, isolated from wild-type *P. pastoris* strain Y-11430, has been incorporated in expression plasmids (14) and is the most commonly used selectable marker in transformations of the histidine auxotroph, GS115. The *HIS4* gene from *S. cerevisiae* was found to complement the histidine mutation in strain GS115 and was used in the initial characterization of this strain. It was also used as a selectable marker in the first generation of *P. pastoris* vectors.

Sreekrishna et al. demonstrated that the *S. cerevisiae* invertase gene could be used as a dominant selectable marker in *Pichia* (16). *Pichia* yeast transformed with autonomous plasmids carrying the invertase gene grows on sucrose, whereas wild-type *Pichia* cannot. It was found that a stable Suc$^+$ phenotype was obtained only after chromosomal integration of the gene. A two-step transformation procedure was required, however, because of utilization of components of the spheroplast regeneration medium as carbon sources and subsequent cross-feeding of nontransformed cells.

The kanamycin-neomycin phosphotransferase gene of *Tn903* was used to convey antibiotic G418 resistance to *P. pastoris* (W. R. McCombie, personal communication). *P. pastoris* is sensitive to G418 at about 300 μg/ml (minimum inhibitory level). *P. pastoris* transformed with plasmids carrying the gene became resistant to G418, with individual isolates showing a range of resistances from less than 300 μg/ml up to 1200 μg/ml. Clonal variability in resistance levels made direct selection on kanamycin difficult. This problem could perhaps be overcome by the use of a yeast promoter for control of gene expression.

D. Expression Vectors

Vectors developed for the expression of heterologous proteins in *P. pastoris* contain DNA sequences for selection and maintenance in *E. coli*, as well as sequences derived from the genome of *P. pastoris* yeast for the control of heterologous gene expression, selection, and, in some instances, chromosomal integration. Two basic types of expression vectors have been developed: those for autonomous replication and those for site-directed integration into the primary alcohol oxidase locus (*AOX1*) of the yeast genome. Examples of each are shown in Fig. 1.

Both types of vectors include the ampicillin resistance gene and the origin of replication from plasmid pBR322 for maintenance and selection in *E. coli*. The only other DNA sequences of bacterial origin contained in the vectors are short, noncoding segments less than 300 bp in length which join the assembled *P. pastoris* sequences.

Most heterologous proteins have been expressed in *P. pastoris* yeast using control sequences isolated from the alcohol oxidase (*AOX1*) gene, although

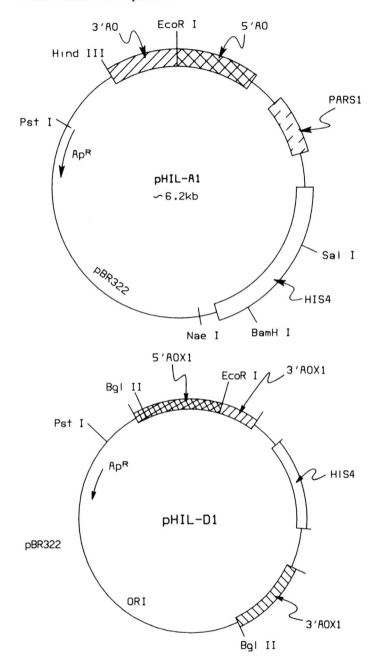

Figure 1 Autonomous and integrative *P. pastoris* expression vectors pHIL-A1 and pHIL-D1, respectively.

Tschopp et al. showed that the *DAS* promoter was also effective in the methanol-regulated production of β-galactosidase (17). The *AOX*1 terminator region, isolated from the *Pichia* genome as a 300-bp fragment immediately downstream from the structural gene, is included for proper termination of transcription and translation in *Pichia* yeast. A synthetic linker providing a unique *Eco*RI-cloning site for the insertion of heterologous genes is present between the *AOX*1 promoter and terminator sequences. Variations of these basic vectors with multilinkers inserted at the unique *Eco*RI site have also been developed to facilitate insertion of genes with various ends (W. R. McCombie, personal communication).

Unique to autonomous vectors, such as pHIL-A1, is an autonomous replicating sequence, PARS1, isolated from the *P. pastoris* genome (14). Plasmids containing the PARS1 element exist autonomously for several generations of growth, followed by spontaneous integration into the *Pichia* genome. This has been found to be an addition-type insertion, with no deletion of DNA at the insertion site (18). Such integrants can be quite stable, as had been found in the expression of β-galactosidase (17) and streptokinase (19). Alternatively, the autonomous vectors can be linearized at a unique *Stu*I or *Sal*I restriction site within the *HIS*4 gene for site-directed integration at the chromosomal *HIS*4 locus (17). Such transformants display a high degree of stability.

The generalized integrative vector pHIL-D1 is a second-generation *P. pastoris* expression vector designed specifically for site-directed integration of the expression cassette at the *AOX*1 chromosomal locus with displacement of the *AOX*1 gene (18). It differs from autonomous vectors by the presence of approximately 1000 bp of *AOX*1 3' DNA required for homologous recombination at the *AOX*1 locus and by the absence of a PARS sequence. The expression cassette, which contains the *AOX*1 controlling sequences, the foreign gene, the HIS4 gene for selection, and a larger flanking *AOX*1 3' region, can be released from the circular vector by digestion with *Bgl*II. *P. pastoris* host strain GS115 is readily transformed to histidine prototrophy by introduction of the linear cassette.

Between 5 and 35% of the His$^+$ transformants exhibit a methanol-slow phenotype, indicative of replacement of the *AOX*1 gene with the expression cassette. The remaining transformants are phenotypically His$^+$, methanol-normal and have the cassette integrated elsewhere in the genome. In the *AOX*1-disrupted strains, the ability to sustain low levels of growth on methanol is due to the presence of a second alcohol oxidase gene (*AOX*2) which contributes 10–15% of the AO activity in wild-type cells. Proteins and mRNAs from the two *AOX* genes have been characterized by Cregg and Madden (20).

Spontaneous integration of multiple copies of the *Bgl*II integrative expression cassette at the *AOX*1 locus was first noted in studies on TNF expression (K. Sreekrishna, personal communication). It was discovered that from 3 to

30% of the methanol-slow transformants produced levels of TNF in excess of 30% of the soluble protein. Southern blots revealed a copy number variation in individual transformants with the high producers containing multiple copies of the integrative cassette. To facilitate screening for multicopy integrants, a colony hybridization procedure was developed for *P. pastoris* (M. J. Hagenson, unpublished), based on the procedure reported by Hinnen et al. (21). Transformants with stable, multicopy integrations of the expression cassette have been found for nearly all heterologous genes; however, the enhancement in expression level obtained is gene-dependent.

IV. INTRACELLULAR PRODUCTION OF HETEROLOGOUS PROTEINS

A. β-Galactosidase

Using promoter fusions with the *E. coli lacZ* gene, Tschopp et al. demonstrated that the alcohol oxidase and *DAS* promoters could be used to drive the production of heterologous proteins in *P. pastoris* (17). To ensure stability, autonomous plasmids were integrated at the *HIS4* gene. The effect of various carbon sources on gene expression for each of the promoter-*lacZ* fusions is shown in Fig. 2. Both the *AOX1* and *DAS* promoters regulate heterologous gene expression in the same manner in which the native proteins are regulated. Each is repressed during growth on glucose or a glucose-methanol mixture. The *AOX1* promoter was derepressed in response to carbon starvation, with small levels of β-galactosidase detected, while the *DAS* promoter did not show derepression. Each promoter was induced 50- to 100-fold by the addition of methanol to the media. When induced cells were switched onto a glucose-methanol mixture, rapid reduction in both alcohol oxidase and β-galactosidase activities were seen.

β-Galactosidase produced in *P. pastoris* was 115,000 Da in size, was enzymatically active, and composed up to 20% of the soluble protein in induced cells. In the GS115 host the yield was about $5-13 \times 10^3$ U/mg compared to $70-80 \times 10^3$ in the *AOX1*-defective host, KM71.

B. Hepatitis B Surface Antigen

The use of the *P. pastoris* expression system for the production of hepatitis B surface antigen was described by Cregg and coworkers (22). To be useful as a vaccine, the 23-kDa monomers must be assembled into particles resembling the 22-nm Dane particle found in blood serum of infected humans, with about 100 monomers included in each assembled particle. The alcohol oxidase promoter and termination sequences were used for expression of the S form of HBsAg. Expression cassettes released as linear fragments from *AOX1*-integrative

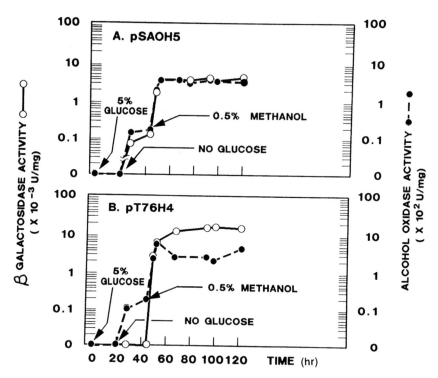

Figure 2 Effect of carbon source on the regulated expression of β-galactosidase and alcohol oxidase in *P. pastoris* transformed with autonomous plasmids utilizing the *AOX*1 promoter (A) or the *DAS* promoter (B) for expression of β-galactosidase. (From Ref. 17, with permission of the authors and IRL Press, Eynsham, Oxford.)

plasmids were used to direct chromosomal integration. Transformants with methanol-slow (*AOX*1-disrupted integration) as well as methanol-normal (addition-type integration) phenotypes were evaluated for the production levels of the 23,000-Da monomer and the extent of particle assembly.

During growth on methanol, the amount of monomer produced was about 3–4% of the soluble protein for each type of integration, but significant differences in the amount of monomer assembled into particle were observed. In the *AOX*1-disrupted strains, nearly 100% of the monomer was assembled into particles while about 10% was assembled in *AOX*1 wild-type strains.

Slow growth rate seems to be crucial for efficient particle assembly. Cregg postulates that posttranslational events with slow kinetics may be responsible. Other possible explanations are that particle assembly is subject to growth rate regulation or is dependent on the available lipid pool in the cell (R. E. Torregrossa, personal communication). The lipid content of *P. pastoris* HBsAg particles was examined by isopycnic and velocity sucrose gradient centrifugations.

It was concluded that the particle produced in this yeast had a protein-lipid content very similar to the natural 22-nm particles, with sucrose sedimentation densities of 1.16 g/ml.

Production of HBsAg particles in methanol-slow strains was found to scale up linearly from the shake flask to 400-liter fermentation at several different densities (Fig. 3). It was calculated that a 240-liter batch fermentation at high density yields about 90 g of hepatitis particle.

The production of the S form of HBsAg and particle assembly was examined in *E. coli* (23) and in *S. cerevisiae* (3). It was found that the bacterial host does not support proper formation of particles. In *S. cerevisiae*, the levels of expression of monomer were reportedly quite similar to those obtained in *P. pastoris*, being about 2–3% of the soluble protein. However, only about 10% of the monomers are assembled into particles, in contrast to nearly 100% in *AOX*1-disrupted strains of *P. pastoris*.

C. Streptokinase

Streptokinase is a potent plasminogen activator with widespread clinical use as a thrombolytic agent. It is naturally secreted by several strains of hemolytic streptococci for which gene transfer mechanisms have not been described. The

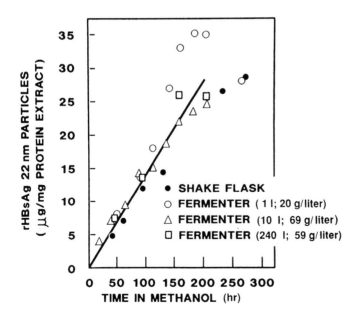

Figure 3 Production of HBsAg particles with time in *P. pastoris* after shift to methanol-containing medium in shake flasks and at three different fermentation densities. (From Ref. 22, used by permission of *Biotechnology*, 1987.)

gene encoding the 47,000-Da streptokinase from *S. equisimilis* was expressed in *E. coli* (24) and in *S. sanguis* (25) under control of its native promoter. The authors observed gene instabilities in *E. coli* (J. J. Ferretti, personal communication) and posttranslational cleavage of the streptokinase molecule in *S. sanguis* (25).

As an alternative to bacterial expression, the gene encoding the 47,000-Da streptokinase gene minus the signal sequence was inserted in an autonomous vector for expression in *P. pastoris* (19). The gene was stripped of all noncoding upstream DNA sequences and a synthetic oligonucleotide was used to provide a methionine initiation codon and to fuse the "mature" coding sequence to the *AOX*1 promoter. Because of gene instabilities encountered in *E. coli*, ligated DNA was transformed directly into *P. pastoris* GS115 where the desired recombinant plasmid was recovered. After induction on methanol, the vector spontaneously integrated into the *Pichia* genome where it was stably maintained.

Expression of streptokinase was tightly controlled by the available carbon source, with no production during growth on glycerol or glucose. Low levels of production were detected during carbon limitation, with rapid induction upon switching to methanol medium. Several methods of cell breakage were evaluated to find optimum lysis conditions for release of streptokinase, including French press and agitation with glass beads, both with and without various concentrations of Triton X-100. Optimum yield was obtained for cells broken by agitation for 3 min in the presence of 0.5-mm glass beads using a Minibead beater (Biospec Products, Bartlesville, OK) in a breaking buffer consisting of 0.2 M Tris-HCl, pH 7.4, and 0.1% BSA.

Production levels were first determined for cells grown and induced in shake flasks followed by small-scale continuous fermentation. Production levels scaled up directly with density of the culture. In the fermentor, the level of production of streptokinase was evaluated over an 11-day continuous fermentation (10-liter fermentor with 2-liter working volume). Cells were brought up to the desired cell mass on glycerol (59 g/liter dry wt, equivalent roughly to an OD_{600} of 300) and then induced on methanol. Gene expression was rapidly induced with full production levels reached within 14 hr. Samples taken during the 256-hr continuous, steady-state fermentation indicated steady production of streptokinase with a yield of approximately 80 mg/liter (0.5 −1% of soluble protein).

The streptokinase produced in *P. pastoris* was determined to be 47,000 Da, the same size as that secreted from the native bacterial host, and it reacted with antibody raised against native streptokinase. Plasminogen activation ability was detected on casein plasminogen plates and in solution using synthetic esterolytic substrates. Based on quantitation by Western blots and the bioassays, the specific activity of the streptokinase produced in *P. pastoris* appeared to be the same as that of the natural protein. The streptokinase expression plasmid was transformed into *P. pastoris* AOX1-disrupted host KM71 and the expression of streptokinase evaluated in shake flasks (M. J. Hagenson and G. M. Whited, unpublished results). Evaluation of nine independent isolates in shake flasks revealed that each produced streptokinase at levels about twice those obtained using the AOX1 wild-type host GS115 (Fig. 4).

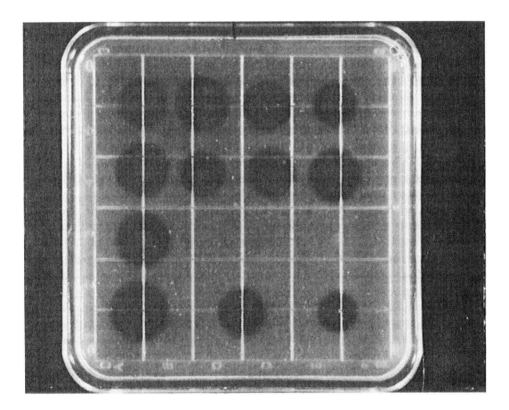

Figure 4 Clearing zones in casein plasminogen agar produced by streptokinase in lysates of shake flask cultures of *P. pastoris* KM71 transformed with streptokinase expression plasmid, pHTskc25. One microliter of each of the following samples was spotted onto the agar surface and the plates were incubated at 37°C. From left to right: row 1, KM71/pHTskc25 isolates 1–4; row 2, KM71/pHTskc25 isolates 5–8; row 3, KM71/pHTskc25 isolate 9, blank, KM71 transformed with vector not containing the streptokinase gene (negative control); row 4, serial dilutions of standard streptokinase containing 1 mg/ml, 0.1 mg/ml, and 0.01 mg/ml of streptokinase protein.

D. TNF and TNF Analogs

Tumor necrosis factor (TNF) is a 17,000-Da antitumor protein secreted by macrophages (26,27). Sreekrishna and coworkers inserted the human TNFα gene into an expression vector pA0804 (herein referred to as pHIL-D1) and evaluated production of TNF in *P. pastoris* host strain GS115 (28). The gene was trimmed of all 5' noncoding DNA and synthetic linkers were used to join it to *AOX*1 controlling sequences at the unique *Eco*RI site of the vector. A segment of noncoding sequence derived from the human cDNA, 221 bases in length, was retained at the 3' end of the gene. The plasmid was treated with *Bgl*II to release the expression cassette prior to transformation of strain GS115.

Transformants were selected on histidine-deficient media prior to selection of methanol-slow isolates. Southern blots confirmed that methanol-slow isolates had undergone an integration of the cassette, displacing the *AOX*1 gene. Isolates displaying normal growth kinetics on methanol were also examined and were found to have genomic integration of the cassette without disruption of the *AOX*1 locus. The Southern analysis also revealed a variation in hybridization signal which indicated a copy number variation among the isolates tested (K. Sreekrishna, personal communication).

Lysates of methanol-induced cultures were analyzed by SDS-PAGE and densitometer scans were used to estimate the amount of TNF in the lysate. Production levels of TNF were greater than 30% of soluble protein for the isolates having multicopy insertions of the expression cassette. TNF had the expected molecular weight of 17,000 Da. Batch fermentation of a high-producing strain yielded TNF in excess of 30% of the soluble protein and productivity of 6–10 g TNF/liter broth at cell densities of 56 and 80 g/liter dry wt, respectively. Continuous fermentations of the same strain gave 17% TNF with a productivity of 1.08 g/liter/hr at a cell density of 25 g/liter dry wt.

A three-step process was developed for the isolation of TNF from *P. pastoris* cultures broken with glass beads (28). Greater than 95% purity was obtained, with a recovery of 78% (based on activity). Analysis of purified TNF revealed correct processing of the NH_2 terminal initiator methionine. Some heterogeneity of the 5' end was observed with 80% of the material analyzed yielding the expected amino terminal valine, while the remainder began with the penultimate arginine residue. Bioactivity of the purified material did not seem to be affected by the NH_2-terminal heterogeneity.

Characterization of the nondenatured molecule by gel filtration and by measurements of sedimentation equilibrium and velocity indicated that the TNF produced in *P. pastoris* has a trimeric structure within the cell with a molecular weight of 51,000 Da. This is consistent with reports of trimeric TNF (29, 30).

Analogs of TNF were constructed for studies of structure–function relationships and purification (31). Four analogs were studied, two with internal deletions corresponding to the coding regions for amino acids 135–153 (TNF9), and amino acids 7–47 (TNF16) and two with carboxy terminal amino acid extensions. Analog TNF14 has a C-terminal extension of eight amino acids which was expected to increase hydrophobicity and analog TNF15 provides divalent metal ion-chelating ability to TNF by the addition of an eight-residue histidine-alanine-rich tail. Each of the four proteins was expressed in *E coli* and each was present in the insoluble fraction. When expressed in *P. pastoris*, all four were produced at 30% levels; however, only the TNF14 analog was soluble. This analog was determined to retain full biological activity. Whether or not the analog retains the C-terminal extension remains to be established.

V. SECRETION OF HETEROLOGOUS PROTEINS FROM *Pichia pastoris*

A. Invertase

Invertase is naturally produced by *S. cerevisiae* and is secreted into the periplasmic space, thereby enabling this yeast to utilize sucrose as a source of carbon (32). The *SUC2* gene which encodes invertase has been cloned and sequenced (33).

Secretion of invertase from *P. pastoris* and comparisons to invertase secreted from wild-type *S. cerevisiae* and the *sec*18 mutant strain of *S. cerevisiae* have been done by J. Tschopp and coworkers at SIBIA (34). The 532-amino-acid coding sequence of invertase and its natural secretion signal was inserted into an autonomous *P. pastoris* expression vector. The *AOX1* promoter was used for gene expression in both GS115 and KM71 host strains. Integration of single-copy integrants either at the *AOX1* or the *HIS4* gene of the *P. pastoris* genome was confirmed by Southern blots. Expression of the invertase gene was found to be tightly controlled by carbon source, with repression on glucose or glycerol, and induction on methanol.

It was found that up to 2.5 g/liter of invertase was secreted from the KM71 host strain. Over 80% of the invertase was secreted through the cell wall and into the medium, while the remaining fraction was retained in the periplasmic space. Secreted invertase composed 80–90% of the protein in the culture medium. This is the contrast to *S. cerevisiae* where the invertase is found in the periplasmic space.

On SDS-polyacrylamide gels, invertase from *S. cerevisiae* migrated as a heterogeneous population with a molecular weight ranging from 100 to 140 kDa, while invertase from *P. pastoris* was more homogeneous in size and migrated as an 85- to 90-kDA band. This indicated a higher level of glycosylation in the *S. cerevisiae* yeast and more variability in the size of the mannose side chains. Upon EndoH treatment, invertase from each yeast migrated as a 58-kDa nonglycosylated protein.

Cesium chloride equilibrium density gradient measurements of invertase from *S. cerevisiae* wild-type cells, the *sec*18 mutant of *S. cerevisiae*, and *P. pastoris* (Fig. 5) were made to compare the relative carbohydrate content of each. Invertase secreted from wild-type *S. cerevisiae* had a broad distribution with a peak around 1.46 g/ml, reflecting the heterogeneity of the mannose side chains and the overall high carbohydrate content. In contrast, invertases from the *sec*18 mutant and *P. pastoris* have a more homogeneous carbohydrate content with a peak around 1.40 g/ml. The *sec*18 mutant of *S. cerevisiae* is

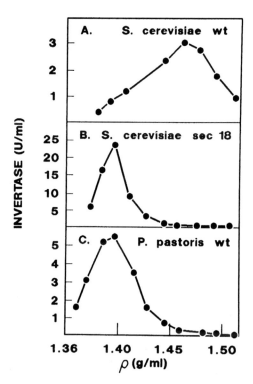

Figure 5 Cesium chloride equilibrium density distribution of invertase activity in *S. cerevisiae* wild-type (A), *S. cerevisiae sec*18 mutant (B); and *P.* pastoris wild-type strains (C). (From Ref. 35, used by permission of *Biotechnology*, 1987.)

known to have N-linked oligosaccharide containing eight mannose residues (35). Based on size of the side chains, the *P. pastoris* glycosylation pattern resembles that of higher eukaryotes.

B. Secretion Using Mammalian Pre-Pro Signals

Several proteins have been expressed and secreted in *P. pastoris* using their native mammalian pre-pro signals. Using the alcohol oxidase promoter for gene expression, bovine lysozyme has been secreted at levels of 200–400 mg/liter of broth (36). Amino terminal sequencing revealed proper processing of the mammalian signal sequence. The secreted bovine lysozyme had full enzymatic activity and comprised 50–70% of the protein in the broth.

γ-Interferon was secreted from *P. pastrois* using its native pre-pro signal under control of the alcohol oxidase promoter (G. Thill, personal communication). Secretion efficiency was about 40%, with levels in the broth of 15–20 mg/liter. Discrete mannose glycosylation was observed. The γ-interferon remaining inside the cell was of molecular weight 20–25 kDa, while the secreted protein was in the 35- to 38-kDa dimeric size range. After EndoH treatment, 16 and 32-kDa molecular weight forms of the protein were obtained, corresponding to monomeric and dimeric forms of the molecule.

C. N-Linked Oligosaccharides in *P. pastoris*

The N-linked oligosaccharides of the invertases from *P. pastoris* and *S. cerevisiae* were examined with respect to their size distribution and structural features. *P. pastoris* utilizes an average of 9 or 10 of the 14 available sites for N-asparagine-linked glycosylation of invertase (34). It was recently found that two to four chains of *P. pastoris* invertase are $Man_{8,9}GlcNAc_2$, five or six chains are in the range of $Man_{10-14}GlcNAc_2$, and one chain is $Man_{>30}GlcNAc_2$ (37). In contrast, invertase synthesized in *S. cerevisiae* utilizes 13 of 14 potential glycosylation sites with 9 of the 13 chains being short, $Man_{8-14}GlcNAc$ species and four being long chains of greater than 50 mannose residues ($Man_{>50}GlcNAc$) each (38,39). Using incorporated mannose label, Grinna and Tschopp showed that *P. pastoris* invertase incorporated 40–70% of the label in the small oligosaccharide fraction, compared to 5–10% for invertase from *S. cerevisiae*.

In the same series of experiments, Grinna and Tschopp also examined N-asparagine-linked oligosaccharides isolated from soluble *P. pastoris* periplasmic glycoproteins (37). Oligosaccharides released by EndoH treatment fell into two major size groups. Smaller oligosaccharides, $Man_{8-14}GlcNAc$, constituted a significant portion of the total oligosaccharides synthesized by *P. pastoris*.

VI. DISCUSSION

Pichia pastoris is an industrial methylotrophic yeast which has proven to be an excellent recombinant host for the production of heterologous proteins. The *P. pastoris* expression system utilizes tightly controlled, methanol-inducible promoters for control of gene expression. Growth on glucose or glycerol results in total repression of the *AOX*1 and *DAS* regulatory regions. Derepression during carbon limitation is seen with *AOX*1 but not for *DAS*. Both are fully induced on methanol.

Autonomous and *AOX*1-disruptive vectors using these promoters have been developed using the *P. pastoris HIS*4 gene for selection. Autonomous plasmids undergo spontaneous, stable integration either at the *HIS*4 or *AOX*1 loci, or can be directed to the *HIS*4 chromosomal locus by prior linearization with restriction endonucleases which cut near the center of the *HIS*4 gene contained on the plasmid. Disruptive vectors, such as pHIL-D1, have been designed to release a linear expression cassette which replaces the *AOX*1 gene upon chromosomal integration. In practice, this replacement event, as determined by growth rate on methanol, occurs from 5 to 35% of the time. Spontaneous multicopy insertions have been detected. The number of tandem copies at the *AOX*1 locus ranges from several to as many as 20–30, as seen for TNF. The multicopy inserts appear to be stably maintained.

Regulation of key methanol utilization enzymes has been found to differ between the methylotrophs *P. pastoris* and *H. polymorpha*. In *H. polymorpha*, up to 70% of the alcohol oxidase activity of methanol-grown cells is detected in cells grown on glycerol (40). Recent isolation of the alcohol oxidase (*MOX*) (41) and *DAS* (42) genes and their regulatory regions will facilitate studies of gene expression in this yeast.

Fermentation of recombinant *P. pastoris* production strains is done using simple, defined minimal media consisting of salts, biotin, ammonia gas for pH control and nitrogen source, and a carbon source. *P. pastoris* can be grown in batch or continuous fermentation. Most typically, cells will be grown in the repressed state to the desired cell density (about 70 g/liter dry wt in most cases) on glycerol and then induced by the addition of methanol. Host strains with *AOX*1 disruptions are generally grown in the batch mode because of the extremely slow generation time of these strains on methanol (~30 hr). Production strains with the expression cassette integrated at the *HIS*4 site can be grown in the continuous mode, which may be advantageous in some situations.

Expression of a variety of intracellular and secreted proteins in *P. pastoris* has been described. Expression levels have ranged from <1% to as high as 35% for various proteins. In almost all cases, good mRNA levels were obtained for heterologous proteins. Production levels were found to scale up linearly with cell density from the shake flask to small-scale fermentation for most proteins, reducing process development time. In most cases, intracellular proteins were readily solubilized with simple buffers and were biologically active.

Secretion of heterologous proteins in the gram per liter range have been obtained. Several mammalian pre-pro signals have been used successfully to secrete proteins into the culture medium and all were correctly processed. In addition, proteins are secreted through the wall and into the medium, rather than retained in the periplasmic space. *P. pastoris* secretes few indigenous proteins, thereby simplifying purification of recombinant proteins from the culture broth.

Pichia pastoris has been found to process some proteins differently from *S. cerevisiae*. For example, nearly 100% of HBsAG monomers were assembled into particles when an *AOX1*-disrupted *Pichia* host strain was used, in contrast to 10% assembly into particles in *S. cerevisiae*. And, as discussed in detail with invertase, *P. pastoris* secretes the protein into the media, whereas *S. cerevisiae* secretes invertase into the periplasmic space.

Significant differences have been observed between *P. pastoris* and *S. cerevisiae* in the extent of glycosylation of secreted proteins. Many potential pharmacological proteins are glycosylated as are most, if not all, proteins of the immune system. Carbohydrate side chains appear to be involved in a protein's serum half-life, its solubility and stability, and its interaction with receptors. The pattern of glycosylation obtained in a given recombinant expression system and the carbohydrate structure of the side chains is thus a growing concern. Data accumulated to date for glycoproteins secreted from *P. pastoris* reveal a more discrete pattern of glycosylation than that obtained in other yeast, with most of the side chains being short, mannose 8–14 units. This pattern of glycosylation more closely resembles the mammalian glycosylation pattern (43) and may have important implications for the use of *P. pastoris* for the production of pharmaceutical glycoproteins.

ACKNOWLEDGMENTS

I thank my colleagues at Phillips Petroleum Company and at the Salk Institute Biotechnology/Industrial Associates, Inc. (SIBIA) for many helpful discussions and for sharing unpublished results. I also thank Drs. K. Sreekrishna, W. W. Fish and G. H. Wegner for reviewing this manuscript, and Shirley Corley for her ever excellent secretarial skills.

REFERENCES

1. Hitzeman RA, Hagie FE, Levine HL, Goeddel DV, Ammerer G, Hall BD. Expression of a human gene for interferon in yeast. Nature 1981; 293:717-722.

2. Tuite MF, Dobson MJ, Roberts NA, King RM, Burke DC, Kingsman SM, Kingsman AJ. Regulated high efficiency expression of human interferon-α in *Saccharomyces cerevisiae*. EMBO, 1982; 1:603-608.
3. Valenzuela P, Medina A, Rutter WJ, Ammerer G, Hall BD. Synthesis and assembly of hepatitis B virus surface antigen particles in yeast. Nature 1982; 298:347-350.
4. Mellor J, Dobson MJ, Roberts NA, Tuite MF, Emtage JS, White S, Lowe PA, Patel T, Kingsman AJ, Kingsman SM Efficient synthesis of enzymatically active calf chymosin in *Saccharomyces cerevisiae*. Gene 1983; 24:1-14.
5. Urdea MS, Merryweather JP, Mullenbach GT, Coit D, Heberlein U, Valenzuela P, Barr PJ. Chemical synthesis of a gene for human epidermal growth factor urogastrone and its expression in yeast. PNAS 1983; 80:7461-7465.
6. Rietveld K, Bakhuis JG, Jansen in de Wal NJ, van Leen RW, Noordermeer ACM, van Ooyen AJJ, Schaap A, van den Berg JA. *Kluyveromyces* as a host for heterologous gene expression. Yeast 1988; 4:S163.
7. Janowicz ZA, Merckelbach A, Eckart M, Weydemann U, Roggenkamp R, Hollenberg CP. Expression system based on the methylotrophic yeast *Hansenula polymorpha*. Yeast 1988; 4:S155.
8. Ogata K, Nishikawa H, Ohsugi M. A yeast capable of utilizing methanol. Agr Biol Chem 1969; 33:1519-1522.
9. Harder W, Trotsenko YA, Bystryka LV, Egli T. Metabolic regulation in methylotrophic yeasts. In: Van Verseveld, HW, Duine, JA eds. Microbial growth on C_1 compounds: Proceedings of the 5th international symposium. Boston: Martinus Nijhoff, 1987:139-149.
10. Veenhuis M, Van Dijken JP, Harder W. The significance of peroxisomes in the metabolism of one carbon compound in yeast. Adv Micro. Physiol. 1983; 24:2-81.
11. Wegner GH, Harder W. Methylotrophic yeasts—1986. In: Van Verseveld HW, Duine JA, eds. Microbial growth on C_1 compounds: Proceedings of the 5th international symposium. Boston: Martinus Nijhoff, 1987:131-138.
12. Cauderc R, Baratti J. Oxidation of methanol by the yeast *Pichia pastoris*: Purification and properties of alcohol oxidase. Agr Biol Chem 1980; 44:2279-2289.
13. Ellis SB, Brust PF, Koutz PJ, Waters AF, Harpold MM, Gingeras TR. Isolation of alcohol oxidase and two other methanol regulatable genes from the yeast *Pichia pastoris*. Mol Cell Biol 1985; 5:1111-1121.
14. Cregg JM, Barringer KJ, Hessler AY, Madden KR. *Pichia pastoris* as a host system for transformations. Mol Cell Biol 1985; 5:3376-3385.
15. Cregg JM. Genetics of methylotrophic yeast. In: Van Verseveld HW, Duine JA. eds. Microbial growth on C_1 compounds: Proceedings of the 5th international symposium. Boston: Martinus Nijhoff, 1987:157-167.
16. Sreekrishna K, Tschopp JF, Fuke M. Invertase gene (*SUC2*) of *Saccharomyces cerevisiae* as a dominant marker for transformation of *Pichia pastoris*. Gene 1987; 59:115-125.
17. Tschopp JF, Brust PF, Cregg JM, Stillman CA, Gingeras TR. Expression of the *lacZ* gene from two methanol-regulated promoters in *Pichia pastoris*. Nucl Acids Res 1987; 15:3859-3876.
18. Thill G, Davis G, Stillman C, Tschopp JF, Craig WS, Velicelebi G, Greff J, Akong M, Stroman D, Torregrossa R, Siegel RS. The methylotrophic yeast *Pi-*

chia pastoris as a host for heterologous protein production. In: Van Verseveld HW, Duine JA. eds. Microbial growth on C_1 compounds: Proceedings of the 5th international symposium. Boston: Martinus Nijhoff, 1987:288-297.
19. Hagenson MJ, Holden KA, Parker KA, Wood PJ, Cruze JA, Fuke M, Hopkins TR, Stroman DW. Expression of streptokinase in *Pichia pastoris* yeast. Enzyme Microb Technol 1989; 11:650–656.
20. Cregg JM, Madden KR. Development of yeast transformation system and construction of methanol-utilization-defective mutants of *Pichia pastoris* by gene disruption. In: Stewart GG, Russell I, Klein RD. Hiesch RR, eds. Biological research on industrial yeasts. Vol. 2. Boca Raton: CRC Press, 1988:1-18.
21. Hinnen A, Hicks JB, Fink GR. Transformation of yeast. Proc Natl Acad Sci USA 1978; 75:1929-1933.
22. Cregg JM, Tschopp JF, Stillman C, Siegel R, Akong M, Craig WS, Buckholz RG, Madden KR, Kellaris PA, Davis GR, Smiley BL, Cruze J, Torregrossa R, Velicelebi G, Thill GP. High level expression and efficient assembly of hepatitis B surface antigen in the methylotrophic yeast, *Pichia pastoris*. Biotechnology 1987; 5:479-485.
23. Edman JC, Hallewell RA, Valenzuela P, Goodman HM, Rutter WJ. Synthesis of hepatitis B surface and core antigens in *E. coli*. Nature 1981; 291:503-506.
24. Malke H, Ferretti JJ. Streptokinase: Cloning, expression and excretion by *Escherichia coli*. Proc Natl Acad Sci USA 1984; 81:3557-3561.
25. Jackson KW, Malke H, Gerlach D, Ferretti JJ, Tang J. Active streptokinase from the cloned gene in *Streptococcus sanguis* is without the carboxyl-terminal 32 residues. Biochemistry 1986; 25:108-114.
26. Old LJ. Tumor necrosis factor (TNF). Science 1985; 230:630-632.
27. Gray PW, Aggarwal BB, Benton CV, Bringman TS, Henzel WJ, Jarrett JA, Leung DW, Moffat B, Ng P, Svedersky L.P, Palladino MA, Nedwin GE. Cloning and expression of cDNA for human lymphotoxin, a lymphokine with tumor necrosis activity. Nature 1984; 312:721-724.
28. Sreekrishna K, Nelles L, Potenz R, Cruze J, Mazzaferro P, Fish W, Fuke M, Holden K, Phelps D, Wood P, Parker K. High level expression, purification and characterization of recombinant human tumor necrosis factor synthesized in the methylotrophic yeast *Pichia pastoris*. Biochemistry 1989; 28:4117–4125.
29. Lewit-Bentley A, Fourme R, Kahn R, Prange T, Vachette P, Travernier J, Haquier G, Fiers W. Structure of tumor necrosis factor by X-ray solution scattering and preliminary studies by single crystal X-ray diffraction. J Mol Biol 1988; 199: 389-392.
30. Smith RA, Baglioni C. The active form of tumor necrosis factor is a trimer. J Biol Chem 1987; 262:6951-6954.
31. Sreekrishna K, Potenz RHB, Cruze JA, McCombie WR, Parker KA, Nelles L, Mazzaferro PK, Holden KA, Harrison RG, Wood PJ, Phelps DA, Hubbard CE, Fuke M. High level expression of heterologous proteins in methylotrophic yeast *Pichia pastoris*. J Basic Microbiol 1988; 28:265-278.
32. Carlson M, Botstein D. Two differentially regulated mRNAs with different 5' ends encode secreted and intracellular forms of yeast invertase. Gene 1982; 28:145-154.

33. Taussig R, Carlson M. Nucleotide sequence of the yeast *SUC1* gene for invertase. Nucl Acids Res 1983; 11:1943-1954.
34. Tschopp JF, Sverlow G, Kosson R, Craig W, Grinna L. High-level secretion of glycosylated invertase in the methylotrophic yeast, *Pichia pastoris*. Biotechnology 1987; 5:1305-1308.
35. Esmon PC, Esmon BE, Schauer IE, Taylor A, Schekman R. Structure, assembly and secretion of octameric invertase. J Biol Chem 1987; 262:4387-4394.
36. Cregg JM, Tschopp JF, Digen ME, Siegel B, Craig B, Velicelebi G, Thill G. High-level expression and secretion of heterologous proteins from the methylotrophic yeast, *Pichia pastoris*. In: Brew K, Ahmad F, Bialy H, Black S, Fenna R, Puett D, Scott W, Van Brunt J, Voellmy RW, Whelan WJ, Woessner JF. eds. Advances in gene technology: Protein engineering and production. ICSU Short Reports Volume 8. Oxford: IRL Press, 1988:48-49.
37. Grinna LS, Tschopp JF. Size distribution and general structural features of N-linked oligosaccharides from the methylotrophic yeast, *Pichia pastoris*. Yeast 1987; 5:107-115.
38. Reddy VA, Johnson, RS, Biemann K, Williams RS, Ziegler FD, Trimble RB, Maley F. Characterization of the glycosylation sites in yeast external invertase. I. N-linked oligosaccharide content of the individual sequons. J Biol Chem 1988; 262:6978-6985.
39. Ziegler FD, Maley F, Trimble RB. Characterization of the glycosylation sites in yeast external invertase. II. Location of the endo-β-N-acetylglucosaminidase H-resistant Sequons. J Biol Chem 1988; 262:6986-6992.
40. Eggeling L, Sahm, H. Derepression and partial insensitivity to carbon catabolite repression of the methanol dissimilatory enzymes in *Hansenula polymorpha*. Eur J Appl Microbiol Biotechnol 1978; 5:197-202.
41. Ledeboer AM, Edens L, Maat J, Visser C, Bos JW, Verrips CT, Janowicz Z, Eckart M, Roggenkamp R, Hollenberg C. P. Molecular cloning and characterization of a gene coding for methanol oxidase in *Hansenula*. Nucl Acids Res 1985; 9:3063-3082.
42. Janowicz ZA, Eckart MR, Drewke C, Roggnekamp RO, Hollenberg CP, Maat J, Ledeboer AM, Visser C, Verrips CT. Cloning and characterization of the *DAS* gene encoding the major methanol assimilatory enzyme from the methylotrophic yeast, *Hansenula polymorpha*. Nucl Acids Res 1985; 9:3043-3062.
43. Kornfeld R, Kornfeld S. Assembly of asparagine-linked oligosaccharides. Ann Rev Biochem 1985; 54:631-664.

10
Purification of Monoclonal Antibodies

Tom C. Ransohoff* and Howard L. Levine

Xoma Corporation
Berkeley, California

I. INTRODUCTION

Monoclonal antibodies were first introduced in 1975 when Köhler and Milstein (1,2) established a technique for producing unlimited quantities of homogeneous, monospecific antibodies of preselected specificity. Mouse myeloma cells and spleen cells immunized against sheep red blood cells were fused using inactivated Sendai virus to produce new, hybrid cells called hybridomas. These hybridomas, capable of growing in continuous culture, were found to secrete antibody against sheep red blood cells. In the 13 years since these pioneering experiments, monoclonal antibodies have been used as important reagents in research in medicine, biochemistry, microbiology, and immunology. Today, monoclonal antibodies play a significant role in many commercial in vitro diagnostic tests. In addition, these antibodies and their derivatives are being widely tested in cancer diagnosis and therapy (3-5), transplant rejection (6), and septic shock (7). One monoclonal antibody, OKT-3 (Ortho), has been approved for therapeutic use, and several more will probably be licensed in the near future.

As the use of monoclonal antibodies for both diagnostic and therapeutic indications increases, so too will the need for efficient and economical production methods. Attention will focus on the reproducibility of production processes and on the purity, potency, and safety of the final product. Most of the

---

*Present affiliation: Dorr-Oliver, Inc., Milford, Connecticut.

monoclonal antibodies currently being studied and purified on a large scale are murine-derived immunoglobulins. While this may change as the technology to produce human monoclonal antibodies develops, this chapter focuses on the large-scale purification of murine monoclonal antibodies

II. PHYSICAL CHARACTERIZATION OF ANTIBODIES

Antibodies consist of two types of polypeptide chains held together by disulfide bonds. The heavy chain, molecular weight approximately 55,000, and the light chain, molecular weight approximately 25,000, are each further subdivided into a constant and variable region. The N terminus of each chain, consisting of approximately 100 amino acids, is called the variable region and contains the residues that make contact with the antigen. The C-terminal constant portions of each chain fall into a limited number of sequences and define the antibody subtype. The five human isotypes of antibodies—IgM, IgD, IgG, IgA, and IgE—are determined by the five different heavy-chain types—μ, δ, γ, α, ϵ, respectively. There are also two types of light chain, κ and λ, for all imunoglobin classes. IgG (the major immunoglobulin in serum), IgD, and IgE are composed of two heavy and two light chains. Structurally, IgM is a pentamer of IgG-like molecules while IgA is a dimer. Murine antibodies of the IgG and IgM subtypes, the most common monoclonal antibodies in use today, are structurally similar to their human counterparts.

Despite many similarities, each monoclonal antibody is a unique protein and presents its own challenge in purification. The diversity of amino acids found in the variable regions and the variable degree of glycosylation of antibodies results in a wide range of isoelectric points for these proteins. A recent study shows that the pI of murine monoclonal antibodies ranges from 4.9 to 8.2 (8). Similarly, the charge density and hydrophobicity of each antibody is different. Also, since antibodies are glycoproteins, the nature and quantity of the carbohydrate moiety contributes to the protein's physical properties. Therefore, each antibody must be considered as a separate protein for purposes of purification process design.

III. PRODUCTION OF MONOCLONAL ANTIBODIES

The traditional method for producing monoclonal antibodies involves injecting hybridoma cells into the peritoneal cavity of histocompatible mice (9,10), where the cells grow and secrete antibody. Sometimes an ascites tumor forms. The ascitic fluid containing the secreted monoclonal antibody can be collected from the peritoneal cavity. The concentration of antibody and volume of ascites collected are related to the biological behavior of the particular cell line. As shown in Table 1, the volume of fluid generated ranges from 3 ml/mouse to as high as 14 ml/mouse while the titer of immunoglobulin in the fluid ranges from approximately 2 mg/ml to about 10 mg/ml (11).

Table 1 Volume and Monoclonal Antibody Yields in Lots of Ascites Generated by Different Cell Lines

Cell line	ml/Mouse[a,b]	mg Ig/ml[b]	mg Ig/mouse[c]
AA	4.7	1.8	8.4
DD	5.0	1.8	9.0
P	4.0	2.0	8.0
NN	5.0	2.4	12.0
K	3.6	2.7	9.7
BB	5.5	2.8	15.4
LL	4.9	2.9	14.2
J	3.1	3.3	10.2
E	5.6	3.5	19.6
W	6.4	3.5	22.4
H	3.4	3.8	12.9
B	4.5	4.1	18.0
Z	3.8	4.1	15.6
V	4.2	4.1	17.2
JJ	4.3	4.3	18.5
KK	6.1	4.4	26.8
U	3.9	5.0	19.5
HH	3.5	5.4	18.9
MM	3.9	5.7	22.2
X	3.4	6.2	21.1
CC	4.6	6.9	31.7
GG	4.2	7.4	31.7
T	2.4	7.5	18.0
Y	4.4	7.5	33.0
D	7.2	7.8	51.8
II	3.8	8.0	30.4
L	4.8	8.0	38.4
EE	3.8	8.1	30.8
G	14.1	8.1	114.2
I	5.5	0.0	49.5
Q	4.5	9.1	40.9
N	4.9	9.6	47.0
FF	5.9	9.6	56.6
O	4.3	10.0	43.0
M	4.3	10.8	46.4
R	4.2	13.3	55.8
A	8.4	13.4	115.9

Note: Mice were inoculated intraperitoneally with hybridoma cells. Antibody concentration was determined by cellulose acetate electrophoresis.
[a]ml/mouse injected.
[b]Average monoclonal antibody concentration of lot or lots of mice injected.
[c]ml/mouse X mg Ig/ml ascites.
Source: Reprinted from (11) with permission.

More recently, in vitro production methods have been developed for the large-scale production of monoclonal antibodies (12). These include suspension methods (such as stirred fermenters, airlift fermenters, and porous beads) and perfusion methods (such as hollow-fiber cartridges and ceramic matrices). In vitro production of monoclonal antibodies offers many advantages over the in-vivo (ascites) method. Because of the small volume of fluid collected per mouse, thousands of mice and several manipulations per animal are required to produce gram or kilogram quantities of monoclonal antibodies in vivo. Once operational, in vitro cell culture production offers a simpler, more efficient production scheme.

Regardless of the source, monoclonal antibody purification can be divided into three phases: clarification, isolation, and final polishing and formulation. During clarification, cellular debris and other particulates are removed from the crude ascitic fluid or cell culture fluid. Soluble impurities such as lipids, endotoxins, and nucleic acids may also be removed in this phase. The main objective of clarification is to produce a clean, particulate-free feedstock for subsequent purification. In the isolation phase of antibody purification, the antibody product is purified to near-homogeneity, usually via a series of adsorption-desorption column separations. In the final polishing and formulation phase of monoclonal antibody purification, the last traces of impurities are removed from the protein and the antibody is formulated for its intended use.

IV. CLARIFICATION

When monoclonal antibodies are produced in ascites tumors in mice, the ascitic fluid must be clarified to remove red blood cells, cellular debris, and fibrin clots collected with the fluid. In the laboratory, this clarification is often accomplished by high-speed centrifugation. For example, Stanker et al. (13) centrifuged ascitic fluid at 20,000g for 30 min to clarify it before loading a hydroxylaptite column. Bruck et al. (14) recommend centrifugation at 1000g for 5 min to remove cells and then further centrifugation at 100,000g for 30 min to remove smaller debris. These high-speed centrifugations often produce a pellet of cellular debris and a layer of lipid which floats on top of the aqueous, antibody-containing phase. These clarification methods are difficult to perform on a large scale and should be avoided if possible in the production and purification of monoclonal antibodies.

Neoh et al. (15) attempted to simplify the removal of lipids and other solids from ascitic fluid by adding silica gel to the crude ascitic fluid. Addition of 30g Cab-*O*-Sil (Union Carbide) per liter of ascitic fluid followed by a moderate-centrifugation (2000g, 20 min) yielded a lipid-free, clear solution. The yield of both IgG and IgM antibodies by this process was quantitative, and subsequent purification of the antibodies was unaffected by the treatment. Other solid absorbents, such as Biocryl Bioprocessing Aids (TosoHaas) and Cell Debris Remover (Whatman) also show promise as ascitic fluid clarification aids on a small scale. However, use of these adsorbents in systems of greater than 10 liters has not been reported and their use on a large scale may be difficult. Another approach to filtering ascitic fluid is to use a combination of depth filters and cascading membrane filters to remove lipids and other colloidal material, as well as solid material, from the solution.

While filter aids and prefilters coupled with moderate-speed centrifugation or dead-end filtration can facilitate the clarification of ascitic fluid, tangential flow filtration (TFF) offers an alternative to these methods (16). In tangential flow filtration, liquid flows parallel to the filter surface rather than perpendicular to it as in conventional dead-end filtration. This parallel flow provides a continual sweeping of the membrane surface, eliminating the buildup of a filter cake and thereby preventing filter plugging.

TFF is frequently used to clarify cell culture fluids because of its ease of operation and high yields. Shiloach et al. (17) present data on tangential flow filtration of mammalian (RBL-2H3) cells. In large-scale experiments, 30- and 64-liter samples of cell culture fluid were clarified using a 0.2-μm hollow-fiber cartridge (CFP-2-D-4, A/G Technology). In each case, the initial cell concentration was approximately 8×10^5 cells/ml, and a concentration factor of 30:1 was achieved with no decrease in cell viability. For both clarifications, an average inlet pressure of 4 psig was used with a recycle flow rate of 270 liters/hr. Average permeate fluxes for the 30- and 64-liter experiments were 245 liters/m^2/hr and 1.5 liters/m^2/ hr, respectively. In both cases, the flux decreased continually throughout the run, which indicates significant membrane fouling.

To minimize fouling, Shiloach et al. (17) suggest increasing membrane area during TFF scale-up by increasing the number of fibers instead of the fiber length. This will help to avoid the excessive buildup of pressure resulting from pumping cell culture fluid at the high cross-flow rates necessary for rapid processing. This suggestion is consistent with the results of Reismeier et al. (18), who demonstrated that during TFF of microbial cell suspensions, the height of the fouling gel layer is proportional to the transmembrane pressure. Under similar operating conditions (i.e., velocity and system back pressure), long hollow fibers will produce higher average transmembrane pressures and greater fouling than short ones. Therefore, in systems with equivalent surface area, a large number of short hollow fibers will be more efficient for clarifying cell culture fluid than a small number of long fibers.

Maintenance of high fluxes with TFF is not always possible. Concentration polarization and the buildup of a fouling layer at the membrane surface can minimize the effectiveness of TFF in clarifying cell culture fluid (19). The challenge in TFF system design and operation is to minimize irreversible protein adsorption onto the membrane surface and other fouling while retaining high fluxes. In order to preserve antibody conformation and activity, relatively low shear stress should also be maintained throughout the clarification. These challenges can be met by optimizing membrane material, system design, and operating conditions.

The choice of membrane material is particularly important in reducing the nonspecific adsorption of proteins. High protein binding on membrane surfaces may cause not only undesirable product losses during clarification but also increased fouling and reduction in permeate flow rates. Hydrophilic membranes such as polysulfone, cellulose acetate, nitrocellulose, and polyvinylidene difluoride (PVDF) are most effective at limiting protein binding (20). New membranes constructed from ceramic metal oxides (e.g., Norton Ceraflo) and carbon composites (e.g., GFT Murielle System) are currently being tested in bioprocessing applications, but their protein-binding characteristics are not yet well established. Passage through the membrane may affect a protein's conformation as well as binding to the membrane surface. Circular dichroism studies have shown that filtration of IgG antibodies through a nylon membrane severely alters the molecule's conformation. Passage of the IgG through polysulfone or poly-HPA-PVDF, however, had little or no effect on the antibody conformation (21). The exact cause of these changes is unknown, but they are not believed to be shear-related. While every system is different, cellulose acetate, polysulfone, and PVDF (the most frequently used membrane materials for TFF of hybridoma cell culture fluid) appear to provide low nonspecific antibody binding and have no detectable effect on antibody conformation.

Operational conditions can strongly influence the effectiveness of a tangential flow clarification. Hanisch (22) reviewed the effect of operating parameters on the performance of a tangential flow filtration system and concluded that increasing the shear rate at the membrane surface dramatically increases permeate flux. In most systems, a trade-off between the buildup of a gel layer and the shear imposed on the proteins and cells in the feed dictates the optimal recirculation velocity. While evidence suggests that mammalian cells in a stable biochemical environment are less shear-sensitive than originally feared (23-25), very high shear rates have been shown to damage mammalian cells (25). Table 2 (26) compares shear rates in various systems: the shear rates generated in TFF systems, with the exception of ceramic membrane systems, are generally no greater than those experienced in other systems. Minimizing the mass average shear (total residence time at given shear rate) (21) as well as the so-called "secondary" effects, such as the presence of many gas bubbles that could lead to interfacial denaturation (24,27), are also important in TFF. Finally, effective cleaning procedures (28) and periodic backflushing during clarification (29) can profoundly affect system performance and lifetime.

Figures 1–3 show optimization of a tangential flow clarification of a hybridoma cell culture (30). A 0.22-μm Durapore (Millipore) membrane was used to determine the effect of flux and time of operation on transmembrane pressure (TMP). A constant flux was maintained over time by varying the TMP of the

Table 2 Wall Shear Rates in Tangential Flow Filtration and Other Systems

System	Velocity U (cm/s)	Diameter d (cm)	Re^a	Wall shear $\Gamma_w (s^{-1})^c$
TFF systems:				
Plate and frame	90	0.1	900	5,400
Hollow-fiber	65	0.063	410	8,300
Ceramic membrane	540	0.25	$14,000^b$	110,000
Rotary annular gap^d	250	0.2	$5,000^b$	**$*^e$
Recirculation^f	100	2.5	$25,000^b$	3,100
Other systems:				
Human aorta	18	2.5	$4,500^b$	150
Syringe transfer (18 ga.)^g	21	0.1	210	1,700
Syringe transfer (23 ga.)^g	240	0.03	710	64,000
HPLC loading	51	0.025	130	16,000

^{a}The Reynolds number, $Re = Ud/v$, where v is the kinematic viscosity of the fluid, assumed to be equal to the value for water, 0.01 cm^2/s.
^{b}Reynolds numbers over 2100 indicate turbulent flow.
^{c}Formulas used for calculating the wall shear rate, Γ_w, are:
  (1) For pipe and tube flow (laminar): $\Gamma_w = 8U/d$
   (turbulent): $\Gamma_w = U^2 f/2v$,
   where f is the Fanning friction factor.
  (2) For plane poiseuille flow (laminar): $\Gamma_w = 6U/d$
^{d}Assumes no retentate recirculation.
^{e}Because turbulent couette flow in an annular gap is a very complex problem, the shear rate is not easily calculated for this system, but we estimate it to be in the range 1000 to 5000 s^{-1}.
^{f}Assumes 30 liters/min flow in 1-in. tubing.
^{g}Assumes 10 ml/min injection through a needle of the indicated gage thickness.

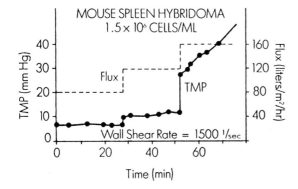

Figure 1 Effect of flux and time on transmembrane pressure. (Reprinted with permission from Ref. 30)

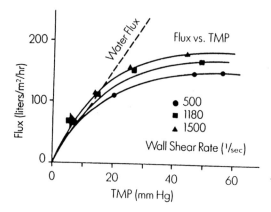

Figure 2 Effect of wall shear rate and TMP on membrane flux. Reprinted with permission from Ref. 30.)

system during total recycle of the permeate back into the feed tank (Fig. 1). At low fluxes, the TMP remained stable over time. As the flux was increased to 160 liters/m^2/hr, a steady time-dependent rise in TMP was noted, indicating the formation of a gel layer. In a separate set of experiments, the effects of reciruculation rate and TMP on flux was determined (Fig. 2). In tests with recirculation rates that generated wall shears of 750–1800 sec^{-1} and fluxes of 70–180 liters/m^2/hr, cell counts and cell viability measurements showed that for a given shear rate, viability decreased as the flux increased.

Based on the results obtained from the experiments shown in Fig. 1 and 2 (30), an operating flux of 85 liters/m^2/hr at a wall shear rate of 1500 sec^{-1} was chosen for the clarification. Figure 3 shows that not only did transmembrane pressure and cell viability remain stable over a 6-hr processing period but, more importantly, 90% passage of IgG antibody was demonstrated over this same period.

V. PRECIPITATION

After clarification, the monoclonal antibody can be partially purified by ammonium sulfate precipitation. At ammonium sulfate levels above 50% saturation, most IgG and IgM antibodies will precipitate from solution. Manil et al. (31) purified a variety of IgG1 and IgG2 antibodies from ascites fluid by ammonium sulfate precipitation. In all cases, greater than 90% of the monoclonal antibody was recovered in the solid phase following addition of ammonium sulfate to a final concentration of 50% saturation. Two-dimensional electrophoresis of the precipitated antibodies revealed that ammonium sulfate precipitation preferentially removed high molecular weight and acidic proteins from the ascites. This technique also effectively isolates and concentrates monoclonal antibodies from in vitro cell culture (32). However, ammonium sulfate precipitation can lead to denaturation of immunoglobulins, and disposal of large quantities of ammonium sulfate can be difficult and expensive.

An alternative precipitation technique for isolating IgM antibodies is euglobulin precipitation. This technique takes advantage of the fact that most IgMs are insoluble at an ionic strength less than that of 0.15 M NaCl (33). Dilution of either ascitic fluid or cell culture fluid with water causes the IgM to precipitate. Unfortunately, as in ammonium sulfate precipitation, euglobulin precipitation may also denature IgM antibodies. In addition, the low concentration of antibody in cell culture fluid makes the volume of liquid involved in euglobulin precipitations quite large and quantitative yields difficult to obtain.

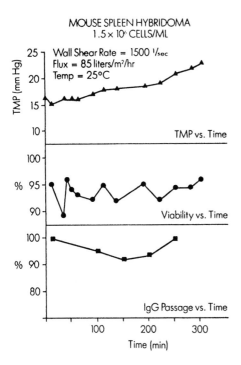

Figure 3 IgG harvesting by TFF. TMP (▲), cell viability (●), and IgG passage (■) were measured vs. time for a large-scale cell culture clarification. (Reprinted with permission from Ref. 30.)

VI. ION EXCHANGE SEPARATIONS

Ion exchange separations have been used extensively for purifying monoclonal antibodies as well as other proteins in general. Antibodies have been purified from ascitic fluid by adsorptive separation on anion exchange media such as DEAE cellulose (Whatman, 34) or DEAE Affi-gel Blue (Bio-Rad, 35). Bruck et al. (35) found that at pH 7.2, both IgG1 and IgG2 immunoglobulins bind to DEAE Affi-gel Blue, which is actually a bifunctional media containing immobilized Cibacron Blue on an ion exchange backbone. These antibodies were eluted with a gradient of 30–50 mM NaCl. The purified antibodies were recovered in >80% yield and were free of contaminating transferrin, albumin, and proteases.

Because most immunoglobulins have high pI values relative to ascites or serum proteins, cation exchange separations are often more effective in purifying monoclonal antibodies than anion exchange separations. With cation exchange separations, a pH can be chosen to minimize media binding of contaminants in the cell culture fluid or ascitic fluid. Carlsson et al. (8) showed that most IgG antibodies will bind to the cation exchange media SP-Sephadex C50 (Pharmacia) at pH 4.0. Ostlund et al. (36) found that when a cell culture supernatant containing an anti-tPA antibody was applied to an S-Sepharose Fast-Flow (Pharmacia) column at pH 5.5, all of the antibody bound to the media whereas 52% of the contaminating proteins flowed through the column. Selective elution of the antibody from the cation exchange column with 10 mM sodium citrate, 75 mM sodium chloride, pH 5.5, yielded an antibody that was greater than 60% pure.

Scott et al. (34) presented a large-scale cation exchange process for the purification of an unspecified IgG antibody. In this case, cell culture fluid was diluted with water, adjusted to pH 6.5 with acetic acid, and loaded onto an S-Sepharose Fast-Flow column equilibrated with 10 mM sodium phosphate, 10 mM sodium chloride, pH 6.5. The purified antibody was then eluted from the column with 10 mM sodium phosphate, 100 mM sodium chloride, pH 6.5 (Fig. 4). Under these conditions, the capacity of the ion exchanger was 15 g IgG/liter of media. With optimization of the pH and ionic strength of an ion exchange separation, as shown in these examples, monoclonal antibodies can typically be purified 25- to 100-fold (32,37). Furthermore, the high capacity, high flow rates, and chemical stability of ion exchange media such as Pharmcia's Sepharose Fast-Flow exchangers permit high antibody throughput at a reasonable cost (37). For this reason, ion exchange separations are attractive for process scale production of monoclonal antibodies.

The ion exchange purification of IgM antibodies has been more difficult than that of IgG-type immunoglobulins. Most IgM isolation procedures involve an initial precipitation step, which can denature the antibody and cause at least a partial loss of biological activity (38,39). Clezardin et al. developed a pro-

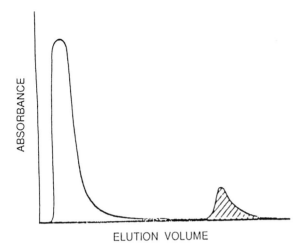

Figure 4 S-Sepharose Fast-Flow cation exchange purification of an IgG antibody. Five liters of concentrated cell culture fluid containing 6.5 g of monoclonal antibody was diluted 11-fold with water, adjusted to pH 6.5 with acetic acid; and loaded onto a 11.5 x 10 cm column at 27 liters/hr. The column was equilibrated in 10 mM sodium phosphate, 10 mM NaCl, pH 6.5 was eluted with 10 mM sodium phosphate, 100 mM NaCl, pH 6.5. (Reprinted with permission from Ref. 32.)

cedure for purifying IgM antibodies from either ascitic fluid (40) or cell culture fluid (41) by anion exchange separation on Mono Q (Pharmacia), which did not reduce biological activity. In either case, the starting antibody-containing solution was applied to a Mono Q column equilibrated with 20 mM L-histidine, pH 6.0. After nonimmunoglobulin-related material was washed off the column, the antibodies were eluted with a linear gradient of 0–0.5 M NaCl in 20 mM L-histidine, pH 6.0. The antibodies were recovered from the Mono Q column in 80% yield with complete retention of immunoreactivity.

VII. PROTEIN A AFFINITY SEPARATIONS

Affinity separations using immobilized protein A are well documented for the small-scale purification of antibodies (42,43). Protein A, a 42,000 Da cell wall protein isolated from *Staphylococcus aureas* (44), interacts with human immunoglobulins as well as most mouse antibodies (45). Mouse IgG2a, IgG2b, and IgG3 bind tightly to protein A while IgG1 and IgM bind weakly, if at all (42,46). The binding of antibodies to immobilized protein A is pH-dependent, with the strongest binding generally occurring at pH 7.0–8.5. The bound immunoglobulin is eluted from the affinity media using either a chaotropic agent (46) or solutions of low pH (42).

The large-scale purification of an antimelanoma IgG2a monoclonal antibody on protein A-Sepharose CL-4B (Pharmacia) was described by Lee (47). This media consists of protein A covalently coupled to Sepharose CL-4B using the cyanogen bromide method. Concentrated cell culture fluid containing 3.2 g/liter antibody was applied to a column preequilibrated with 0.05 M potassium phosphate, 0.2 M NaCl, pH 8.4. To protect the expensive affinity column, a guard column of Sepharose CL-4B was used to adsorb cell debris from the cell culture concentrate. After binding, the column was washed with equilibration buffer and the antibody was eluted with 0.05 M monobasic potassium phosphate, 0.2 M NaCl, pH 4.5. The yield of antibody from this column varied from 50 to 100%, and the recovered antibody was greater than 90% pure. However, the antibody did contain significant levels of endotoxin, bovine IgG (derived from the serum added to the original cell culture medium), nucleic acid, and protein A. For reuse, the column was regenerated with 3 M KSCN before reequilibration.

The example described above demonstrates the advantages and disadvantages of affinity separations using immobilized protein A. The simplicity of the process, its high purification power, and its relatively high yield make the process an attractive one for scale-up. Indeed, for in vitro studies, the antibody obtained by a single-step purification on a protein A affinity column might be sufficiently pure. However, for clinical use the antibody would require further processing to remove residual endotoxin, contaminating proteins, and nucleic acids. An additional ion exchange or gel filtration step might remove the residual impurities.

The leakage of protein A from the protein A–Sepharose CL-4B media represents a serious regulatory barrier. Using cyanogen bromide, the protein A is coupled to Sepharose via a N-substituted isourea. This linkage is chemically unstable (50), and hydrolysis of the isourea leads to a small but constant leakage of the ligand from the solid matrix (48,49). In the example cited above, two lots of antibody contained 3.7 ng protein A/mg antibody and 0.3 ng protein A/mg antibody. Furthermore, protein A was detected in all column fractions, indicating a steady release of protein A from the media.

To produce a more chemically stable linkage, many other activating agents have been developed for preparing affinity media (50,51). Activated gels ready for ligand coupling are commercially available in epoxy, aldehyde, vinyl sulfone, carbonyldiimidazole, and alkyloxypyridinium salt-activated forms. With protein A, many of these activated gels from a stable secondary amine between the ϵ-amino groups of lysine residues on the protein and the activated media.

VIII. DYE LIGAND AFFINITY SEPARATIONS

Johnson et al. (52) used immobilized Cibacron Blue F3GA to purify two IgM monoclonal antibodies. This chemical dye interacts with a variety of proteins, permitting affinity like separations. Crude ascitic fluid was loaded onto an Affi-gel Blue (Bio-Rad) column equilibrated with 0.05 M sodium acetate, pH 5.0. Most of the contaminating proteins either flowed through the column or were eluted with up to 2 M sodium chloride. The IgM was then eluted in a biologically active form with 2 M NaCl. The remaining protein impurities—haptoglobulin, transferrin, and trace quantities of albumin—were then removed from the IgM by gel filtration.

While the dye ligand affinity column effectively purifies the IgM antibodies, the regulatory issues of ligand leakage from the media remain. As with immobilized protein A, the use of this media in the manufacture of monoclonal antibodies for therapeutic use will require validation of its stability.

Despite the efforts of many researchers to develop better affinity ligands and more stable linkages, affinity purification of monoclonal antibodies remains a difficult and expensive process to use on a commercial scale. The regulatory issues of media reuse the ligand leakage may make this type of affinity separation unattractive for the production of monoclonal antibodies for therapeutic use.

IX. HYDROXYLAPATITE SEPARATIONS

Adsorptive separations on hydroxylapatite present an alternative to affinity separation as a powerful one-step technique for monoclonal antibody purification. The technique uses hydroxylapatite crystals as a medium for adsorption. Adsorption of proteins on hydroxylapatite results from an interaction between the divalent calcium ions of the hydroxylapatite and phosphate groups associated with the proteins (53). The strong binding of IgG and IgM antibodies to hydroxylapatite, coupled with the weak or nonexistent binding of contaminating proteins commonly found in ascitic and cell culture fluid, such as albumin and transferrin (13,54), make this separation technique a powerful one for antibody purification.

Stanker et al. (13) present a procedure for a single-step purification of IgG or IgM antibodies from ascitic fluid using hydroxylapatite. After clarification by centrifugation, the ascitic fluid was applied to a column of HTP grade hydroxylapatite (Bio-Rad). The column was equilibrated in 0.01 M sodium phosphate, 0.02% sodium azide, pH 6.8. After loading, elution was performed using a 5.6 column volume (CV) linear gradient of 0.01–0.3 M sodium phosphate. Regeneration was effected using successive washes (3 CV) of 0.5 M sodium phosphate buffer and 1 M NaCl. Approximately 80–90% of the initial

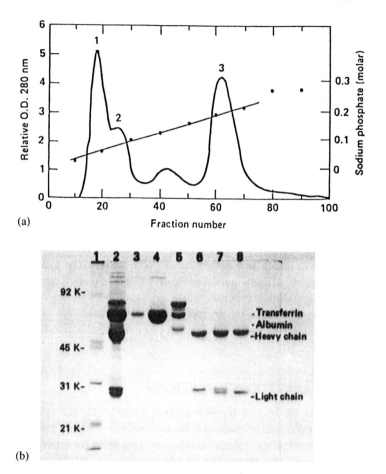

Figure 5 Hydroxylapatite purification of IgG antibodies. (a) The OD_{280} profile obtained following hydroxylapatite fractionation of ascites fluid. The column was equilibrated in 0.01 M sodium phosphate, pH 6.8. Bound proteins were eluted with a linear gradient of 0.01–0.3 M sodium phosphate, pH 6.8. (b) SDS-PAGE of hydroxylapatite purified immunoglobulins. Lane 1: MW standards—phosporylase B, 92, 500; bovine serum albumin, 66,200; ovalbumin, 45,000; carbonic anhydrase, 31,000; soybean trypsin inhibitor, 21,500. Lane 2: Crude ascites fluid. Lane 3: Column fraction 16. Lane 6: Column peak 1. Lane 5: Column peak 2. Lane 6: Column peak 3. Lane 7: Column peak 9. Lane 8: Column peak 5. (Reprinted with permission from Ref. 13.)

antibody was recovered in this procedure. The authors provided no quantitative purity data for the purified antibodies, but SDS-polyacrylamide gels (Fig. 5) indicated that the recovered antibody was highly purified.

Hydroxylapatite represents a less expensive alternative to protein A for large-scale antibody production. While the yield and purity of the recovered antibody are not always as high as those for affinity separations using immobilized protein A (13), the technique is far less expensive and could be used successfully to supplement an ion exchange column in a large-scale monoclonal antibody purification. In the examples cited above, no excessive media compression or system backpressures were reported at linear velocities of 10–40 cm/hr (13,54). Furthermore, the improved forms of the media currently available should be amendable to large-scale purification of monoclonal antibodies.

X. HYDROPHOBIC SEPARATIONS

Hydrophobic interaction chromatography (HIC) is emerging as a purification technique for monoclonal antibodies and other proteins. It offers a selectivity very different from that of ion exchange separations and, therefore, like hydroxylapatite, provides an alternative to affinity separations as an adjunct to an ion exchange process. HIC takes advantage of the fact that most antibodies have similar underlying hydrophobic properties. In HIC, proteins are bound or "precipitated" onto a hydrophobic medium, usually a hydrocarbon bound to the support matrix, from a mobile phase with a high salt concentration. The bound proteins are then eluted with progressively lower salt concentrations.

An example of the purification of an IgG monoclonal antibody by HIC is given in Fig 6. (55). Anion exchange purification of this antibody was not possible because the acidic pI of the antibody caused it to coelute with serum albumin. In addition, the antibody tended to undergo hydrophobic denaturation; high concentrations of ammonium sulfate destabilized the antibody and thus could not be used to remove albumin. The HIC media alkoxy Superflow (Sterogene) effectively separated the immunoglobulin from albumin. Approximately 1 liter of mouse ascitic fluid was loaded onto a 2-liter column of alkoxy Superflow preequilibrated with 0.9 M ammonium sulfate, 0.9 M glycine, 25 mM sodium phosphate, pH 7.2. After thorough washing of the column with equilibration buffer, the bound antibody was eluted with a low-salt buffer. Both IgG and IgM monoclonal antibodies have been separated from major serum contaminants including albumin and transferrin by HIC (55). HIC-purified antibodies were further purified to homogeneity by a subsequent ion exchange step purification.

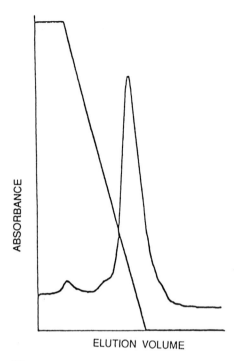

Figure 6 Purification of an IgG on alkoxy Superflow. An alkoxy Superflow column (2-liter bed volume) was equilibrated with 0.9 M ammonium sulfate, 0.9 M glycine, 25 mM sodium phosphate, pH 7.2. Ascites fluid, diluted 10:1 with this buffer, was applied to the column at a linear velocity of 200 cm/hr. After loading, the antibody was eluted from the column with 25 mM sodium phosphate, pH 7.2.

XI. MIXED-MODE SEPARATIONS

The recently introduced "mixed-mode" media, such as Matrex Silica PAE-300 (Amicon) and Bakerbond ABx (J. T. Baker), permit several different types of interactions with proteins. These media are prepared by coating a silica support with a hydrophilic polymer, which is then modified to provide multiple ligand types for interaction with proteins (56,57). In a procedure presented by Nau (57,58) for purifying of antibodies using the ABx media, concentrated cell culture fluid was loaded into a column of ABx (40-μm particle size) preequilibrated with 10 mM MES, pH 5.6. Under these conditions, albumin, transferrin, various proteases and indicator dyes did not bind to the ABx media. Antibodies were then eluted from the column with a linear gradient from the equilibration buffer to 1 M sodium acetate, pH 7.0 over approximately 10 column volumes (Fig. 7). Nau reports purities ranging from 70 to 99+%, with good recovery, using this technique.

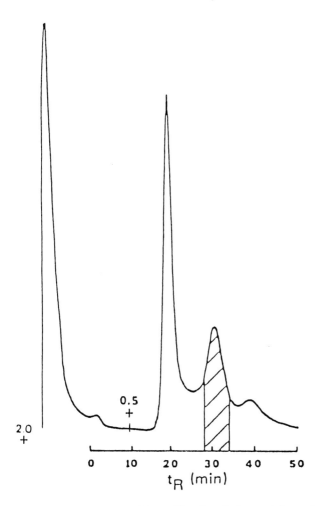

Figure 7 Chromatographic profile of an IgG containing cell culture fluid on 40-μm. Bakerbond ABx. Concentrated cell culture fluid was diluted with 3 volumes of 10 mM MES, pH 5.6 and applied to a 7.75 × 100 mm column equilibrated with 10 mM MES, pH 5.6. After loading, the column was eluted with a linear gradient from equilibration buffer to 1 M sodium acetate, pH 7.0. The void volume peak contained albumin, transferrin, proteases, and phenol red. The cross hatched peak contained the immunoglobulin. Following elution of the void peak, the UV montior sensitivity was increased four-fold (+) to better visualize the antibody peak. (Reprinted with permission from Ref. 57.)

The advantages of mixed-mode media such as ABx are their physical stability, low resistance to flow, and high binding capacity. In the example discussed above, a linear velocity of 120 cm/hr generated a backpressure of only 3 psig on a 0.78 × 10 cm column, which suggests that high throughputs are possible when this media is used in production. In addition, the binding capacity of the media is reportedly as high as 150 mg IgG/G media (57). However, because the media is relatively new, the performance of ABx compared to conventional ion exchange media, as well as its chemical stability, is unknown.

XII. GEL FILTRATION CHROMATOGRAPHY

Chromatography is the term that is commonly but incorrectly used to describe the column-based separation techniques described above. A more precise term for practical purposes would be reversible adsorptive separations. These separations are generally run by strongly binding a protein of interest to a solid adsorbent during loading and then eluting the protein from the same adsorbent using different conditions. Therefore, instead of a true chromatographic, or differential migration, separation, the result is an adsorption-desorption or "on–off" type of separation, which is inherently more easily scaled up than true chromatography (37). Because separation by size is not amenable to any currently available adsorptive separation techniques, gel filtration or size exclusion chromatography is an example of true chromatography (i.e., differential migration through a porous media) that is used for large-scale purification of monoclonal antibodies.

Gel filtration chromatography is most frequently used as a finishing or polishing step in the production of monoclonal antibodies because it is highly effective at removing trace contaminants of differing molecular weight (8,36). Processes using gel filtration as a polishing step often produce antibody solutions that are greater than 95% pure. Furthermore, from an operational standpoint, the most practical place for a gel filtration column is at the end of a process, where the produce is relatively pure and concentrated. Typically, gel filtration media do not support flow rates as high as ion exchange or affinity media. Also, because gel filtration is a true chromatography technique, the volume of material loaded onto the column is limited to a certain percentage of the column volume. These operating characteristics often preclude the use of gel filtration on a large scale except at the end of the process as a finishing step.

Ostlund et al. (36) reported on a large-scale process for isolating monoclonal antibodies from cell culture fluids that incorporated a preparative Superose 6 (Pharmacia) column as the final, polishing step. The antibody pool from an ion exchange separation was loaded onto a 5 × 80 cm Superose 6 column, equilibrated in 20 mM sodium phosphate, 500 mM NaCl, pH 8.0. To maintain high resolution, the load volume was limited to 1.7% of the total column volume, and the linear velocity of the mobile phase was limited to 5 cm/hr. Because of the low load and velocity of the mobile phase was limited to 5 cm/hr. Because of the low load and velocity constraints, four cycles were required on the Superose 6 column to process one batch of antibody, and the time for gel filtration was greater than the time required for all of the other three process steps combined (desalting, ion exchange, and ultrafiltration). Nevertheless, the final product was 95% pure IgG, and the yield on the gel filtration step was 89%.

XIII. REMOVAL OF NONPROTEIN IMPURITIES

The preparation of monoclonal antibodies for therapeutic use requires the removal not only of contaminating proteins but other impurities as well. The FDA published a guide for the development and use of monoclonal antibodies as therapeutic agents which states that the antibody should be "as free as possible of extraneous, immunoglobulin and non-immunoglobulin contaminants" (59). Since monoclonal antibodies are products of transformed cells, it is essential to design a purification process for effective separation of viruses and nucleic acids. A separate step, specific for one or both of these agents, may or may not be necessary so long as the overall process reduces them to below an acceptable maximum.

Finally, it is important that the purified antibody be sterile and pyrogen-free. Sterility is usually accomplished by aseptic filtration of the purified antibody through a 0.2-μm membrane filter. Bacterial endotoxins, or pyrogens, may be present in the crude antibody either from the media used in bioreactors or through bacterial infection of the mice producing the ascitic fluid. In either case, the purification process should be designed to reduce endotoxin levels and no single step should permit introduction of significant levels of endotoxin into the process stream.

Monoclonal antibodies appear to have a bright future as both diagnostic and therapeutic agents. To date, most of the published information on the purification of monoclonal antibodies relates to small-scale laboratory purification methods. From these data it is becoming increasingly apparent that the molecular diversity among immunoglobulins makes a single-step generic purification process highly unlikely. The challenge in large-scale antibody purification is to develop simple, efficient, and economical processes which reproducibly yield a pure, stable, and biologically active product.

REFERENCES

1. Köhler G, Milstein C. Continuous cultures of fused cells secreting antibody of predefined specificity. Nature 1975; 256:495-497.

2. Köhler G, Howe SC, Milstein C. Fusion between immunoglobulin-secreting and nonsecreting myeloma cell lines. Eur J Immunol 1976; 6:292-295.
3. Dillman RO, Royston I. Applications of monoclonal antibodies in cancer therapy. Br Med Bull 1984; 40:240-246.
4. Blair AH, Glose TI Linkage of cytotoxic agents to immunoglobulins. J Immunol Meth 1983; 59:129-143.
5. Spitler LE, del Rio M, Khentigan A, Wedel NI, Brophy NA, Miller LL, Harkonen WS, Rosendorf LL, Lee HM, Mischak RP, Kawahata RT, Stoudemire JB, Fradkin LB, Bautista EE, Scannon PJ. Therapy of patients with malignant melanoma using a monoclonal antimelanoma antibody-ricin A chain immunotoxin. Cancer Res 1987; 47:1717-1723.
6. Kernan NA, Byers V, Scannon PJ, Mischak RP, Brochstein J, Flomenber N, Dupont B, O'Reilly RJ. Treatment of steroid-resistant graft vs. host disease by in vivo administration of an anti-T-cell ricin A chain immunotoxin, J Am Med Assoc 1988; 259:3154-3157.
7. Harkonen S, Scannon P, Mischak RP, Spitler LE, Foxall C, Kennedy D, Greenberg R. Phase I study of a murine monoclonal anti-lipid A antibody in bacteremic and non-bacteremic patients. Antimicrob Agents Chemother 1988; 47:710-716.
8. Carlsson M, Hedin A, Inganas M, Harfast M, Blomber F. Purification of in vitro produced mouse monoclonal antibodies. A two step procedure utilizing cation exchange chromatography and gel filtration. J Immunol Meth 1985; 79:89-98.
9. Kennett RJ, McKearn TJ, Bechtal KB. (eds.) Monoclonal antibodies—hybridomas: A new dimension in biological analyses. New York: Plenum Press, 1980.
10. Langone JL, van Vunakis H. (eds.) Methods in enzymology. Volume 92. New York: Academic Press, 1983.
11. Chandler JP. Factors influencing monoclonal antibody production in mouse ascites fluid. In: Seanes SS. ed. Commercial production og monoclonal antibodies: A guide to scale-up. New York: Marcel Dekker, 1987:75-92.
12. Seaver SS. (ed.) Commercial production of monoclonal antibodies: A guide to scale-up. New York: Marcel Dekker, 1987.
13. Stanker LH, Vanderlaan M, Juarex-Salinas H. One step purification of mouse monoclonal antibodies from ascites fluid by hydroxylapatite chromatography. J Immunol Meth 1985; 76:157-169.
14. Bruck C, Drebin JA, Glineau C, Portelle D. Purification of mouse monoclonal antibodies from ascitic fluid by DEAE Affi-Gel Blue chromatography. In: Langone JL, van Vunakis H. eds. Methods in enzymology. Volume 121. New York: Academic Press, 1986:587-596.
15. Neoh SH, Gordon C, Potter A, Zola H. The purification of mouse antibodies from ascitic fluid. J Immunol. Meth 1986; 91:231-235.
16. McGregor WC. (ed.) Membrane separations in biotechnology. New York: Marcel Dekker, 1986.
17. Shiloach J, Kaufman JB, Kelly RM. Hollow fiber microfiltration for recovery of rat basophilic leukemia cells (RBL-2H3) from tissue culture media. Biotechnol Progr 1986; 2:230-233.
18. Riesmeier B, Kroner KH, Kula MR. Studies on secondary layer formation and its characterization during cross-flow filtration of microbial cells. J Membr Sci 1987; 34:245-266.

19. Blatt WF, Dravid A, Michaels AS, Nelson L. Solute polarization and cake formation in membrane ultrafiltration: Causes, consequences, and control techniques. In: Flinn JE. ed. Membrane science and technology. New York: Plenum Press, 1970:47-97.
20. Pitt AM. The nonspecific protein binding of polymeric microporous membranes. J Parenter Sci Technol 1987; 41:110-113.
21. Truskey GA, Gabler R, Dileo A, Manter T. The effect of membrane filtration upon protein conformation. J Parenter Sci Technol 1987; 41:180-121.
22. Hanisch W. Cell harvesting. In: McGregor WC. ed. Membrane separations in biotechnology, New York: Marcel Dekker, 1986:61-88.
23. Augenstein DC, Sinskey AJ, Wang DIC. Effect of shear on the death of two strains of mammalian tissue cells. Biotechnol Bioeng. 1971; 13:409-418.
24. Virkar PD, Narendranathan TJ, Hoare M, Dunhill P. Studies of the effects of shear on Globular proteins: Extension to high shear fields and to pumps. Biotechnol Bioeng 1981; 22:425-429.
25. Gabler FR. Tangential flow filtration for processing cells, proteins, and other biological components. ASM News 1984; 50:299-304.
26. Burnett MB, McGregor WC. unpublished results.
27. Ransohoff TC, Kawahata RT, Levine HL, McGregor WC. The use of surface tension measurements in the design of antibody-based product formulations; in preparation.
28. McGregor WC. Selection and use of UF membranes. In: McGregor WC. ed. Membrane separations in biotechnology. New York: Marcel Dekker, 1986:1-36.
29. Ripperger S, Schultz G. Microporous membranes in biotechnical application. Bioprocess Eng 1986; 1:43-49.
30. Anon. Harvesting mammalian cells with higher recoveries and higher yields. Downstream Process Update, Millipore Systems Division, 1986.
31. Manil L, Motte P, Pernas P, Troalen F, Bohuon C, Bellet D. Evaluation of protocols for purification of mouse monoclonal antibodies. J Immunol Meth 1986; 90:25-37.
32. Scott RW, Duffy SA, Moellering BJ, Prior C. Purification of monoclonal antibodies from large-scale mammalian cell culture perfusion systems. Biotechnol 1987; 3:49-56.
33. Deutsch HF, Morton JI. Human serum macroglobulins and dissociation units. J Biol Chem 1958; 231:1107-1118.
34. Parnham P, Androlewicz MJ, Brodsky FM, Holmes NJ, Ways JP. Monoclonal antibodies: Purification, fragmentation and application to structural studies of class I MHC antigens. J Immunol Meth 1982; 53:133-173.
35. Bruck C, Portetelle D, Glineur C, Bollen A. One-step purification of mouse monoclonal antibodies from ascitic fluid by DEAE Affi-gel Blue chromatography. J Immunol Meth 1982; 53:313-319.
36. Ostlund C, Borwell P, Malm B. Process scale purification from cell culture supernatants: Monoclonal antibodies. Dev Biol Standard 1985; 66:367-375.
37. Levine HL. High performance adsorption separations. In: Sikdar S., Bier M., Todd P. eds. Frontiers in bioprocessing. Boca Raton, FL: CRC Press, 1990:303–319.
38. Jehanli A, Dough D. A rapid procedure for the isolation of human IgM myeloma proteins. J Immunol Meth 1981; 44:199-204.

39. Bovet JP, Pires R, Pillat J. A modified gel filtration technique producing an unusual exclusion volume of IgM: A simple way of preparing monoclonal IgM. J Immunol Meth 1984; 66:299-305.
40. Clezardin P, Bougro G, McGregor JL. Tandem purification of IgM monoclonal antibodies from ascites fluid by anion exchange and gel fast protein liquid chromatography. J Chromatog 1986; 354:425-433.
41. Clezardin P, Hunter NR, McGregor IR, McGregor JL, Pepper DS, Dawes J. Tandem purification of mouse monoclonal antibodies produced in vitro using anion exchange and gel fast protein liquid chromatography. J Chromatogr 1986; 358:209-218.
42. Ey PL, Prowse SJ, Jenkins CR. Isolation of pure IgG1, IgG2a and IgG2b Immunoglobins from mouse serum using protein A Sepharose. Immunochemistry 1978; 15:429-436.
43. Golding JW. Use of *Staphylococcal* protein A as an immunological reagent. J Immunol Meth, 1938; 20:241-253.
44. Sjoquist J, Movitz J, Johansson IB, Hjelm H. Localization of Protein A in the bacteria. Eur J Biochem 1972; 30:190-194.
45. Forsgren A, Sjorquist J. Protein A from *S. aureus*, J Immunol 1966; 97:822-827.
46. MacKenzie MR, Warner NL, Mitchell GF. The binding of murine immunoglobulins to *staphylococcal* protein A. J Immunol 1978; 120:1493-1496.
47. Lee S. Affinity purification of monoclonal antibody from tissue culture supernatant using protein A-Sepharose 4B. In: Seaver SS. ed. Commercial production of monoclonal antibodies. New York: Marcel Dekker, 1987:199-216.
48. Wilcheck M, Miron T. Polymers coupled to agarose as stable and high capacity spacers. In: Jacoby WB, Wilcheck M. eds. Methods in enzymology. Volume 34. New York: Academic Press, 1978:72-76.
49. Wilcheck M. Oka T, Topper YJ, Structure of a soluble super-active insulin is revealed by the nature of the complex between cyanogen bromide activated Sepharose and amines. Proc Natl Acad Sci USA 1975; 72:1055-1058.
50. Dean PDG, Johnson WS, Middle FA. (eds.) Affinity chromatography—a practical approach. Oxford: IRC Press, 1985.
51. Lasch J, Janowski F, Leakage stability of ligand-support conjugates under operational conditions. Enzyme Microb Technol 1988; 10:312-314.
52. Johnson E, Miribel L, Arnaud P, Tsang KY, Purification of IgM monoclonal antibody from murine ascitic fluid by a two-step chromatography procedure. Immunol lett 1987; 14:159-165.
53. Bernardi G. Chromatography of proteins on hydroxyapatite. In: Hirs CHW, Timashoff SN. eds. Methods in Enzymology. Volume 27. New York: Academic Press, 1973:471-479.
54. Bukovsky J. Kennett R. Simple and rapid purification of monoclonal antibodies from cell culture supernatants and ascites fluids by hydroxylapatite chromatogrpahy on analytical and preparative scales. Hybridoma 1987; 6: 219-228.
55. Henner J. personal communication, 1988.
56. Anon. Matrex Silica PAE-300 Chromatography Media, Publication I-295, Amicon Corp., 1987.

57. Nau DR. ABx: A novel chromatographic matrix for the purification of antibodies. In: Seaver SS. ed. Commercial production of monoclonal antibodies. New York: Marcel Dekker, 247-275.
58. Nau DR. Procedures for Using ABx to Purify Antibody. In 1987: Seaver SS. ed. Commercial production of monoclonal antibodies. New York: Marcel Dekker, 1987:305-313.
59. Points to consider in the manufacture and testing of monoclonal antibody products for human use (1987).

IV

RECENT TRENDS IN THE AREA OF RECOMBINANT PROTEIN PURIFICATION AND ANALYSES

11
Engineering Proteins to Enable Their Isolation in a Biologically Active Form

Stephen J. Brewer and Barry L. Haymore

Monsanto Company
St Louis, Missouri

Thomas P. Hopp and Helmut M. Sassenfeld

Immunex Corporation
Seattle Washington

I. INTRODUCTION

Protein engineering is the application of scientific knowledge to the practical design and construction of proteins. The first applications of protein engineering have changed the properties of proteins to enable their isolation in a biologically active form. Continued development of this technology has improved the downstream processing properties of proteins so that high-purity proteins can be isolated at large scale with low cost. This chapter will describe how basic principles of separations science and protein chemistry can be used to engineer proteins with these properties.

Recombinant DNA technology allows the construction and cloning of a gene encoding the sequence of amino acids of any desired protein. Expression of this gene in a living organism leads to the synthesis of a linear sequence of amino acids. This sequence of amino acids encodes the folding pathway which leads to the protein acquiring its characteristic structure and function. The knowledge required to accurately predict the effect of sequence changes on the structure and function of a protein is in its infancy. However, by applying basic principles of protein chemistry and making small sequence changes, it is possible to modify structure and function in a predictable manner.

II. ENGINEERING PROTEINS

Native proteins may be engineered by changing amino acids within their sequence or by adding polypeptide extensions (fusions) to the C or N terminus. Modifications within the sequence will be permanent but fusions may be permanent or temporary. Permanent changes are possible if the desired properties of the natural protein are either improved or unaltered by the modification. For example, provided the modification does not reduce the catalytic activity or specificity of an enzyme, it may be used as a diagnostic reagent or in a chemical transformation. If, however, the product is to be used as a pharmaceutical, the modification must not introduce any additional toxicity. Therefore, if the fusion is to be temporary, then it is necessary to introduce cleavage sites which will allow the native protein to be produced free from any additional amino acids.

A. Introducing Cleavage Sites

There are many protein chemical techniques which allow specific peptide bonds to be hydrolyzed. These methods are either enzymatic or chemical. Proteases which cleave peptide bonds within polypeptide chains (endopeptidases) or sequentially digest amino acids from the C or N terminus (exopeptidases) and chemical methods can be used (Table 1). Therefore, by engineering a protein with a cleavage peptide which directs hydrolysis to the junction of a polypeptide fused at the N and/or the C terminus, the structure of the native protein can be restored (Fig. 1). In this manner, reversible fusions can be made.

Since each protein has a unique sequence, the design of a cleavage site depends on the specificity of the cleavage reaction. Chemical reactions or proteases which hydrolyze a peptide bond formed with a single amino acid can be used to release small peptides from larger precursors. For example, trypsin, which cleaves the peptide bond after lysine or arginine, and cyanogen bromide, which cleaves after methionine, have long been used to specifically digest proteins into smaller peptides for sequence analysis. Therefore the absence of methionine in insulin allows the use of cyanogen bromide to cleave peptide fusions and release the A and B chains from a fusion protein precursor (1). When enzymes are used to cleave peptides, selectivity may also be determined by the folding of the protein. Urogastrone contains three peptide bonds susceptible to trypsin but the protein is highly resistant to this protease. Thus trypsin only hydrolyzes the C- and N-terminal fusions from a precursor protein leaving the urogastrone intact (Fig. 2). In certain circumstances, the selectivity of proteases may also be enhanced by chemical modification of reactive amino acids in proteins. Therefore, the five lysines in β-endorphin can be blocked with citraconic anhydride. Trypsin will then selectively cleave after an arginine which links this polypeptide to a β-galactosidase fusion protein (2). Mild hydrolysis can then be used to produce the authentic polypeptide.

ENGINEERING PROTEINS FOR ISOLATION

Table 1 Cleavage Peptides Used in Fusion Proteins[a]

Method	Specificity	Product
	Single Amino Acid	
Chemical		
Cyanogen bromide (4)	N-Met-C	N-Hse & C
Endopeptidase		
Trypsin (6)	N-Lys/Arg-C	N-Lys/Arg & C
V8-Protease (23)	N-Glu/Asp-C	N-Glu/Asp & C
	Multiple Amino Acid	
Chemical		
Acid (24)	N-Asp-Pro-C	N-Asp & Pro-C
Endopeptidase		
Collagenase (25)	N-Pro-X-Gly-Pro-C	N-Pro-X & Gly-Pro-C
Clostropain (26)	N-Lys-Arg-C	N-Lys-Arg & C
Enterokinase (27)	N-Asp-Asp-Lys-C	N-Asp-Asp-Lys & C
Factor Xa (28)	N-Ile-Glu-Gly-Arg-C	N-Ile-Glu-Gly-Arg & C
Exopeptidase		
Aminopeptidase I (5)	Glu-Ala-Glu-C	Glu-Ala-Glu & C
Carboxypeptidase B (29)	N-(Lys)n/(Arg)n	n(Lys)/n(Arg) & N

[a]The specific hydrolysis of peptide bonds within a protein can be achieved using chemical or enzymatic methods. The amino acid sequences above (where X indicates any amino acid) have been used in fusion proteins to allow a specific hydrolysis reaction. Depending on the hydrolysis method, these cleavage peptides may be used with fusion polypeptides linked at either or both their amino and carboxy termini. These are indicated as N and C polypeptides, respectively. Chemicals or endopeptidases will generally hydrolyze accessible cleavage sequences anywhere in a fusion protein. Exopeptidases, however, are a special group of proteases which digest individual amino acids or peptide sequences from either the N terminus (aminopeptidase) or the C terminus (carboxypeptidase). The cyanagen bromide reaction converts Met to homoserine (Hse) while cleaving the peptide chain.

Selectivity can also be increased by using hydrolysis reactions which require two or more amino acids. For example, Asp-pro bonds are particularly susceptible to acid hydrolysis and there are a large number of proteases specific for sequences of two or more amino acids. Finally, there is a class of enzymes which can remove repeating sequences of amino acids from the N or C terminus. These enzymes are particularly useful for removing short polypeptide fusions from either end of a protein.

Therefore, a knowledge of enzymatic and chemical methods for selective peptide hydrolysis combined with rDNA technology allows the design and synthesis of a protein with a reversible modification. Applications of this technology have enabled a number of rDNA-derived proteins to be purified.

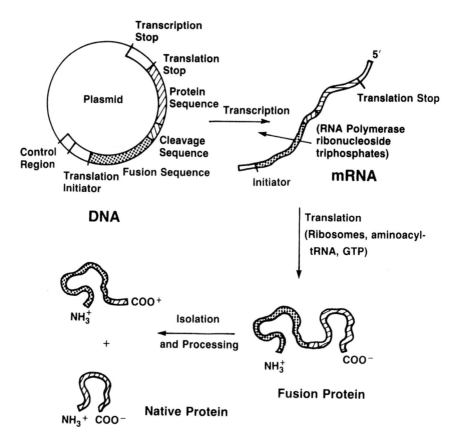

Figure 1 Synthesis and isolation of fusion proteins produced by bacteria.

III. ENGINEERING TO ENABLE ISOLATION OF ACTIVE PROTEINS

Although the transfer of genetic materials between eukaryotic and prokaryotic organisms can be readily achieved, the proteins are expressed in radically different environments. The higher organizational level of mammalian cells means that proteins may be expressed in specialized areas, such as the Golgi complex, and hydrolytic enzymes compartmentalized in microsomes. Furthermore, the rate of protein synthesis is much slower than that in bacteria, which also do not glycosylate proteins. It is impossible to predict what will happen when a mammalian protein is expressed in a bacterium such as *E. coli*, but in many cases proteins either are rapidly degraded or aggregate into inclusion bodies. In the form of inclusions, these proteins are stable but require refolding in order to attain the native structure and function. The refolding process is com-

ENGINEERING PROTEINS FOR ISOLATION

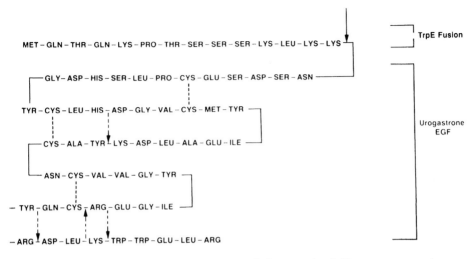

Figure 2 Trypsin cleavage of β-urogastrone fusion protein. β-Urogastrone contains many internal arginine and lysine residues which could be cleaved by trypsin (⇢) but are not because of steric effects. Therefore, an N-terminal fusion of 14 amino acids could be selectively removed by trypsin digestion (→) to produce authentic human β-urogastrone with no N-terminal modifications.

plex and with certain proteins not readily achieved. Examples will now be given where protein engineering has been used to overcome these stability problems to allow the isolation of a product with a native structure and/or function.

A. Production of Inclusions

Bacterial cell cultures often express genetically engineered proteins in the form of inclusion bodies. These inclusions may offer an attractive route for protein purification if they can be readily separated from the bulk of intracellular proteins by homogenization and centrifugation. In addition, proteolysis is particularly sever when polypeptides and small proteins are expressed in bacteria. Proteins which are in inclusions, however, are largely resistant to proteolysis.

The reason why recombinant DNA-derived proteins form inclusions is unclear but it appears to be related to the inability of the protein to fold into the correct structure in the physiological conditions of the bacterial cell, combined with a high rate of expression. Denatured proteins have a low solubility product and as this is exceeded the protein precipitates (3). Bacterial proteins can be made to form inclusions by expressing truncated proteins in large amounts. Therefore, by making fusions of unstable proteins and peptides with *E. coli* proteins or fragments of proteins, stable inclusions may be formed (Table 2).

Table 2 Protein Fusions for Stability[a]

Polypeptide	Mechanism	Protein
β-Galactosidase (4)	Inclusions	Somatostatin
β-Galactosidase (1)	Inclusions	Insulin α and β chains
β-Galactosidase (2)	Inclusions	β-Endorphin
TrpE (30)	Inclusions	Urogastrone
Polyarginine (30)	Inclusions	Urogastrone
Secretion signal (24) and protein A	Extracellular	Insulin-like growth factor (IGF-1)
Secretion signal (8)	Periplasmic	Urogastrone

[a]Small proteins and peptides expressed in *E. coli* are rapidly digested by proteases. They can, however, be stabilized by forming fusions with other polypeptides which encourage the formation of inclusion bodies. In addition, the protein may be fused to a polypeptide which causes secretion into the periplasmic space of *E. coli* or into the extracellular medium. The secretion signal is removed by enzymes in the cell envelope during export.

The expression of a β-galactosidase/somatostatin fusion protein was the first example of a polypeptide being stabilized by forming an inclusion body (4). Stabilization can also be achieved by using much smaller polypeptide fusions. Small basic polypeptides stabilize urogastrone by encouraging the formation of inclusions (Table 3). The ability to stabilize polypeptides with small fusions is important where high culture productivity is required. The larger the fusion, the more production capacity is wasted in making an ultimately un-

Table 3 Stabilization of Urogastrone by Inclusion Bodies Formed with Basic Polypeptide Fusions[a]

Fusion Polypeptide	Soluble		Insoluble	
	%	Half-life (min)	%	Half-life (min)
None	>90	6	—	—
N-Terminal-TrpE	67	25	33	Stable
C-Terminal-Polyarg	61	19	39	Stable
TrpE and Polyarg	83	36	17	Stable

[a]The effect of the addition of small, basic polypeptide fusioins on the stability of urogastrone when expressed in *E. Coli* was studied using pulse-chase experiments (30). The percentage of the total urogastrone expressed as soluble and insoluble and the corresponding half-lives are shown. The results demonstrate that insoluble proteins in inclusions are stable.

wanted product. Furthermore, large fusions will almost certainly need to be removed because of interferences with the efficacy of the protein product. This in turn increases the complexity of downstream processing.

B. Production of Native Proteins

Bacterial expression systems normally add methionine at the N terminus of proteins as part of the translation initiation step. This N-terminal methionine may also be formylated to a variable extent. Thus human growth hormone produced by *E. coli* is not identical to the human product. Furthermore, incomplete N-terminal formylation results in a heterogeneous product. By engineering human growth hormone with an N-terminal sequence cleaved by aminopeptidase I the authentic growth hormone can be isolated (5). Similarly, a lysine cleavage site engineered into a trypE/urogastrone fusion protein may be used to produce authentic human urogastrone. Although the urogastrone contained four potential trypsin cleavage sites, steric effects directed hydrolysis to the N-terminal fusion (6).

Although the production of rDNA-derived proteins as inclusion bodies has advantages for purification, chaotropic agents (e.g., urea, guanidine, and SDS) have to be used to dissolve them. This means that the proteins need refolding. For some proteins this is a major problem. Protein engineering can provide two solutions: enable the soluble but unstable product to be rapidly isolated, or change the primary structure of the protein to improve refolding efficiency. An example of the former was the use of a polyarginine purification fusion (see below) combined with preparative HPLC to isolate an unstable, soluble γ-interferon. In this case, the majority of the γ-interferon was produced in inclusion bodies and was stable. However, it could not be refolded to the correct specific activity. By allowing the rapid isolation of the small amount of soluble protein, the specific activity of the rDNA-derived protein was shown to be similar to the native, glycosylated protein (Table 4). β-interferon was also produced in an inclusion body and could not be refolded to the correct specific activity because it preferred to form the wrong disulfide bond pair. In this case, protein engineering was used to replace the miss pairing cysteine with a serine. This allowed the recombinant protein to be isolated with the correct activity (7).

IV. ENGINEERING TO ENABLE SIMPLIFIED PURIFICATION

The examples above use protein engineering to overcome problems encountered when bacterial expression systems are used for mammalian protein or polypeptide production. However, changes in the expression systems may also solve these problems. For example, extracellular expression systems allow

Table 4 Purification of Polyarginine-Tailed γ-Interferon by Ion Exchange HPLC[a]

Sample	Volume (ml)	Protein (mg)	Interferon Activity Units	Units/mg
Applied	34	272	8.5×10^7	3.1×10^5
Nonadsorbed	38	240	0.6×10^7	—
Elution pool	1.5	1.2	4.2×10^7	4.2×10^5

[a]The soluble fraction of γ-interferon/polyarginine fusion protein was rapidly purified from homogenates of *E. coli* using the affinity for an ion exchange resin. Typical of affinity chromatography, the product has a 135-fold increase in specific activity and is concentrated 11-fold (30).

urogastrone, which is unstable when expressed intracellularly, to be expressed in a stable, soluble form (8). However, protein engineering can also be used to improve purification properties. This application makes a significant addition to the existing separations technology available for purification process development.

A. Engineering Improved Chromatography Properties

Purification is achieved by exploiting differences in charge, size, hydrophobicity, and chemical reactivity between proteins and other impurities. Separation methods can be based on all four characteristics either independently or in combination using affinity methods. Chromatography is the most powerful separation method in terms of resolution and versatility. This is because different columns and solvents allow all these physicochemical interactions to be used.

Liquid chromatography uses differential partitioning between liquid and surface phases in a selective binding and/or a dynamic mode. For example, immunoaffinity chromatography uses the selective binding of a protein to an immobilized antibody by its unique antigenic site. Ion exchange chromatography selectively binds classes of proteins which have the required charge distribution. In addition, by adjusting solvent conditions, proteins of similar structure can be resolved by dynamic partitioning between the mobile solvent phase and the stationary surface phase (Fig. 3). Finally, size exclusion is based entirely on a dynamic partitioning effect between solvent and a porous stationary phase (Table 5).

In the examples described below, the major emphasis is to engineer sites which improve the selective adsorption of the protein onto immunoaffinity, metal affinity, and ion exchange chromatography media. However, these methods may also improve the resolution of liquid phase methods and these applications will be briefly reviewed.

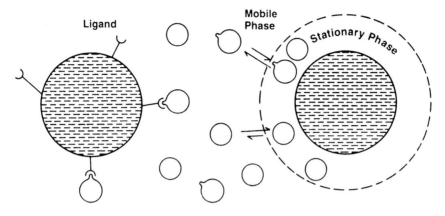

Figure 3 The two modes of protein chromatography: selective adsorption and dynamic partitioning. Protein chromatography occurs through two distinct mechanisms. In selective adsorption, a specific structure on the surface of the protein binds specifically to an immobilized ligand. Only molecules which have this structure are bound and selectively removed from the stationary phase. After impurities are removed, the protein can be released in a purified form. In a partitioning mode, the protein is in a dynamic equilibrium between a stationary phase and the mobile phase. Separation depends on different equilibrium constants between proteins. Those species which spend more time in the mobile phase are eluted before those which spend most of their time in the stationary phase.

B. Engineering for Immunoaffinity Chromatography

An eight-amino-acid sequence has been designed (9) which combines a four-residue antigenic site, with an overlapping five-residue enterokinase cleavage site (Fig. 4). This enables the purification of a recombinant DNA-derived protein engineered with this peptide at the N terminus by immunoaffinity chromatography. After purification, enterokinase can be used to cleave the immunoaffinity sequence at the C terminus of the peptide to liberate the authentic protein product, which is free of extraneous amino acids (Fig. 5).

To design a peptide with such generic utility for the purification of proteins, a polymer of eight amino acids was chosen as the optimum which satisfies the specificity of enterokinase with minimal steric effects on enzyme/substrate binding. In addition, antibodies require at least six amino acids for good binding, and the aromatic amino acid flanked by charged amino acids appears to be important for antigenicity (10). The peptide was also extremely hydrophilic, which ensures that it is readily available to aqueous solvents for antibody binding and minimizes its effect on protein folding and biological

Table 5 Selective Adsorption and Dynamic Partition in Chromatography[a]

Chromatography	Adsorption	Partitioning
Affinity	+	−
Covalent	+	−
Dye ligand	+	±
Metal affinity	+	±
Ion exchange	+	+
Reverse phase	+	+
Size exclusion	−	+

[a]With the exception of size exclusion, all chromatography methods are able to bind proteins selectively onto the surface of the media. Binding is by a small patch on the surface of the protein. These methods are able to distinguish proteins which have differences in this adsorption site, but not on other parts of its surface (31). However, under certain circumstances, ion exchange and reverse phase media are able to utilize the effects of dynamic partitioning which allow these related proteins to be resolved.

activity. Finally, the peptide was such that even if the protein was denatured, the immunological activity of the peptide would be maintained and used to detect the protein in small amounts.

This purification fusion removes several factors which have previously limited the application of immunoaffinity chromatography to the purification of rDNA-derived proteins. A unique antibody does not need to be produced for each product because the same monoclonal antibody can be used to purify any

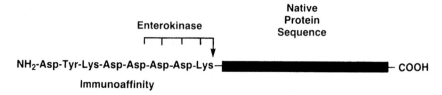

Figure 4 An N-terminal peptide designed for immunoaffinity purification and enterokinase cleavage. A purification fusion has been designed which has a sequence of four amino acids which are recognized by a specific antibody. This antibody can be used for both immunoaffinity purification and for the assay of any rDNA-derived protein which has this sequence at the N terminus. In addition, the peptide has a cleavage sequence which is recognized by the endopeptidase enterokinase. This can be used to release the native protein from the fusion protein precursor.

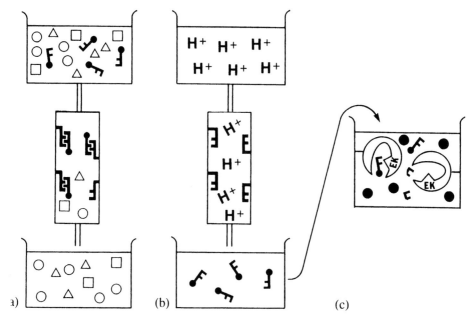

Figure 5 Principle of the affinity purification fusion. When an extract containing a mixture of proteins (○△□) and a rDNA protein with an N-terminal immunoaffinity purification fusion (F) is applied to an immunoaffinity column, it is selectively adsorbed and the contaminants pass through. Purified protein is eluted by acids or chelating agents. Immobilized enterokinase (EK) cleaves the purification fusion to produce the native purified protein (●).

rDNA protein with this N-terminal sequence. Considerable care has been given to the selection of an antibody which shows optimal purification characteristics. Thus the antibody selected for this purification will dissociate under physiological conditions by simply removing calcium (on which binding is dependent) from the medium or with mild acid. This allows the immunoaffinity column to be eluted without resorting to denaturing conditions of low pH which are often required to disrupt antibody–antigen interactions.

The highly specific recognition of an antibody for its antigen makes immunoaffinity chromatography a powerful method for purifying proteins from dilute and impure solutions. Purification factors of several orders of magnitude can easily be obtained. This method is therefore particularly suitable for producing research quantities of poorly expressed or dilute proteins. This is illustrated by an example where a colony-stimulating factor could be purified by 300-fold and concentrated 800-fold from cultures of recombinant yeast (Table 6). The selectivity of the antibody for the short fusion peptide means

Table 6 Purification of Colony-Stimulating Factor with an N-Terminal Immunoaffinity Fusion[a]

Sample	Volume (ml)	Protein (mg)	Units (x10−5)	Activity (unit/mg)
Supernatant	1000	150	5.5	3.7×10^3
nonadsorbed	1050	148	1.8	—
pH 7 Wash	50	0.2	1.3	—
pH 3 Elution (peak)	1.5	0.6	7.0	1.2×10^6

[a]The purification scheme shows the advantages of affinity chromatography in its ability to concentrate and purify dilute proteins. Not only does the specific activity increase by over 300-fold, but the concentration increases by over 800-fold (9).

that any protein with this sequence will be copurified. This is particularly useful when attempting to keep the biological activity of variably glycosylated proteins together. The ready accessibility of the N-terminal peptide allows enterokinase to cleave the purification fusion and release the native protein (Fig. 6).

C. Engineering for Ion Exchange Chromatography

In this second example, a protein is engineered to have a small sequence of charged amino acids at the C terminus. This makes the protein unusually acidic or basic in nature, which facilitates purification by ion exchange chromatography. This polymer extends away from the protein because such amino acids are hydrophilic. Consequently, this fusion has minimal effect on protein refolding and biological activity. In the special case of C-terminal arginine or lysine fusions, a generic method for removing the purification fusion is available using the enzyme carboxypeptidase B. This enzyme specifically digests C-terminal arginines and lysines and stops when any other amino acid is reached (Fig. 7). Therefore, provided the C-terminal amino acid is not lysine or arginine, the polypeptide of Arg/Lys can be specifically digested away to leave the native protein. Therefore, a three-step process has been developed which uses two cation exchange steps with an intervening digestion with carboxypeptidase B to yield a highly purified protein (Fig. 8). If the C-terminal amino acid is an arginine or lysine which is required in the product, a polyglutamic/aspartic fusion can be used and digested away with carboxypeptidase A. This enzyme has a complementary specificity to carboxypeptidase B and can digest these acidic amino acids, but stops at arginine or lysine.

Of the chromatography methods, ion exchange is highly favored because differences in binding properties are a function of charged amino acids which,

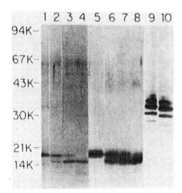

Figure 6 SDS-PAGE illustrating the purification and enterokinase cleavage of affinity-labeled proteins. Three proteins with N-terminal purification fusions were purified to homogeneity in a single pass of affinity chromatography: lane 1, purified interleukin 2 (IL-2); lane 5, purified granulocyte-macrophage-stimulating factor (GM-CSF); lane 9, macrophage colony-stimulating factor (M-CSF) exhibiting multiple carbohydrate molecular weight forms. These preparations were incubated with increasing concentrations of bovine enterokinase at 37°C for 16 hr and analyzed for changes in molecular weight using SDS-PAGE. This enzyme specifically cleaves this fusion to produce a protein with lower molecular weight. Lanes 2–4: IL-2 fusion cleaved to yield authentic interleukin 2. Lanes 6–8: GM-CSF cleaved to yield authentic protein. Lane 10: a heterogeneous population of M-CSF cleaved to produce the corresponding authentic proteins with lower molecular weight.

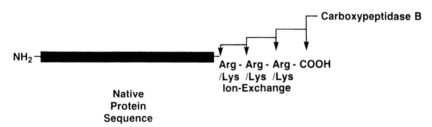

Figure 7 A C-terminal peptide designed for ion exchange purification and carboxypeptidase B digestion. A purification fusion has been designed which has a sequence of the basic amino acids arginine and lysine at the C terminus. This imparts a strong basic charge on the protein which can be purified by cation exchange chromatography. The peptide can then be specifically digested with the enzyme carboxypeptidase B which only hydrolyzes arginine or lysines from the C terminus of proteins to release the native protein from the fusion protein precursor (29).

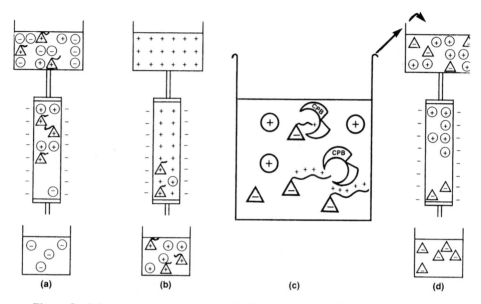

Figure 8 Principle of the Polyarginine Purification Fusion. When an extract containing a mixture of basic (⊕), acidic (⊖), and polyarginine-fused (△) proteins are applied to a cation exchange column, only basic proteins bind. The basic protein contaminants and polyarginine fused proteins can then be eluted with salt or acid. Carboxypeptidase B (CPB) digests the polyarginine purification fusion to produce the native protein which becomes acidic (△). The basic protein contaminants still bind to the cation exchange column while the native protein passes through the column and is purified.

being hydrophilic, will tend to be on the protein's surface. This minimizes denaturing effects which may occur when proteins bind to the hydrophobic surfaces used in reverse phase chromatography. Protein elution is also mild and involves changing the ionic strength or pH of an aqueous buffer. Ion exchange media have been made from particulate cellulose to micrometer-sized spherical beads. The former is suitable for processing proteins in bulk where low cost is paramount and the latter can be used for research preparations or to manufacture clinical quality proteins.

The binding to ion exchange media can be modulated in a precise way by varying the number of charged residues in the polymer. For example, each arginine residue requires an additional 125 mM sodium chloride to elute polyarginine-tailed proteins from a cation exchange column. However, with some proteins, local salt bridging may occur which will mask the binding properties of the first few arginines (Fig. 9). Optimal effects can be achieved with four to seven amino acids and there is some evidence that an Arg/Lys copolymer may have advantages over polyarginine by resisting proteases in *E. coli*. In bacterial extracts, the binding of nucleic acids to basic fusions can prevent adsorption to cation exchange resins. This effect can be overcome by

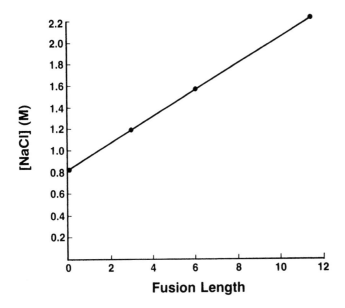

Figure 9 Effect of polyarginine fusions on the selective adsorption of γ-interferon to cation exchange media. The sodium chloride concentration required to elute γ-interferon from an analytical cation exchange column is proportional to the lengths of polyarginine fused to its C terminus.

using precipitating agents such as polyethylenimine or by selecting a high-efficiency ion exchange media. The purification of poly(arg/lys)/aminotransferase fusion protein from *E. coli* (Fig. 10) demonstrates how effective and simple this method can be for isolating a recombinant protein produced by *E. coli*.

The disadvantage of only using selectivity to purify proteins is that any degradation products with the correct affinity will also be copurified. Thus, a truncated form of soluble γ-interferon was copurified with γ-interferon because both species were selectively adsorbed as a dimer. However, when a salt gradient was used to elute the proteins from a preparative HPLC column, the truncated species are resolved (Fig. 11). This is consistent with the model proposed for chromatography in which ion exchange can operate in both selective adsorption and dynamic partitioning modes. The latter mode cannot be used for immunoaffinity chromatography but can be used by ion exchange to resolve closely related species.

D. Engineering for Metal Affinity Chromatography

This final example illustrates the use of protein engineering to introduce strong metal-binding sites which can be used to purify proteins by metal affinity

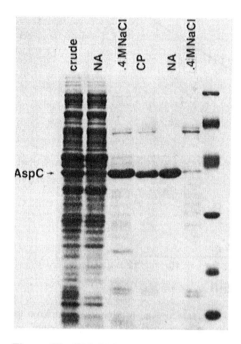

Figure 10 SDS-PAGE illustrating the purification of aminotransferase. Lanes from left to right contained the following samples: Crude; applied crude lysate. NA; unbound protein, first ion exchange column. 0.4 M NaCl; eluate first ion exchange column. CP; carboxypeptidase B-digested eluate. NA, digested, unbound aminotransferase from second ion exchange column. 0.4 M NaCl; eluate second ion exchange column. Last track; molecular weight standards. AspC indicates position of the aminotransferase (30).

chromatography. Instead of using C- or N-terminal polypeptide fusions, the affinity site is engineered within the primary sequence of the protein and takes advantage of its natural secondary structure. Engineering first requires the identification of suitable structural elements within the protein. This is achieved using secondary structure prediction algorithms. Although internal sequence changes are generally irreversible and lead to a variant protein, the metal-binding site is introduced with only one or two single-amino-acid modifications. Such small changes usually do not affect the biological activity of the protein or its antigenic characteristics.

1. Principle of Metal Affinity

The transition metals used in metal affinity chromatography are general Lewis acids which form four to six coordination bonds with suitable Lewis bases

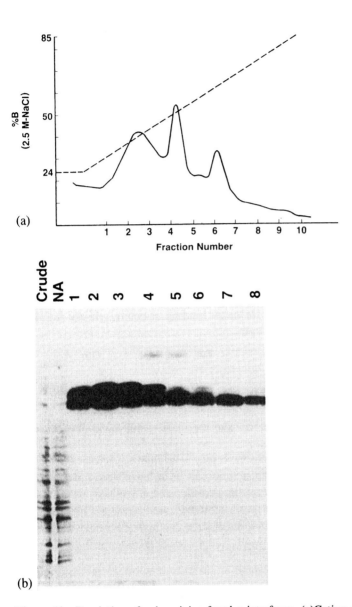

Figure 11 Resolution of polyarginine-fused γ-interferon. (a) Cation exchange chromatography used to purify γ-interferon with a C-terminal polyarginine fusion from the soluble fraction of *E. coli*. (b) SDS-PAGE analysis of fractions. Tracks labeled from left to right contained the following samples: crude; applied crude lysate. NA; unbound protein. Samples 1-9 fractions from column. If the column is eluted in a bind release mode typical of affinity chromatography, a degraded species is copurified with the γ-interferon. However, by applying a gradient, the ion exchange column resolves γ-interferon from the truncated species. Thus ion exchange can use a combination of selective adsorption to the polyarginine fusion and dynamic partitioning to resolve closely related species (31).

such as thiols, carboxylates, and amines. These bases are called ligands. If a single ligand contains two or more properly oriented binding sites, both atoms can simultaneously bind a single metal in a cooperative manner, and favorable enthalpy and entropy effects (due to conformational constraints and selective solvation) cause the metal-binding strength to increase. Such ligands are called chelates and this enhanced binding of chelates is called the chelate effect, which can amount to several orders of magnitude (11,12). Iminodiacetic acid (IDA) is a tridentate ligand which forms stable, facially bonded metal complexes with Cu^{2+}, Zn^{2+}, Ni^{2+}, and Fe^{3+}. These metal-IDA complexes can still form one to three additional coordination bonds depending on the metal. Several amino acid side chains found in proteins can form these bonds, and metal-IDA complexes which are properly immobilized on biocompatible supports will bind amino acids, peptides, and proteins. This effect can be used to purify proteins and is called metal affinity chromatography or ligand exchange chromatography (13).

The amino acids responsible for metal binding at neutral pH are ranked in the following order of bond strength:

$$Cys \geq His >> Asp; Glu > Trp, Tyr, Lys, Met, Asn, Gln, Arg$$

In order to bind to an immobilized metal, these amino acid residues must be situated on the surface of the protein. Although thiol groups usually bind most strongly to metals, exposed cysteine residues with free -SH groups are uncommon in stable peptides and proteins, and the most useful metal-protein interactions involve histidines.

Metal affinity matrices usually have relatively high capacities for binding proteins (25-125 g/liter). Protein elution can be achieved using a variety of bases or acids which compete with the peptide for immobilized metal or vice versa. Imidazole is probably the most commonly used base; elution with chelating bases such as glycine or EDTA usually gives inferior results. In addition, NaCl or another appropriate salt is added to buffers in order to overcome the ion exchange properties of the metal affinity matrix. Of the common metals used, Cu^{2+} is more tightly bound to an IDA column than Ni^{2+}, Zn^{2+}, or Co^{2+}. Owing to metal leakage, it is a good practice to strip a column with EDTA, clean, and reload with fresh metal before each run in order to obtain good chromatographic reproducibility.

2. Designing Proteins with Chelation Sites

Surface histidines play a major role in determining the strength of protein and peptide binding to metal-affinity matrices. The strength of the interaction is determined by the number of histidine residues and the degree to which these residues are exposed. However, if two histidines are held rigidly in the correct spacial alignment on a fully exposed surface of a protein, there is a dramatic increase in the strength of the metal binding. This occurs when the two histidines themselves combine to form a metal-chelating site. A few metal-binding sites with this geometry are known in nature. For example, thermolysin has a His73/His77 zinc-binding site (14), and superoxide dismutase has a His44/His46 copper-binding site (15). However, neither of these sites is on the surface of the protein and would not be useful for metal affinity chromatography.

Engineering a metal affinity site for purification requires the identification of a stable structure on the protein surface. The two most common stable structures in proteins are the β sheet and α helix. To meet the geometrical demands of a metal chelate, histidines in β sheets must be separated by one amino acid residue (Fig. 12) and those in α helix should be separated by three residues (Fig. 13). Knowing the primary structure, it is then possible to predict

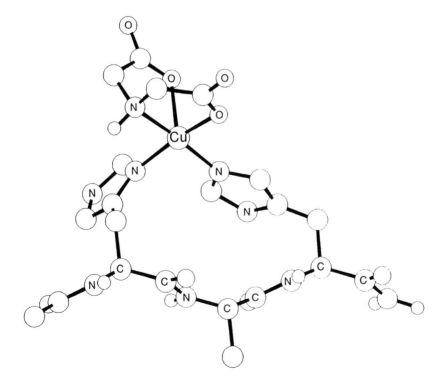

Figure 12 Metal-ligand chelate formed with histidines in a β sheet. The model shows iminodiacetatocopper(II) bound to Ac-His-Ala-His-NH$_2$ in the form of a β-sheet structure ($\phi = -139°$, $\omega = +135°$). Aliphatic and imidazoyl hydrogen atoms are omitted for clarity.

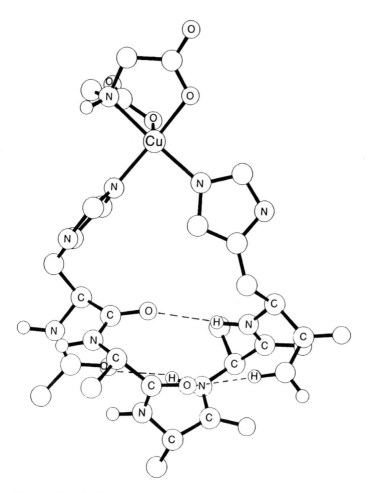

Figure 13 Metal-ligand chelate formed with histidines in an α-helix. The model shows iminodiacetatocopper(II) bound to Ac-His-Ala-Ala-Ala-His-NH$_2$ in the form of an α-helix structure ($\phi = -57°$, $\omega = -47°$). Aliphatic and imidazoyl hydrogen atoms are omitted for clarity.

many of these secondary structure features with a reasonable degree of accuracy (16). This information can be augmented by spectroscopic data from nuclear magnetic resonance, circular dichroism, and Raman spectroscopy. Knowing the parts of these structures which are on an exposed protein surface is also important. The hydropathy of the α and β structures provides useful information in this regard (17). An X-ray crystal structure provides, of course, the highest degree of structural information but analogies with known protein structures can help to establish a short list of potentially useful surface sites. Site-directed mutagenesis is then used to introduce the changes in amino acid sequence and the resultant proteins are screened for the desired effect. The application of these principles will be illustrated by the engineering of chelating sites into porcine and bovine growth hormone (somatotropins).

3. Engineering Somatotropins with Chelation Sites

Human growth hormone is required for the normal development of children; the bovine and porcine hormones can be used to improve animal productivity (18-20). The protein is required in large quantities at high purity. Porcine and bovine somatotropins have three endogenous histidines at positions 19, 21, and 169. Spectroscopic studies indicate that there are large amounts of α-helical structure in the proteins. Structural calculations suggest that there are four α-helical sequences and that all three histidines were in these regions (Fig. 14). Furthermore, hydropathy calculations predicted that two of the three histidine residues are appropriately exposed including the His19 in the first helix and His169 in the fourth helix (Fig. 15).

Based on these data, variants were made with the HisX3His Sequence and the resulting proteins were isolated and tested for strong binding affinity to metal affinity matrices. Changing Leu15 to His15 and utilizing the endogenous His19 yielded an engineered somatotropin which displayed very strong binding to a Cu affinity column (Fig. 16). This single change produced an effect approximately equal to six isolated, fully exposed histidine residues. This result was only consistent with the presence of a chelating metal-binding site on the protein's surface. The introduction of yet another histidine at position 11 resulted in a protein able to form two tandem chelation sites which doubled the affinity of the protein for the immobilized copper. Other single-histidine substitutions at surface exposed sites which did not meet the structural requirements for chelation (residues 16, 149) only incrementally increased the binding to immobilized metals (Haymore, manuscript in preparation).

The behavior of these variants is consistent with the reported structure of porcine somatotropin (21). The orientation of the first α-helix was such that the chelate site formed by the natural His19 and the new His15 was exposed to the solvent. The proteins with engineered affinity sites had similar biological activity to the native proteins in several different biological assays and allowed highly purified protein to be isolated in a single chromatography step from crude bacterial cell lysates (Table 7).

E. Engineering for Low-Cost Purification

The increase in processing steps to remove a purification fusion reduces the application of this technology in cases where cost is a major driving force.

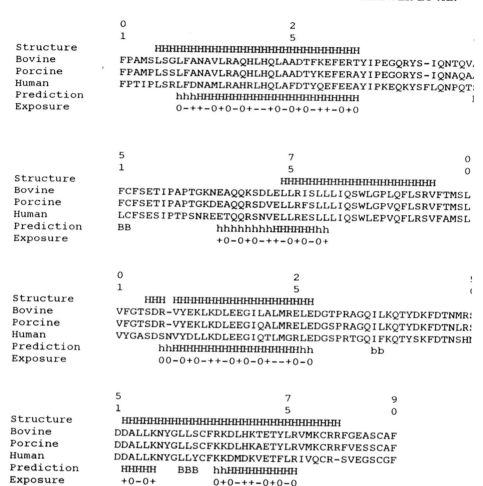

Figure 14 Secondary structure predictions for somatotropins. The sequences of three somatotropins (bovine, porcine, and human) are shown with helical regions (H = helix) obtained from the X-ray structure of porcine somatotropin (17). The joint prediction of secondary structure based solely on the sequences of seven different somatotropins (cow, pig, human, monkey, horse, rat, and mouse) is also shown (H = strong helix, h = moderate helix, B = strong β, b = moderate B) (13). The surface exposure of the helical residues was predicted from calculations of the periodicity of the hydropathic moment and the results are given on the final line (+ = exposed, − = buried, 0 = intermediate). Of the five predicted helical regions, 152–156 was considered too short to be useful and 65–80 possessed the weakest hydropathic moment and the poorest internal consistency in the prediction calculations. The following sites were predicted to be potentially useful for designing an exposed His-X3-His metal-binding site: 11–15, 15–19, 22–26, 26–30, 29–33, 110–114, 113–117, 117–121, 166–170, and 169–173.

However, the small, hydrophilic purification fusions described above are "silent" in terms of their effects on the structure and function of the proteins. These developments will increase the application of this technology to the production of bulk, low-cost proteins and enzymes.

Free-zone electrophoresis has many advantages for large-scale purification of proteins but lacks resolution. Using the Biostream system, the urogastrone-polyarginine-fused protein migrates toward the cathode and is purified to near homogeneity from a crude cell extract. This method is particularly useful for purifying proteins from turbid solutions which would cause the fouling of chromatography columns (C. Dickerson, personal communication).

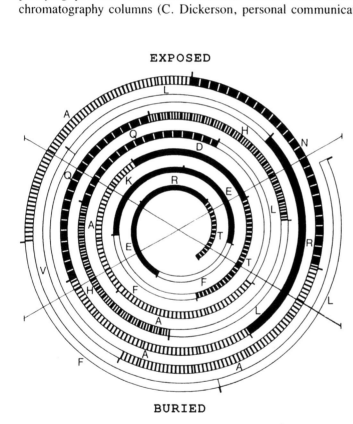

Figure 15 Prediction of exposed surface of somatotropin α-helix. A helical wheel representation of the first α-helix (leu9-thr34) of bovine somatotropin shows the predominant orientation of hydrophilic residues to one surface. This surface would be predicted to be exposed in water. This was obtained using published hydropathic indices (14). The center of the hydrophilic side was calculated to be oriented 245° clockwise from the center of the first residue, Leu9. A helix pitch of 100°/residue was used. The hydrophobic yet fully exposed Leu15 is the only residue which appears to be out of place by this technique.

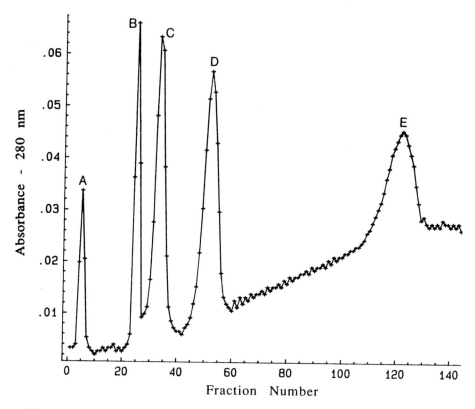

Figure 16 Metal affinity chromatography of engineered somatotropins. A copper-loaded metal affinity matrix (iminodiacetic acid immobilized to Trisacryl GF2000M by a diethylene glycol diglycidylether linker; 43 umol-Cu^{2+}/ml, 10 × 130 mm column) was equilibrated in buffer (50 mM NaH_2PO_4, pH 7.0). Proteins were loaded and eluted in this buffer and eluted with N-α-acetylhistidine (1–97 mM linear gradient 0–500 min, constant 97 mM 500–576 min). Fractions (2 ml) were collected and the adsorbance at 280 nm measured. Tuna cytochrome C (A, a nonbinding reference protein); ser169-porcine somatotropin (B, one His less than native, 1 mg); native bovine somatotropin (C, 1.9 mg); his149-bovine somatotropin (D, 2 mg) and His15-bovine somatotropin (E, 1.6 mg).

Solution methods using liquid-liquid partitioning and selective precipitation have considerable advantages for large-scale applications but lack the resolution of column chromatography methods. By engineering proteins with unique selectivities, powerful generic methods may be developed for isolating many different protein species. This is being investigated by engineering a β-galactosidase-polyaspartate fusion protein. The polyaspartate fusion is at the C terminus and contains a methionine cleavage site and a tyrosine alanine residue

Table 7 Purification of His15 Bovine Somatotropin Variant by Metal Affinity Chromatography[a]

Sample	Volume (ml)	Protein (mg)	His15-Somatotropin (mg)	% protein
Applied	790	9,600	20.1	0.21
Nonadsorbed	1030	9,400	<0.5	—
Elution pool	200	19.4	18.8	97

[a] The protein from 100 g of *E. coli* cell paste in 4 M urea/borate (pH 9.0) was purified on a copper-loaded chelating column by applying a linear imidazole gradient similar to that described in Fig. 16. Typical of affinity chromatography product was purified by 460-fold to a purity of 97% as determined by reverse phase HPLC.

to facilitate the in vivo determination of tail length (22). This charged fusion protein is being used to investigate the possibility of using polyelectrolyte precipitation to produce low-cost proteins in bulk quantities.

F. Engineering for Improved Analysis

The development processes for the isolation of proteins is often limited by the absence of a suitable assay. Ideally, the analysis should be accurate, precise, and selective for the desired product. The latter is particularly important during fermentation development and early stages of purification where samples are extremely impure. Any protein engineering used to improve purification can also form the basis of an HPLC or immunological analytical method (Table 8). Therefore, analytical ion exchange HPLC can provide a convenient and specific assay for polyarginine-tailed proteins and immunoaffinity fusion proteins can be assayed by highly sensitive ELISA or dot-blot methods.

V. CONCLUSION

General procedures are available which use protein engineering to facilitate the isolation of rDNA-derived proteins from microorganisms. Proteins with enhanced purification properties can be made by making small changes to the primary sequence of proteins. Stability can also be improved by the use of peptide extensions which encourage the formation of inclusion bodies. If the protein is modified by peptide extension, these can be removed by engineering selective hydrolysis sites into the fused protein. The refolding of proteins denatured during production in recombinant organisms is also a key process in the isolation of rDNA-derived proteins (32). In a limited number of cases,

Table 8 Comparison of HPLC and Immunological Methods for Protein Assay[a]

Assay Parameter	Assay method		
		Immunological	
	HPLC	ELISA	Blot
Precision	High	Moderate	Low
Accuracy	High	Moderate	High
Sensitivity	Low	High	Moderate
Selectivity	High	Moderate	Moderate

[a]HPLC methods have the advantage of high precision and accuracy, but are not able to detect low levels of product. Immunological methods are highly sensitive, but because antibodies recognize only a small portion of the protein's surface, they may also recognize partially degraded species.

improvements in refolding may be made by changes to the primary sequence. However, there are currently no general procedures for achieving this goal and this is the major challenge for protein chemists isolating biologically active proteins from genetically engineered microorganisms.

REFERENCES

1. Goeddel DV, Kleid DG, Bolivar F, Heyneker HL, Yansura DG, Crea R, Hirose T, Kraszewski A, Itakura K, Riggs AD. Expression in *Escherichia coli* of chemically synthesized genes for human insulin. Proc Natl Acad Sci USA 1979; 76:106-110
2. Shine J, Fettes I, Lan NCY, Roberts JL, Baxter JD. Expression of cloned β-endorphin gene sequences by *Escherichia coli*. Nature 1980; 285:456-461.
3. Mitraki A, King J. Protein folding intermediates and inclusion body formation. Biotechnology 1989; 7:690-697.
4. Itakura K, Hirose T, Crea R, Riggs AD, Heyneker HL, Bolivar F, Boyer HW. Expression in *Escherichia coli* of a chemically synthesized gene for the hormone somatostatin. Science 1977; 198:1056-1063.
5. Dalbøge H, Dahl HHM, Pedersen J, Hansen JW, Christensen T. A novel enzymatic method for the production of authentic hGH from an *Escherichia coli* produced hGH-precursor. Biotechnology 1987; 5:161-164.
6. Smith J, Cook E, Fotheringham I, Pheby S, Derbyshire R, Eaton MAW, Doel M, Lilley DMJ, Pardon JF, Patel T, Lewis H, Bell LD. Chemical synthesis and cloning of a gene for human-β-urogastrone. Nucleic Acids Res 1982; 10:4467-4482.
7. Mark DF, Lu SD, Creasey AA, Yamamoto R, Lin LS. Site-specific mutagenesis of the human fibroblast interferon gene. Proc Natl Acad Sci USA 1984; 81:5562-5666.

8. Oka T, Sakamoto S, Miyoshi K, Fuwa T, Yoda K, Yamasaki M, Tamura G, Miyake T. Synthesis and secretion of human epidermal growth factor by *Escherichia coli*. Proc Natl Acad Sci USA 1985; 82:7212-7216.
9. Hopp TP, Prickett KS, Price VL, Libby RL, March CJ, Cerretti DP, Urdal DL, Conlon PJ. A short polypeptide marker sequence useful for recombinant protein identification and purification. Biotechnology 1988; 6:1204-1210.
10. Hopp TP. Protein surface analysis. Methods for identifying antigenic determinants and other interaction sites. J Immunol Meth 1986; 88:1-18.
11. Frusto da Silva JJR. The chelate effect redefined. J Chem Educ 1983; 60:390-392.
12. Drey CNC, Fruton JS. Metal chelates of bis-imidazole. Biochemistry 1965; 4:1258-1263.
13. Sulkowski E. Purification of proteins by IMAC. Trends Biotechnol 1985; 3:1-7.
14. Holmes MA, Matthews BW. Structure of thermolysin refined at 1.6 angstrom resolution. J Mol Biol 1982; 160:623-639.
15. Trainer JA, Getzoff ED, Richardson JS, Richardson DC. Structure and mechanism of copper, zinc superoxide dismutase. Nature 1983; 306:284-287.
16. Stenkamp RE, Sieker LC, Jensen LH. Adjustments of restraints in the refinement of methemerythrin and azidomethemerythrin at 2.0 angstrom resolution. Acta Crystallogr 1983; B39:697-703.
17. Kidera A, Konishi Y, Oka M, Ooi T, Scheraga HA. Statistical analysis of the physical properties of the 20 naturally occuring amino acids. J Protein Chem 1985; 4:23-55.
18. Asimov GJ, Krouze NK. The lactogenic preparations from the anterior pituitary and the increase of milk yield in cows. J Dairy Sci 1937; 20:289-306.
19. Annexstad RJ, Otterby DE, Linn JG, Hansen WP, Soderholm CG, Eggert RG. Responses of cows to daily injections of recombinant bovine somatotropin BST during a second consecutive lactation. J Dairy Sci 1987; 70:176.
20. Peel CJ, Bauman DE. Somatotropin and lactation. J Dairy Sci 1987; 70:474-486.
21. Abdel-Meguid SS, Shieh HS, Smith WW, Dayringer HE, Violand BN, Bentle LA. Three-dimensional structure of genetically engineered variant of porcine growth hormone. Proc Natl Acad Sci USA 1987; 84:6434-6437.
22. Zhao J, Ford C, Gendel S, Rougvie M, Glatz C. Genetic engineering B-galactosidase with a poly-aspartate tail to enhance down-stream product recovery by polyelectrolyte precipitation. Adv Gene Tech and Protein Eng. Tech. ICSU Short Reports 1988, Vol. 8.
23. Mai M, Bittner MM, Goldberg S, Heeren B, Galluppi G. Expression of atriopeptins in E. Coli. Fed Proc 1986; 45:3049.
24. Nilsson B, Holmgren E, Josephson S, Gatenbeck S, Philipson L, Uhlen M. Efficient secretion and purification of insulin-like growth factor I with a gene fusion vector in Staphylococci. Nucleic Acids Res 1985; 13:1151-1162.
25. Germino J, Bastia D. Rapid purificatioin of a cloned gene product by genetic fusion and site specific proteolysis. Proc Natl Acad Sci USA 1984; 81:4692-4696.
26. Bennett A, Rhind SK, Lowe PA, Hentschel C. U.K. Patent 1984; 2140810A.
27. Mayne NG, Burnett JP, Belegaje R, Hsiung, HM. Eur. Patent Appl. 1983; EP95361.

28. Nagai K, Thøgerson HC. Generation of B-globin by sequence specific proteolysis of a hybrid protein produced by *Escherichia coli*. Nature 1984; 309:810-812.
29. Sassenfeld HM, Brewer SJ. A polypeptide fusion designed for the purification of recombinant proteins. Biotechnology 1984; 2:76-81.
30. Sassenfeld HM. Thesis 1986; University of Liverpool, UK.
31. Brewer SJ, Larsen BR. Isolation and purification of proteins using preparative HPLC. In: Verrall and Hudson. ed. Separations for biotechnology Ellis Horwood, Chichester, 1987; pp 113-126.
32. Sassenfeld HM. Engineering proteins for purification. Trends Biotechnology 1990; 8:88–93.

12
Practical Aspects of Receptor Affinity Chromatography

Pascal S. Bailon, David V. Weber, and John E. Smart

Roche Research Center
Hoffmann-La Roche, Inc.
Nutley, New Jersey

I. INTRODUCTION

Until affinity chromatography became a full-fledged protein purification technology in the 1970s, biomolecules such as enzymes, hormones, antibodies, and receptors were purified on an empirical basis. Affinity chromatography exploits the ability of the biologically active macromolecules in solution to bind specifically and reversibly to matrix-bound substrates and other ligands (1–4). Potentially, affinity chromatography is a facile, rapid, and selective purification method and its applications are virtually limitless. Immunoaffinity chromatography, which utilizes the specificity of the antigen–antibody interaction, has realized its full potential only after the discovery of monoclonal antibody-producing hybrid cell lines by Kohler and Milstein in 1975 (5–8). Immunoaffinity chromatography has been used successfully for the purification of several bioactive molecules, such as interferons of various origin (9–12), interleukin 2 (13,14), blood-clotting factors (15), membrane antigens (16), and interleukin 2 receptor (17).

Previously, the authors reported an alternative affinity purification method termed "receptor affinity chromatography," which exploits the biochemical interactions between an immobilized receptor and its soluble protein ligand (14). The receptor affinity-purified biomolecules are expected to be biochemically and biologically more homogeneous than the immunoaffinity purified material due to the conceptual hypothesis that only a fully active biomolecule in its native conformation should bind to its receptor with high avidity. In contrast, in immunoaffinity chromatography it has been demonstrated that various molecular forms of the antigen with varying degrees of biological activity and renaturation state bind to the antibody column (14).

Recombinant technology has produced a soluble form of interleukin 2 receptor, denoted as IL-2R-Δ-Nae (18), and is on the verge of producing recombinant receptors of other biomolecules such as interleukin 1, tumor necrosis factor, γ-interferon, etc. Hence, we believe that in the near future receptor affinity chromatography will become an established method for the purification of high-value recombinant proteins.

In this chapter, we present an overview of the use of receptor affinity chromatography as a downstream purification tool for recombinant proteins from microbial and mammalian sources. A systematic approach to the development and optimization of receptor affinity purification systems is discussed. To date, only recombinant interleukin 2 (rIL-2) has been purified using this technique (14). Consequently, we will use the receptor affinity purification of rIL-2 as a model system to demonstrate the utility of this approach for the purification of recombinant proteins for clinical use in humans.

II. RECEPTOR AFFINITY CONCEPT

A. Receptor–Ligand Interaction

In most cases, molecular recognition between a receptor and its ligand occurs through the formation of a noncovalently bonded receptor–ligand complex. Receptor affinity chromatography is based on the specificity and reversibility of the interactions between the receptor and its soluble protein ligand. Assuming the simplistic one-to-one binding in the receptor–ligand complex, theoretically one ligand molecule can bind to one receptor molecule. However, this theoretical binding capacity is seldom achieved with immobilized receptors due to a variety of reasons described elsewhere in this chapter.

The unique properties of the receptor–ligand interaction impart the following practical characteristics to receptor affinity chromatography:

1. Rapid formation of a stable but reversible receptor–ligand complex
2. Uniquely high selectivity, which allows the isolation of a single dilute molecule from a complex mixture of components
3. Equivalent applicability for the purification of biomolecules of all sizes
4. High recovery of biological activity after elution with mild desorbing agents

B. Schematic Outline of Receptor Affinity Chromatography

The general receptor affinity purification scheme is given in Fig. 1.

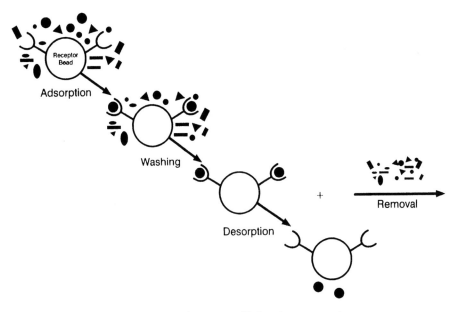

Figure 1 A schematic outline of receptor affinity chromatography.

Receptor covalently attached to an inert polymer support is packed into a chromatographic column. The crude extract containing the soluble ligand is passed over the immobilized receptor column, unadsorbed materials are washed away, and the specifically bound ligand is eluted with mild desorbing agents.

III. DESIGN PARAMETERS IN RECEPTOR AFFINITY CHROMATOGRAPHY

The success of receptor affinity chromatography is very much dependent on establishing conditions that optimize the natural interaction between the receptor adsorbent and the soluble protein ligand. We have used the receptor affinity purification of rIL-2 as a model system to determine the general design parameters in the purification of recombinant proteins for therapeutic uses.

A. Selection of Receptors Suitable for Immobilization

Like the monoclonal antibodies used in immunoaffinity chromatography, the receptors chosen for immobilization should have a high affinity during the adsorption phase of receptor affinity chromatography and the ability to be switched to low affinity by the simple titration of specific amino acid residues such as histidine or lysine (19) under mild pH conditions, thus permitting the release of bound ligand. Although knowing the relative affinities of the receptors can be useful in narrowing down the number of receptors screened for immobilization, the most practical approach is the empirical method, in which the receptor is immobilized on a small scale and its ligand-binding capacity and selectivity, as well as its reversibility with mild desorbing agents, are determined experimentally.

B. Polymer Supports for Immobilization: Physical, Chemical, Mechanical, and Flow Properties

The insoluble carriers chosen for the preparation of receptor adsorbents should be:

1. Inert, hydrophilic, chemically stable, and contain an optimum number of functional groups (e.g., hydroxyl, carboxyl, aliphatic or aromatic amines, hydrazides, etc.) which can be easily activated for efficient protein coupling
2. Rigid beads with high porosity, thereby allowing the rapid passage of potentially viscous fluids at moderate pressure

It should be pointed out that the carrier may alter the microenvironment of the receptor, thus affecting the strength of the ligand interaction with the immobilized receptor. A carefully chosen carrier should facilitate these interactions in the favorable direction by providing a suitable microenvironment in terms of biocompatibility and hydrophilicity.

We evaluated several commercially available support materials in terms of the mechanical rigidity and flow rates that are best suited for industrial scale operations. NuGel 500 A, 50-μm particle size (Separation Industries, Metuchen, NJ), and Sepharose CL-6B (Pharmacia, Piscataway, NJ) were found to meet all or most of these requirements. The NuGel support has the added advantage of providing a durable bed support that allows a fourfold flux over agarose supports (20).

C. Immobilization Methods

Immobilization methods chosen should be simple and allow efficient receptor coupling with maximum recovery of binding capacities. The common strategy for immobilization is to activate the polymer support with a reagent or sequence of reagents to a chemically reactive form which is capable of forming stable covalent bonds when mixed with the protein. Although numerous immobilization methods have been reported in the literature (21), to date there are no reported methods that allow 100% retention of biological activity plus a leach-proof covalent coupling.

All subsequent data in the present communication are derived from the IL-2 receptor covalently coupled to the commercially available N-hydroxysuccinimide (NHS) ester derivative of a polyhydroxy silica gel (NuGel P-AF Poly-N-Hydroxysuccinimide, Separation Industries). These NHS derivatives readily react with amine nucleophiles to yield covalently coupled ligand in aqueous solutions. The unreacted groups undergo spontaneous hydrolysis to carboxyl derivatives. A typical coupling reaction is shown in Fig. 2.

D. Immobilization Procedure

The activated gel is quickly washed with three volumes of ice-cold water in a coarse-sintered glass funnel. The gel is then mixed with an equal volume of coupling buffer (0.1 M potassium phosphate containing 0.1 M NaCl, pH 7.0) containing a known concentration of receptor. The slurry is shaken gently for 4–16 hrs at 4°C in a stoppered Erlenmeyer. The unbound protein is collected by filtering the reaction mixture and the gel is washed with two volumes of phosphate-buffered saline (PBS), pH 7.4. The filtrate and washes are combined and a small aliquot is dialyzed against PBS. The gel is then treated with an equal volume of 0.1 M ehtanolamine-HCL, pH 7.0, to neutralize any remaining activated groups. Following deactivation, the gel is washed with PBS and stored as a suspension in the same buffer in the presence of 0.02% sodium azide at 4°C. The volume of the unbound protein solution is recorded and the protein concentration in the dialyzed aliquot is determined by the method of Lowry et al. (22) and confirmed by quantitative amino acid analysis in a post-column fluorescamine amino acid analyzer (23). From these two values the total unbound protein is calculated. The difference between the starting amount of receptor and the amount of uncoupled receptor in the pooled filtrate and washes divided by the gel volume gives the receptor coupling density (mg/ml gel).

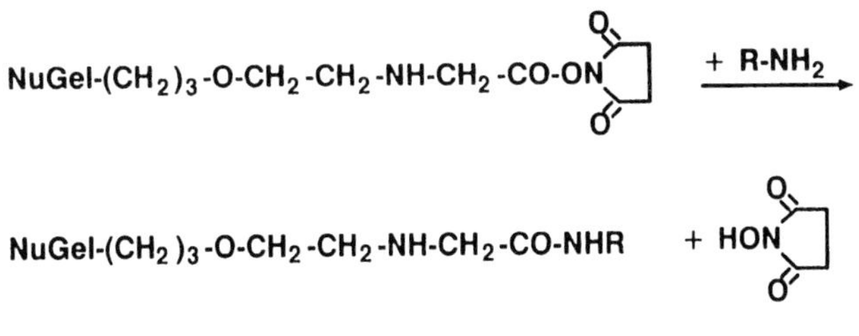

Figure 2 Coupling reaction of NuGel P-AF poly-N-hydroxysuccinimide.

E. Optimum Receptor-Coupling Conditions

1. Residual Ligand-Binding Capacity

The amount of ligand-binding capacity retained by the receptor after immobilization is defined as the residual ligand-binding capacity of the receptor adsorbent. This can be determined experimentally. After proper equilibration, a receptor adsorbent column of known volume (0.5–1.0 ml gel) is saturated with an excess of ligand (purified or crude). The unadsorbed materials are washed out and the specifically bound ligand is eluted with a mild desorbing agent. The number of nanomoles of ligand bound is calculated from the protein content of the eluate and the molecular weight of the protein ligand. The nanomoles of ligand bound per unit volume of gel is defined as the residual ligand-binding capacity of the receptor adsorbent.

2. Factors Affecting Coupling Efficiency and Residual Binding Capacity

Coupling conditions such as pH, activated group density on the matrix, and coupling density contribute to the loss of binding capacity usually accompanied by immobilization.

3. Effect of Coupling pH

At a constant ratio of IL-2R to unit volume of activated matrix (2 mg/ml gel) receptor adsorbents were prepared at various pHs and their IL-2-binding capacities were determined. Results are summarized in Table 1.

4. Effect of Activated Group Density on the Matrix

At very low activated group densities (5–10 μmol/ml gel) coupling efficiencies were very poor while ligand-binding capacities were high. At the medium activated group densities (15–30 μmol/ml gel) both the coupling efficiencies

Table 1 Effect of Coupling pH[a]

Coupling (pH)	IL-2R-coupled		IL-2-binding capacity		
	mg/ml gel	nmol/ml	Calculated nmol/ml	Observed (nmol/ml)	Residual (%)
5.0	0.900	36	72	14	19
6.0	1.150	46	92	36	39
7.0	1.595	60	120	80	67
8.0	1.575	63	126	88	70
9.0	1.625	65	130	84	65

[a]Binding capacities are calculated taking into account the MW of IL-2R and rIL-2 as 25 and 15.5 kDa, respectively. Two IL-2-binding sites (equivalents) are assumed for the IL-2R (24). Coupling efficiencies and IL-2-binding capacities were optimal when the coupling pH was 7–8.

and the residual ligand-binding capacities were optimal. At relatively high functional group densities (50–100 μmol/ml gel) coupling efficiencies were the highest; however, a substantial loss in binding capacity was observed, possibly due to multipoint attachment.

5. Effect of IL-2R Receptor Coupling Density

IL-2R-Δ-Nae was immobilized at various coupling densities and the IL-2-binding capacities of the resultant receptor adsorbents were determined. The results are listed in Table 2.

As observed for antibodies by Comoglio et al. (25), high IL-2R loadings resulted in lower IL-2-binding capacities, possibly due to steric hindrance. The maximum residual binding capacities were observed when the receptor coupling density was 1–2 mg/ml gel.

F. Detection and Prevention of Protein Leaching from Receptor Sorbents

Due to the possibility that the covalent bond formed between the receptor and the matrix during immobilization may not be completely leach-proof, trace amounts of immobilized receptor may leach from the column during the purification process, thereby contaminating the final bulk product.

1. Detection Method

A sensitive bimolecular solid phase receptor-binding assay (18) is used to detect receptor leaching from the affinity sorbent during column operations. The lower limit of the assay's sensitivity is 0.1 ng/ml.

2. Stabilization of Immobilized Receptor

We successfully employed the glutaraldehyde crosslinking method of Kowal and Parsons (26) to stabilize the covalently bonded receptor. At low con-

Table 2 Effect of Coupling Density[a]

IL-2R-coupled		IL-2-binding capacity		
m/ml gel	nmol/ml gel	Calculated (nmol/ml)	Observed (nmol/ml)	Residual (%)
0.830	33	66	45	68
1.505	60	120	80	67
3.610	144	288	128	44
7.780	311	622	174	28

[a]Binding capacities are calculated as in Table 1. It is assumed that there are two IL-2-binding sites (equivalents) on the IL-2R (24)

centrations (0.1–0.5% v/v) of glutaraldehyde and controlled contact time and temperature (60 min at 4°C) receptor leaching is reduced from 0.4 ng/ml to nondetectable levels, without significant loss (<5%) in IL-2-binding capacities.

G. Receptor Affinity Purification Procedures

In this section we focus our attention on the procedures involved in receptor affinity purification of clinical grade recombinant proteins from *E. coli* and cultured mammalian cell supernatants. In his review article, Sharma (27) presented an excellent overview of the recovery of recombinant proteins from *E. coli*. Similarly, Boschetti et al. (28) described the purification of biomolecules from cell culture supernatants.

General receptor affinity purification schemes for the production of pharmaceutical grade proteins from microbial and mammalian sources are given in Fig. 3.

1. Extraction, Solubilization, and Renaturation of Recombinant Proteins

Often recombinant proteins are expressed in *E. coli* in high concentrations in an insoluble form (inclusion bodies). Quite often extraction of the desired protein from *E. coli* requires the use of strong denaturants at high concentrations with or without detergents. The removal of denaturants under conditions optimal for renaturing is of utmost importance. Previous studies (29) showed that optimal refolding of proteins occurs when their concentration is at or below the micromolar range. Consequently, relatively large dilutions of the denatured extracts followed by aging for varying periods of time are required. The results of such an aging or refolding experiment conducted with 7 M guanidine hydrochloride (GuHCl)-extracted rIL-2 after a 40-fold dilution is given in Table 3.

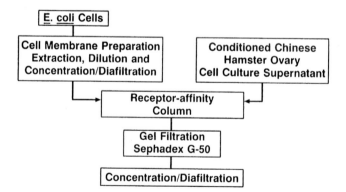

Figure 3 Receptor affinity purification schemes.

Table 3 rIL-2 Refolding Studies

Age of extract (days)	Amount of rIL-2 recovered (mg/kg cells)
1	1918
2	2300
3	2480
4	2604
5	2572

The results indicate that 3–4 days of aging is needed to attain maximum refolding of rIL-2 extracted with 7 M GuHCl.

In general, no special treatments are necessary for the mammalian cell culture supernatants. They can be applied directly on the receptor column.

2. Adsorption

The adsorption stage is one of the most critical aspects of receptor affinity chromatography. During adsorption, the crude material should be in a buffer compatible with maximum adsorption. In order to ensure that no product is wasted during the adsorption phase, sufficient contact time between the soluble ligand and receptor column should be maintained by careful choice of flow rate. The effect of flow rate on capture efficiency of an IL-2 receptor column is shown in Fig. 4. The results indicate no decrease in capture efficiency due to increased flow rate up to 60 ml/min. The linearity of pressure increase suggests no compression of the receptor column bed during increased flow of solvents. Using appropriate buffers, quantitative adsorption of the dilute ligand to the receptor adsorbent can be achieved even at very high flow rates. In order to take full advantage of the selectivity of the receptor sorbent to preferentially bind the fully renatured ligand molecules from a heterogeneous population, the receptor column should be operated at or above its saturation binding capacity.

3. Washing

The purpose of washing the receptor column immediately after adsorption is twofold:

1. To remove the crude materials from within or surrounding the receptor beads
2. To remove materials nonspecifically bound either to the support or to the immobilized receptor

Nonspecific binding to the support can be minimized but is rarely eliminated completely. Electrostatic as well as hydrophobic interactions between

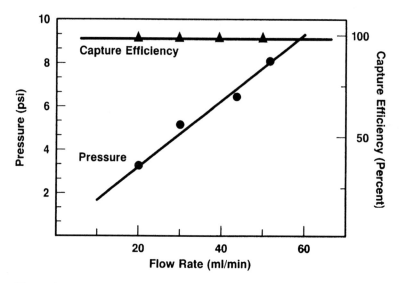

Figure 4 Effect of flow rate on capture efficiency. *E. coli* extract containing 100 mg rIL-2 is applied on the receptor column (4.8 × 6 cm, Kontes, Chromaflex, Vineland, NJ) at various flow rates using a FMI Lab pump (Fluid Metering, Oyster Bay, NY).

the receptor and the extraneous materials in the crude extract are another source of nonspecific binding. These nonspecifically bound contaminants usually can be reduced to an acceptable level by washing extensively with buffers containing salts at neutral or mildly alkaline pH or by inclusion of low concentrations of nonionic detergents in the starting materials and in all buffers used.

4. Elution

The elution of the adsorbed ligand from the receptor column is achieved by causing the dissociation of the receptor–ligand complex. Nonspecific elution methods are commonly used for the desorption of the bound protein ligand. These eluents involve low- or high-pH buffers, protein denaturants such as urea or GuHCl, and chaotropic agents like potassium thiocyanate. If the ligand involved is stable at acid pH and is readily eluted from the receptor sorbent under these conditions, an eluent of choice is a low-pH (<3) buffer.

5. Size Exclusion Liquid Chromatography

Gel filtration as a final step in the purification process is a convenient way of preparing recombinant proteins free of trace high molecular weight contaminants (e.g., pyrogens, oligomers, contaminant proteins of microbial or mammalian origin, etc.) as well as low molecular weight contaminants such as salts, detergents, and denaturants. Quite often gel filtration is a convenient method for exchanging the protein into the final formulation buffer. This step is carried out under aseptic conditions.

H. Concentration, Diafiltration, and Bulking

The gel-filtered product is concentrated by diafiltration using appropriate M_r cut-off membranes under sterile conditions. For example, rIL-2 is concentrated using a 5000 M_r cut-off YM-5 membrane. This step can also be used to exchange the final product into the formulation buffer before bulking and storage.

I. Quality Control Aspects of Pharmaceutical Grade Recombinant Proteins

The FDA's Center for Drugs and Biologics has published several "Points to Consider" bulletins (30,31) regarding the quality control requirements for producing recombinant proteins for medicinal use in humans. For further insight into this subject, the reader is referred to a publication by F. M. Bogdansky (32).

1. Biological Potency

Biological potency data expressed as activity units/mg protein (specific activity) is a good indicator of the nativeness of the purified proteins. Nonnatural disturbances in the secondary or tertiary structure of the intact molecule can render it partially or completely inactive.

2. Protein Structure

The identity, quality, and biological potency of a protein is dependent on its primary, secondary, and tertiary structures. Partial amino acid sequence analysis is useful in determining the primary structure, whereas secondary structure such as the location of intramolecular disulfide bonds can be determined by tryptic mapping under reducing and nonreducing conditions. Full biological activity usually confirms that at least the tertiary structure of the biologically relevant (interacting) domains is correct.

3. Determination of Purity

SDS-PAGE analysis under reducing and nonreducing conditions is usually used as a primary test for purity. Such an analysis for receptor affinity-purified rIL-2s from microbial and mammalian sources is shown in Fig. 5. Fragments, oligomers, and foreign proteins in trace amounts are often detected and separated by a variety of chromatographic procedures. Isoelectric focusing, two-dimensional gel electrophoresis, and Western blotting are good techniques for detecting heterogeneity in the recombinant protein preparation.

4. Analysis of Trace Contaminants in the Final Product

Recombinant proteins used for therapeutic purposes should be analyzed for trace levels of contaminants such as foreign proteins of bacterial or mammalian origin, oligomers, antibiotics, etc. Sensitive customized assays are needed

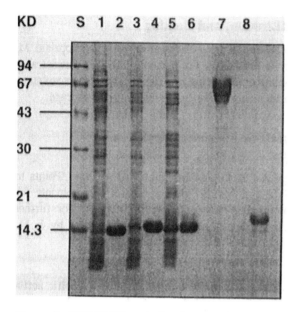

Figure 5 SDS-PAGE analysis of receptor affinity purified IL-2s under nonreducing conditions. Lane S: Standard molecular weight markers; Lanes 1, 3, and 5: *E. coli* extracts; Lane 2: rIL-2 Mutein Des-Ala(1)-Ser(125); Lane 4: rIL-2 Mutein Lys[20]; Lane 6: Wild-type rIL-2; Lane 7: Conditioned Chinese hamster ovary cell culture supernatant; and Lane 8: Mammalian glycosylated rIL-2.

to detect these trace contaminants in the ppm range. Other types of potentially antigenic contaminants are alternate and derivatized N-terminal amino acid (e.g., N-terminal methionine or derivatized N-terminal methionine) which can be detected by sequence analysis. Nucleic acids, which are a potential source of oncogenes, are another potential contaminant in recombinant proteins. Nucleic acids can be detected at the picogram level by hydridization analysis. The final product should be also free of pyrogens. Many of the trace contaminants in the final product can be removed by inclusion of an ion exchange chromatography step followed by size exclusion liquid chromatography.

J. Longevity and Stability of Receptor Adsorbents During Long-Term Use

Receptor columns should be subjected to stability and longevity studies involving at least 100 cycles of operation (automated) to determine their efficacy for long-term use. The qualitative and quantitative recovery of the recombinant proteins are determined by the criteria described above. Receptor sorbents have performed satisfactorily for at least 500 cycles of operation with no significant impairment in the functionality.

K. Scale-up and Automation of Receptor Affinity Chromatography Scale-up Procedures

The design of large-scale receptor affinity purification systems is based on the performance data obtained from small-scale operations. In a true linear scale-up, as the volume of crude material to be processed increases, the size and flow rates of the column employed are increased proportionally. This is seldom achieved in conventional mode operations. In order to process large volumes at increased flow rates, increasing the cross-sectional area is more desirable than increasing the column height. When the column height of soft-gel column support is increased, a pressure drop across the column bed occurs which leads to flow problems. The noncompressible rigid bead supports such as NuGel-PAF (Separation Industries, Metuchen, NJ) and the Superflow column (Sepragen Corporation, Hercules, CA) make linear scale-up a distinct possibility.

Advances made in the recombinant DNA technology for the production of soluble forms of receptor as well as improved purification procedures have provided sufficient quantities of receptors for the construction of receptor columns suitable for large-scale purification (e.g., rIL-2).

1. Automation

The repetitive use of the receptor affinity process can be managed, often unattended, by a microcomputer-based control system. The basic outline of a microcomputer-based receptor affinity purification system is shown in Fig. 6. Basically, it consists of a collection of valves which control the input of fluids to the column bed as well as the output from the column. Included in the system are a pump to control fluid flow, a spectrophotometer to monitor protein levels, and a programmable fraction collector for collecting eluted proteins. A column regeneration step after each cycle is also included in the system.

IV. PROBLEM AREAS OF RECEPTOR AFFINITY CHROMATOGRAPHY

A. Immobilization Methods

During receptor immobilization, loss of significant amounts of ligand-binding capacity occurs for a variety of reasons, which are described elsewhere in this chapter. Improved immobilization methods should allow better coupling efficiencies and residual ligand-binding capacities.

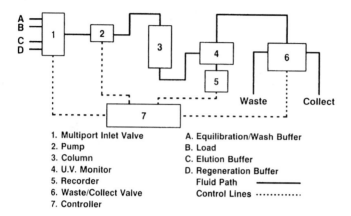

Figure 6 Basic outline of an automated receptor affinity purification system.

B. Receptor Affinity Purification Procedure

Considerable time and effort is spent before the crude microbial extract is ready for chromatography on the receptor column, e.g., centrifugation, addition of strong denaturants for extraction, removal or dilution of denaturants, allowance of adequate time for refolding and reduction of large volumes. Some of these steps are unavoidable due to the origin of the product (e.g., microorganism). At the present time, no general methods are available for the washing and elution stages of the receptor-affinity process. It is anticipated that the second-generation recombinant proteins will be produced by mammalian cells and the product will be secreted into the medium. This should allow the product to be chromatographed directly.

C. Cost Effectiveness

The initial cost of installation of receptor affinity purification systems remains high. The extent to which this technology will be used for the industrial scale production of biomolecules depends on considerable reduction of the production costs of soluble receptors.

D. FDA Regulations

At the present time, the stringent regulations set forth by the FDA make it mandatory to perform extensive tests on the receptor preparations used for immobilization. These tests are often time consuming and expensive.

V. CONCLUSIONS

Receptor affinity chromatography utilizing immobilized receptors is a viable separation technique for the purification of therapeutically useful recombinant proteins for the following reasons:

1. The method is well suited for scaled up continuous operations.
2. Noncompressible, large-pore, rigid-bead supports which allow high flow rates are now available for the preparation of receptor sorbents.
3. High-flow, low-pressure, industrial scale columns and sophisticated automatic control systems are readily available commercially.
4. Efficient immobilization methods have been developed.
5. The cost of producing soluble receptors has been reduced to affordable levels due to recent advances made in recombinant DNA technology.
6. Product recoveries are high and the product quality is superior.

We believe that it is only a matter of time until receptor affinity chromatography will be used routinely for the recovery of high-value recombinant proteins.

ACKNOWLEDGMENTS

We thank Ingrid Newman for the illustrations and Lyn Nelson for her invaluable help in preparing the manuscript. Review of this chapter by Swapan K. Roy is greatly appreciated.

REFERENCES

1. Cuatrecasas P, Wilcheck M, Anfinsen CB. Selective enzyme purification by affinity chromatography. Proc Natl Acad Sci USA 1968; 61:636–643.
2. Cuatrecases P, Anfinsen CB. Affinity chromatography. Ann Rev Biochem 1971; 40:259–278.
3. Lowe CR. Affinity, chromatography: The current status. Int J Biochem 1977; 8:177–181.
4. Katchalski-Katzir E. Some general considerations on the recognition by and of proteins. In: Chaiken IM, Wilcheck M, Parikh I. eds. Affinity chromatography and biological recognition. New York: Academic Press, 1983:7–26.
5. Campbell DH, Luescher E, Lerman LS. Immunologic adsorbents. I. Isolation of antibody by means of a cellulose-protein antigen. Proc Natl Acad Sci USA. 1951; 37:575–578.
6. Gurvich AE, Drzlikh GJ. Use of antibodies on an insoluble support for specific detection of radioactive antigens. Nature 1964; 203:648–649.
7. Silman IH, Katchalski E. Water-insoluble antigen and antibody derivatives. Ann Rev Biochem 1966; 35:896–908.

8. Kohler G, Milstein C. Continuous cultures of fused cells secreting antibody of predefined specificity. Nature 1975; 256:9750–9754.
9. Staehlin T., Hobbs DS, Kung H-F, Lai C-Y, Pestka S. Purification and characterization of recombinant human leukocyte interferon (IFLrA) with monoclonal antibodies. J Biol Chem 1981; 256:9750–9754.
10. Secher DS, Burke DC. A monoclonal antibody for large-scale purification of human leukocyte interferon. Nature 1980; 285:446–450.
11. Novick D, Eshhar Z, Gigi O, Marks Z, Revel M, Rubinstein M. Affinity chromatography of human fibroblast interferon (IFN-B_1) by monoclonal antibody columns. J Gen Virol 1983; 64:905–910.
12. Le J, Barrowclough BS, Vilcek J. Monoclonal antibodies to human immune interferon and their application for affinity chromatography. J Immunol Meth 1984; 69:61–70.
13. Robb RJ, Kutny RM, Chowdhry V. Purification and partial sequence analysis of human T-cell growth factor. Proc Natl Acad Sci USA 1983; 80:5990–5994.
14. Bailon P, Weber DV, Keeney RF, Fredericks JE, Smith C, Familletti PC, Smart JE. Receptor-affinity chromatography: A one step purification of recombinant interleukin-2. Biotechnology 1987; 5:1195–1198.
15. Katzman JA, Nesheim ME, Hibbard LS, Maan KG. Isolation of functional human coagulation factor V by using a hybridoma antibody. Proc Natl Acad Sci USA 1981; 78:162–166.
16. Hermann SH, Mescher ME. Purification of the $H-2K^k$ molecule of the murine major histocompatibility complex. J Biol Chem 1979; 254:8713–8716.
17. Leonard WJ, Depper JM, Crabtree GR, Rudikoff S, Pumphree J, Robb RJ, Kronke M. Svetlik PB, Peffer NJ, Waldman TA, Greene WC. Molecular cloning and expression of cDNAs for the human interleukin-2 receptor. Nature 1984; 311:626–631.
18. Hakimi J, Seals C, Anderson LE, Podlaski FJ, Lin P, Danho W, Jenson JC, Perkins A, Donadio PE, Familletti PC, Pan Y-CE, Tsien W-H, Chizzonite RA, Casabo L, Nelson DL, Cullen BR. Biochemical and functional analysis of soluble human interleukin-2 receptor produced in rodent cells. J Biochem 1987; 262:17336–41.
19. Bartholomew R, Beidler D, David G. Immunoaffinity chromatography with monoclonal antibodies. Protides Biol Fluids 1983; 30:667–670.
20. Roy SK, Weber DV, McGregor WC. High-performance immunosorbent purification of recombinant leukocyte A interferon. J Chromatogr 1984; 303:225–228.
21. Dean PDG, Johnson WS, Middle FA, eds. Affinity chromatography: A practical approach. Oxford: IRL Press Ltd 1985; 16–148.
22. Lowry OH, Rosebrough NJ, Farr AL, Randall RJ. Protein measurement with the folin phenol reagent. J Biol Chem 1951; 193:265–275.
23. Pan Y-CE, Stein S. Amino acid analysis with post column fluorescent derivatization. Shively JE, ed. Methods of protein microcharacterization. Clifton, NJ: Humana Press, 1986; 105–119.
24. Purcell R, et al. Manuscript in preparation.
25. Comoglio S, Massaglia A, Rolleri E, Rosa U. Factors affecting the properties of insolubilized antibodies. Biochim Biophys Acta 1976; 420:246–257.

26. Kowal R, Parsons RG. Stabilization of proteins immobilized on sepharose from leakage by glutaraldehyde crosslinking. Anal Biochem 1980; 102:72–76.
27. Sharma SK. On the recovery of genetically engineered proteins from Escherichia coli. Sep Sc Technol 1986; 21:701–726.
28. Boschetti E, Egly JM, Monsigny M. The place of affinity chromatography in the production and purification of biomolecules from cultured cells. Trends Anal Chem 1986; 5:4–10.
29. Light A. Protein solubility, protein modifications and protein refolding. Biotechniques 1985; 3:298–306.
30. FDA, Points to consider in the production and testing of new drugs and biologicals produced by recombinant DNA technology, Office of Biologics Research and Review Center for Drugs and Biologics, April 10, 1985.
31. FDA, Points to consider in the characterization of cell lines used to produce biological products, Office of Biologics Research and Review, Center for Drugs and Biologics, June 1, 1984.
32. Bogdansky FM. Considerations for the quality control of biotechnology products. Pharmaceut Technol 1987; 11:72–4.

13

Recombinant DNA Technology and Crystallography
A New Alliance in Unraveling Protein Structure–Function Relationships

Alfredo G. Tomasselli, Robert L. Heinrikson, and Keith D. Watenpaugh

The Upjohn Company
Kalamazoo, Michigan

I. INTRODUCTION

Proteins are the molecules that direct the biochemistry of every living organism. As enzymes, proteins catalyze the innumerable reactions indispensable to the survival and replication of cells. Proteins can serve as structural entities, as messengers or receptors for messengers, as regulators of the immune system, and as transporters for oxygen and carbon dioxide, metal ions, etc. Because of their diverse roles in living systems, proteins are the most important targets for drugs. Moreover, their employment as therapeutic agents and as industrial catalysts is becoming more and more the focus of modern medicine and industry. Accordingly, the powerful new methods of genetic engineering have become of pivotal importance in protein research.

In medicine, there are three major areas of interest. One concerns the production, with the tools of biotechnology, of protein drugs, i.e., of scarce proteins, having the same structure and biological activity of their natural counterparts and in quantities sufficient for therapeutic applications (the so-called first-generation products). Furthermore, in many instances, we would like to be able to improve the protein's performance in a predictable fashion (second-generation products) in terms of its solubility in water or blood, stability and protease resistance, molecular weight, thermostability, and pH optimum for activity.

Among the many proteins with therapeutic value are the following: (a) insulin to treat diabetes; (b) growth hormones to regulate the postnatal genetic growth of vertebrates; (c) interferons to treat infectious diseases; (d) interleukins as immunostimulators of T and B lymphocytes (the blood cells responsible for specific immune reactions); (e) blood coagulation factors VIII and XI used to treat the hereditary disorders of hemophilia A and B, respectively; (f) anticlotting and thrombolytic factors such as tissue plasminogen activator (tPA), urokinase, and antithrombin III; (g) α_1-antitrypsin to treat emphysema and related disorders; (h) monoclonal antibodies for cancer therapy diagnosis, tumor imaging, and immunopurification; (i) vaccines to combat diseases; and (j) soluble CD_4 to interfere with the binding of HIV virus gp 120 to cells bearing the CD_4 molecule (e.g., T lymphocytes and cells of the monocyte/macrophage lineage).

A second goal is the development of drug therapeutics based on rational design in which the protein serves as a target for inhibition or destruction, or as a model to delineate the biology of the disease. Here the list is long indeed. It suffices to say that we want to synthesize drugs apt to produce a certain effect by acting specifically on a normal protein, e.g., on renin in order to lower the blood pressure (1), or 3-hydroxy-3-methylglutaryl CoA reductase to lower the level of cholesterol in the blood (2).

A third aim is to develop drugs that can distinguish between a normal cellular protein and the same protein that as a consequence of a genetic mutation has become abnormal and life threatening. This is the case of the body's ubiquitous Ras genes which, upon a single base change at one of a few critical positions, produces a protein with the property of transforming certain normal cells into cancerous cells.

We would also like to develop drugs to inhibit proteins essential for the replication of infectious agents, e.g., HIV reverse transcriptase (RT) and protease, without impairing the host's vital biochemical functions. In the case of RT, we want to find those structural differences (and consequently catalytic differences) from its human counterpart, DNA polymerase, that will point to "inhibitors" that discriminate between the two enzymes. Therefore, these medically oriented goals involve production of recombinant proteins to serve as drugs, as targets of drugs, or as a source of structural analysis in order to provide insights with regard to the control of disease states.

On the industrial front, one of the major goals is to find new catalysts for the safe destruction of toxic products in industrial wastes. Cheese whey is one of the most serious pollutants in the dairy industry (3). Costly waste treatment processes are required to eliminate the lactose portion that accounts for 4–5% of total whey. Sreekrishna and Dickson (4) constructed genes that encode proteins with functions for lactose uptake and utilization, and thus were able to produce strains of *Saccharomyces cerevisiae* that grow on lactose. As pointed out by Russell (5), an organism of this type could be used to produce a useful end product (e.g. fuel ethanol, potable enthanol, biomass, etc.) while disposing of the pollutant whey.

In order to feed a large section of the world population still lacking sufficient food, the agricultural scientist has to improve plant nutritional quality, decrease fertilization requirements, and increase the resistance to environmental stresses and pathogens (6–8). For example, protein-engineering efforts are underway to improve the process of photosynthesis by increasing the carboxylase/oxygenase activity ratio in the enzyme ribulose-1,5-biphosphate carboxylase (9), the most abundant enzyme on earth. This enzyme fixes carbon dioxide in photosynthesis but also uses molecular oxygen as a substrate in photorespiration which leads to about 50% loss of fixed carbon dioxide.

One of the great unsolved problems challenging the scientific community today is to understand how the genetic information encoded in the linear sequence of amino acids is translated into a folded protein (folding problem), and how a folded protein produces specific biological functions (the relationship between structure and function).

Recombinant DNA technology has been fundamental to much of the progress made thus far, and will certainly be crucial for advances hoped for in the improvement of human health. These methods open the door for detailed structural characterization of rare proteins that could never be obtained from natural sources. Indeed, it is increasingly evident that the realization of most of the above goals depends on a detailed knowledge of the three-dimensional (3D) structure of the protein of interest. For example, we want to be able to modify tissue plasminogen activator (tPA) to produce a molecule with increased specificity for fibrin, longer half-life in the blood, and few or no side effects. Though it is relatively easy to change amino acids in proteins by gene manipulation, the predictable improvement or alteration of a desired property of a protein is not an easy task. The rules for rational design are not yet in place and are only learned slowly, step by step. The message is that much basic research is needed in order to understand the relationships among protein structure, function, and physical and chemical properties. X-ray determination of the 3D structures of proteins is a fundamental component of these studies.

II. CRYSTALLOGRAPHIC METHODOLOGY

Crystallography is a mature and powerful technology that has contributed immeasurably to our understanding of biological process through the knowledge of protein and nucleic acid structure and their interaction with each other and with smaller substrates or coenzymes. Nevertheless, it is a technology that is changing as rapidly now as at any time in the past. Following is a brief introduction to the current methodology and areas of future advances.

A. Crystallization of Proteins

The preparation of X-ray diffraction quality crystals can be a major obstacle in protein crystallography. While many standard procedures have been developed for crystallizing proteins, not all proteins crystallize easily by these techniques. Protein purity, solvent, precipitant concentration, pH, trace compounds, and temperature are just a few factors affecting successful crystallization. The underlying principle of protein crystallization is to reach supersaturation of the protein solution slowly while promoting the regular formation of intermolecular hydrogen bonds and electrostatic interactions, followed by gradually decreasing the protein concentration in solution. Supersaturation leads to the formation of ordered aggregates of protein molecules (nucleation phase) that under proper physical and chemical conditions grow into large crystals (growth phase). In order to sustain an ordered and continuous growth, the number of nucleation sites needs to be small to allow adequate protein material to remain for the growth phase. The solubility of the protein depends on its physical and chemical properties, solvent, concentration and nature of the precipitating agent, salt, cofactors, substrates, temperature, etc. Protein purity is of paramount importance in the crystallization of most proteins.

Various methods have been devised to crystallize proteins and microscale techniques have been developed that allow maximal exploration of different conditions with only a few milligrams of material. The methods commonly used include:

(1) Dialysis. The protein solution (10–50 µl at a concentration of 5–25 mg/ml) is confined to a small, shallow, cylindrical depression by a dialysis membrane and very slowly brought to supersaturating conditions by dialysis against the precipitant. The experiment can be planned so that equilibrium is reached in a much slower fashion than with the two techniques described below.

(2) Liquid–liquid diffusion. The protein solution and the precipitant solution are layered on top of one another in a well and allowed to equilibrate by diffusion. Usually the more dense solution is placed at the bottom and the less dense on top of it. Crystal nucleation is favored at the interface of the two solutions because of the high concentrations of both protein and precipitant. Growth of the crystals is sustained while diffusion of the protein with consequent dilution takes place.

(3) Vapor diffusion. The most widely used procedure for crystal growth involves vapor diffusion. In the "hanging drop method," a droplet (less then 10 µl) of a solution containing the protein near its precipitation point is placed on a siliconized 22-mm glass cover slip that is inverted and sealed over a reservoir containing a precipitating solution. The drop will slowly equilibrate with the reservoir solution, precipitating the protein. This method using multiwell

tissue culture plates allows screening of many different crystallization conditions using a minimal amount of protein. Figure 1 describes the method further. Related vapor diffusion methods such as the "sitting drop method" can be used with larger amounts of material.

The production of X-ray quality crystals may require exploration of many different conditions of crystal growth and a considerable amount of protein. In the case of the enzyme adenylate kinase from yeast, for example, crystals were obtained and conditions of growth defined only after innumerable conditions were tested and several hundred milligrams of highly purified enzyme employed (10). However, less than 1 mg has been required in other cases. Precipitants in protein crystallization include salts such as $(NH_4)_2SO_4$, NaCl, sodium citrate, etc., and organic compounds such as polyethylene glycol (PEG). In spite of all the difficulties involved in crystallization of proteins and the subsequent elucidation of their 3D structures, the list of high-resolution structures of soluble proteins is growing at an impressive pace and has passed the 400 mark. On the other hand, the number of 3D structures determined for membrane proteins is relatively small, though significant progress in the field has been made in recent years by the elucidation of the structures of the photosynthetic reaction center from *Rhodopseudomonas viridis* (11) and porin, an integral membrane protein from *E. coli* (12).

In the area of crystallization, new methods are actively being researched to improve crystallization success. For example, McPherson and Shlichta (13) reported nucleating protein crystal growth using mineral surfaces or by seeding with tiny mineral crystals. Crystallization under microgravity conditions on space shuttles and other space flights is being studied (14). Microgravity reduces the effects of density or gravity-driven convection near the crystal sur-

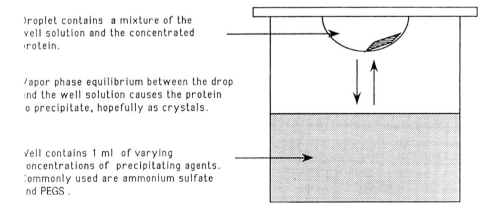

Figure 1 Hanging drop crystallization.

face in crystallization experiments. Because these convection currents cause turbulence around the crystal, their elimination may allow more perfect crystals to grow. Also, because the crystals tend to be suspended rather than settling to the bottom of the drop or container, crystals with much better morphology and less aggregation can be grown. Whether crystals can be grown in space that could not be grown under gravitational conditions has yet to be shown.

B. Crystallographic Data Collection

Crystals with dimensions between 0.3 and 0.8 mm are usually preferred for data collection on standard X-ray sources. Protein crystals are highly hydrated and are usually mounted adhered to the inner wall of capillary tubes by a small amount of the crystallization or storage solution. The capillary tubes are then sealed with plugs of this solution at either end. This is necessary to maintain crystal integrity. X-radiation with a wavelength that approximates the interatomic distance is required to resolve atoms. Conventional sources of X-rays are either sealed-tube anodes or the more powerful rotating-anode generators. The anode, consisting of a metal that can withstand high temperatures, is bombarded with an intense electron beam that ejects electrons from the inner orbitals of the metal atoms producing an excited state. The return to electronic ground state of the metal, while producing X-ray photons over a wide range of wavelengths, also produces more intense radiation at wavelengths characteristic of the metal. With the rotating anode wheel, only a small area is being bombarded with high-energy electrons while the rest is cooling, allowing for nearly 10 times the intensity over sealed-tube X-ray sources. Single crystals are mounted in the center of the X-ray beam and moved through the variety of orientations required to collect the integrated intensities produced by constructive scattering from the planes of atoms in the crystal. The orientation and unit cell parameters of the crystal predict the orientation of the crystal necessary to produce a diffraction point (usually called a reflection), but not its intensity. The measurement of the diffraction data may be accomplished using diffractometers that accurately measure reflections one at a time using a scintillation counter. The rate that reflections can usually be measured on a diffractometer varies from 1000 to 3000 reflections/day. Collecting the data on X-ray-sensitive film cameras is faster, especially on large proteins, but less accurate.

During the data collection, crystal deterioration due to X-ray bombardment, protein instability, and crystal unit cell size influences the choice of data collection methods. Currently, electronic position-sensitive detectors (area detectors) are coming into general use. These detectors can collect multiple reflections at one time like films while approaching the accuracy of diffractometers. The rate at which data can be measured accurately depends on the diffraction efficiency of the crystal. Since the intensity of the diffracted X-rays is proportional to the number of unit cells that make up a crystal, crystals with large unit cells (usually large proteins) or small crystals (less then 0.1 mm) require more intense X-ray beams to produce data above the noise level. Synchrotron X-ray sources at various national laboratories around the world provide X-ray beams many times more powerful than those obtained by conventional sources and have allowed for the study of viruses and other large protein structures.

The measured reflections from the diffraction data are reduced to experimental or "observed" structure factor amplitudes (usually called structure factors). The number of structure factors that exist to a certain resolution is inversely proportional to the cube of the resolution. For example, a protein that contains 1000 nonhydrogen atoms will have around 1250 unique reflections to 4.0 Å resolution, 10,000 reflections to 2.0 Å resolution, and 23,500 to 1.5 Å resolution. For most protein crystals, the data do not extend pass 2.0 Å resolution and in many cases not even that far.

C. Three-Dimensional Structure Solution and Refinement

The diffraction data from a crystal cannot be used directly to create an image of the three-dimensional structure of the molecules that make up the crystal. The diffraction data consist only of the relative structure factor amplitudes of the diffraction pattern and not the relative phases of those amplitudes. In the case of light microscopy, light is refracted from a sample and lenses are used to reconstruct an enlarged image from the diverging light. In electron microscopy, magnetic fields can be similarly used to focus the divergent electron beam. Lenses do not exist for light in the region of the X-ray wavelengths required for determining the atomic structure of molecules. Therefore, the extremely important phase portion of the diffraction data is missing and must be derived by other means. In the case of crystallographic studies of molecules with less than 200 atoms, the approximate phase information can generally be determined directly from a knowledge of the structure factor amplitudes. However, when there are more than 500 atoms in a structure, other means must be used in determining an initial set of approximate phases. A number of good textbooks explain small-molecule crystallography (e.g., Refs. 15 and 16), and Ref. 17 provides a good introduction to protein crystallography. The following is a brief summary of the steps followed in determining a protein structure.

Unique new structures are usually determined by slightly modifying the crystal structure (isomorphous replacement). If a few heavy atom-containing cations or anions can be diffused into discreet positions in a crystal with little disruption of the "native" structure, the changes of the structure factor amplitudes can be used to determine the positions of the heavy atoms in the crystal. Different cations or anions bind to the macromolecules at different positions. Commonly used materials used in these "soaking" experiments are uranyl, mercury, gold and lead salts, lanthanides, and more complicated bulky ions that limit their number of binding sites. Knowledge of these positions, along with the structure factor amplitudes of the perturbed and the native diffraction data sets, is used to obtain a set of very approximate phases. The experimentally measured structure factor amplitudes of the native crystal and the calculated phases can then be used to calculate an electron density map of the unit cell. An initial model of the protein can be constructed from the map.

If knowledge of the structure of a closely related protein already exists, molecular replacement methods may be used to solve the crystallographic structure. These methods consist of generating calculated diffraction data from the known structure that can then be transformed by rotational and translational parameters in attempts to fit the experimentally measured diffraction data of the crystal under investigation. If a satisfactory fit can be determined, the calculated phases from the model can be used along with the structure factor amplitudes to obtain a starting electron density map of the new crystal.

The quality of an electron density map and its interpretability depend on a number of factors. These include accuracy of the data, diffraction angle to which data may be collected, purity of the material used in the crystallization, regularity of the molecular packing in the crystals, isomorphism of heavy-atom derivatives, quality of models for molecular replacement methods, crystal size, and X-ray beam intensity. The crystal quality is paramount in collecting high-resolution data. Crystal size can be in part offset by more intense X-ray beams or greater data collection times. However, crystal damage is increased by increased X-ray exposure. The levels of structure that are interpretable in an electron density map depend on the resolution of the data. At 6–4 Å resolution, the overall shape of the molecule and folding can be ascertained. Helices will appear as fat rods. At 3.5 Å, one should be able to trace the main chain. At 3.0 Å, side groups begin to be resolved and it should be possible to differentiate between bulky side groups and small ones. At 2.5 Å, the torsion angles of the majority of peptides can generally be determined. Aromatic side groups can be differentiated from aliphatic ones. At 2.0 Å, most details of a structure are apparent. Upon refinement of the structure, the accuracy of most atomic positions is probably around 0.2–0.4 Å. Most high-resolution structures have been determined and refined to 2.0 Å resolution. Determination of a structure to 1.5 Å requires twice as many data and most protein crystals do not, in any case, diffract to that resolution anyway. A few structures have been studied at 1.5 Å and a very few to 1.2 Å or better.

The electron density map is used to generate a model of the three-dimensional positions of the nonhydrogen atoms in the protein structure. This is only an approximate model. Accuracy of atomic positions is influenced by the resolution of the data, errors in the phases, disorder of the molecule, high atomic mobility, errors in interpretation, and other factors. Generally, high-performance computer graphics is used to fit the model to the electron density while maintaining reasonable geometry of the model. This proposed structure becomes the starting model in the refinement process. Refinement of a protein model is an iterative process involving both computationally intensive least-squares refinement and interactive computer graphics sessions to correct the proposed model or add new information to it. The course of a refinement can be monitored using the value of the "R factor," defined as:

$$R = \Sigma \, \||F_{obs}| - |F_{calc}|\| / \Sigma |F_{obs}|$$

where the structure factor amplitudes are summed over the data set and $|F_{obs}|$ = absolute value of the observed structure factor amplitude and $|F_{calc}|$ = absolute value of calculated structure factor amplitude.

In the ideal case, this R factor would approach zero but, due to errors in the observed structure factors and the model, the final value is much higher. This R factor is minimized by a least-squares process adjusting the atomic positional and thermal parameters while maintaining a molecular geometry restrained to accepted values. Because errors in the initial model are too great to be eliminated by the least-squares calculations, the process cannot proceed without recalculating new electron density maps. During the refinement process, the errors in the phases and calculated structure factors decrease, allowing for clearer electron and difference electron density maps. Interactive graphics sessions then allow for corrections to be made to the model and least-squares refinement can continue. This cyclic process of least-squares refinement and manual intervention can take from months to years to complete. A beginning model will usually have an R factor of 0.4–0.5. In a highly refined structure, the R factor will be reduced to 0.15–0.2 for 2.0-Å resolution data.

Modeling of new crystal structures in the electron density maps is becoming faster and better using the data bases of the known protein structures. High-performance computer graphics programs are being developed which allow investigators to rapidly survey known structures and to compare consensus structures based on various criteria in order to derive models of new structures under study. Model refinement methods are being improved with the combination of molecular mechanics and dynamics with crystallographic refinement (18,19). These new algorithms may allow trial models to converge on correct structures with less time-consuming interaction by investigators to correct errors in initial interpretations. The process of crystallization, structure solution, and refinement is summarized as a flow chart in Fig. 2.

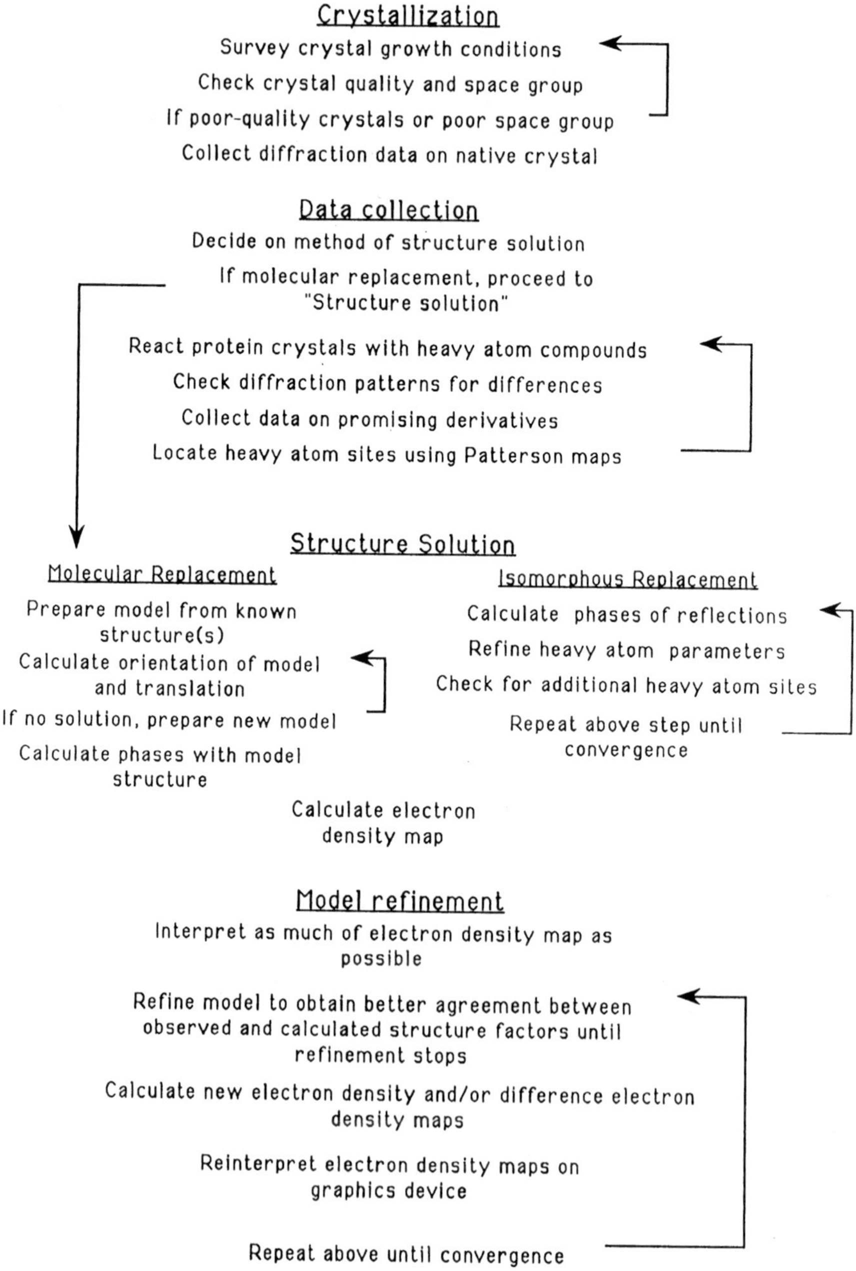

Crystallization
Survey crystal growth conditions
Check crystal quality and space group
If poor-quality crystals or poor space group
Collect diffraction data on native crystal

Data collection
Decide on method of structure solution
If molecular replacement, proceed to "Structure solution"

React protein crystals with heavy atom compounds
Check diffraction patterns for differences
Collect data on promising derivatives
Locate heavy atom sites using Patterson maps

Structure Solution

Molecular Replacement
Prepare model from known structure(s)
Calculate orientation of model and translation
If no solution, prepare new model
Calculate phases with model structure

Isomorphous Replacement
Calculate phases of reflections
Refine heavy atom parameters
Check for additional heavy atom sites
Repeat above step until convergence

Calculate electron density map

Model refinement
Interpret as much of electron density map as possible

Refine model to obtain better agreement between observed and calculated structure factors until refinement stops

Calculate new electron density and/or difference electron density maps

Reinterpret electron density maps on graphics device

Repeat above until convergence

Figure 2 A flow chart describing the basic steps in determining the three-dimensional structure of a protein by X-ray crystallographic methods.

D. Developments of Crystallographic Methods to Study Protein Dynamics

Protein crystallography provides some of the most important knowledge attainable for proteins, but the methodology is not yet advanced to the stage of enabling us to understand the mechanistic details of protein function. Exciting developments to solve limitations associated with the technique are currently underway.

The 3D model of a protein obtained by X-ray analysis represents a static molecule, but proteins in solution are dynamic structures. Enzymes, for example, encounter substrates, process them, and separate them from products; conformational changes, precise contact between atoms, and specific electrostatic and hydrophobic forces are involved in this catalytic process. Elucidating the mechanism of reaction means verifying each step of the reaction from the instant the first substrate encounters the enzyme to the moment the last product separates from it. This involves a molecular level understanding of reaction intermediates, conformation changes, and the contribution of individual amino acids to the catalytic process.

In other words, we have to produce 3D movies of enzymatic catalysis; if we want to do this with the X-ray technique, we must use "catalytic" crystals. Considering that enzymatic reactions are completed in milliseconds and intermediate steps of the reaction, including conformational changes, happen in picoseconds or less, this has been little more than a dream. Conventional X-ray methods of data collection require at least days. One approach to this end is to slow down the reaction several orders of magnitude by cooling down the crystal. The ideal is to have intermediates of the reaction in a "frozen state" for sufficient time to allow collection of diffraction data (20). The applicability of the technique has been limited to the availability of proper cryoprotectants that limit damage of the frozen crystal and preserve the mechanism of the reaction. Recent studies using "shock cooling" show that most protein crystals can be rapidly cooled to liquid nitrogen temperatures (21,22).

A more recent approach uses a technique based on the irradiation of crystals with synchrotron radiation (a high-intensity electromagnetic wave produced during the acceleration of charged particles) and collection of the data on time scales much shorter than those permitted by conventional X-ray methods. There are two different techniques of irradiation; one is the monochromatic technique that uses a single wavelength, and the other is the so-called Laue or white X-radiation technique that uses a broad wavelength range (0.2–2.5 Å) (23). For example, more than 100,000 reflections per second can be obtained with the white X radiation due to its high intensity, its broad effective spectrum, and to the fact that many reciprocal lattice points are located in diffracting positions at the same time. One the other hand, monochromatic X rays still require 0.5–1 hr for data collection. The purpose of the technique is to be able to collect data for each kinetically meaningful step of the reaction. This implies have catalytic crystals, knowing the time scale of the steps under investigation, and being able to tune the data collection time to the life of the event we want to observe (24). Ultrahigh-speed area detectors are being developed that should allow for data collection in milliseconds, and powerful synchroton sources being built in Europe, Japan, and the United States should be completed by the mid-1990s. Using Laue diffraction methods and ultrafast area detectors, entire three-dimensional sets will be collected in a fraction of a second, allowing for time-resolved studies of chemical reaction in crystals.

Promising results have already been obtained by the monochromatic technique with the enzyme phosphorylase b (24,25). In muscle, this enzyme converts glycogen into glucose-1-phosphate, but the enzyme can also slowly convert the natural substrate analog heptenitol (2,6-anhydro-1-deoxy-D-glucohept-1-enitol) into heptulose-2-phosphate. The rate-limiting step in the latter reaction is the interconversion of the ternary complex of enzyme and substrates into products, which means that accumulation of special intermediates occurs preferentially. Johnson and coworkers (25) were able to obtain snapshots of the formation and transformation of the enzyme–substrate complex into the product, heptulose-2-phosphate.

A third approach is also being pursued. It involves crystallizing and determining the 3D structure of the enzyme in the presence of partial sets of substrates, substrate analogs, and inhibitors, to evaluate details concerning the positions of the substrates in noncatalytic situations. This approach is being used for the isozymes of adenylate kinases and their recombinant forms (26).

III. GENETIC ENGINEERING AND PROTEIN CRYSTALLOGRAPHY: THE NEW ALLIANCE

Finally, the combination of recombinant DNA technology and crystallography applied to address mechanistic aspects of enzyme catalysis and assessment of the principles relating protein structure to function is assuming an ascendant role. Among the greatest unsolved problems challenging us today is an understanding of (a) how the genetic information encoded in the linear sequence of amino acids is translated into a folded and biologically active protein molecule; (b) how amino acid sequences that appear to be unrelated correspond to proteins with similar 3D structures and biological properties; (c) the extent to which amino acid replacements affect the structure and properties of a protein; and (d) which mutations are responsible for the functional differences among homologous proteins from different species.

Solutions to these problems are not expected soon, and they will not come by studying only naturally occurring proteins. Amino acid changes made by nature are manifold and distributed randomly over the protein in such a way that a perturbation caused in one site is usually mirrored by a compensatory substitution in another site, often far removed in the sequence from the first. Important but detrimental modifications usually would not be seen in natural mutations. Answers will come from a systematic step-by-step learning process in which a complementary attack by protein engineering and X-ray crystallography will certainly play a major role. Recent advances in determination of 3D structures by X-ray diffraction and nuclear magnetic resonance will allow us to calculate quickly the structures of recombinant proteins differing from each other only by one or a few amino acids. The data will test how perturbations can then be used to engineer new enzymes. Hence, the long journey toward the resolution of the above problems will be traveled slowly in a circuitous path (Fig. 3). At each turn we will learn something useful on which to build our theory. Much work in this area is in progress and exciting results have already appeared. Applying site-directed mutagenesis by taking advantage of the wealth of information on protein 3D structures is delineating the roles of various amino acids in the mechanism of reaction, specificity, and stability of enzymes (27). Here we present examples of X-ray crystallography of engineered proteins as applied to enzyme catalysis, protein structure–function relationships, evolution, and proteins as targets for drugs.

A. Protein Evolution

The work of Nagai et al. (28) on hemoglobin provides an interesting approach to understanding to what extent amino acid replacements affect the structure of

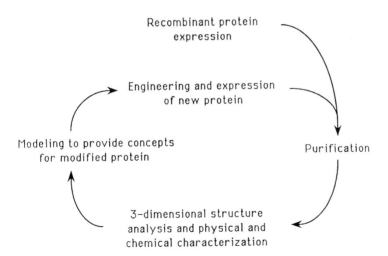

Figure 3 A description of the concerted effort between recombinant DNA technology and X-ray crystallography to solve the problem of how protein structure is related to its function.

hemoglobin and which mutations are responsible for the functional differences among hemoglobins from different species. These investigators (28,29) characterized molecules in which selective amino acids within van der Waals distance from the heme group were modified. Figure 4 shows the region surrounding the heme; of the four amino acids indicated, His-F8, which forms a covalent bond to the heme iron, is the only residue absolutely conserved in all the species (in both hemoglobin and myoglobin). The other three amino acids are mutated in some species, but never as a single mutation. One of the questions addressed by these investigators is what the effect would be of substituting (by site-directed mutagenesis) Val-E11 of the β subunit with Ile, Leu, Met, or Ala (all differing in a single base from the Val codon and all hydrophobic like Val). Val-E11 of the β subunit is substituted by Ile in only one species, but in the α subunit Ile is present at E11 in about 10% of the known hemoglobin sequences. Since several other substitutions take place at the same time, the structural and functional consequences of the Val-E11 → Ile mutation have to be investigated by site-directed mutagenesis. Leu, Met, and Ala have never been found in that position in naturally occurring hemoglobins.

The mutation Val-E11β → Ile resulted in halved O_2 affinity while preserving cooperativity. The side chains of Ile and Val differ in that Ile has an additional methyl group (δ-methyl) attached to the γ_1-methyl group of the Val side chain. Therefore, it is not surprising that the two side chains experience the same rotational restrictions about the bond described by Kendrew for Val-E11β

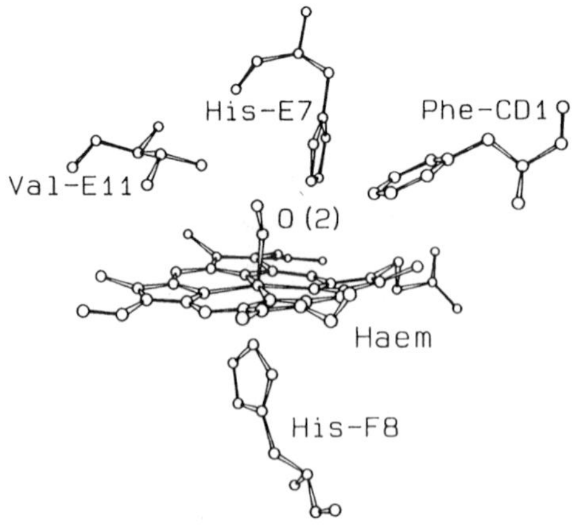

Figure 4 View of four highly conserved amino acid residues surrounding the heme–oxygen complex in the β subunits of hemoglobin. (Modified from Ref. 30. Drawn using the Mosaic system, which was developed in part at the Upjohn Company, and which is based on the MacroModel system developed by Clark Still at Columbia University.)

in myoglobin (31). The reduced affinity for O_2 of the mutant is due to blockage of the oxygen-binding site by the δ-methyl group of Ile as shown by X-ray analysis of the crystalline mutated form. The mutation Val-E11β → Leu preserves both normal O_2 affinity and cooperativity. Electron density maps indicate free rotation of Leu side chains about C_α-C_β and C_α-C_γ bonds without imposing higher steric hindrance than that of the normal Val side chain. The mutation is excluded during evolution because the resulting mutated protein is slightly unstable; for similar reasons of instability, the mutated form Met-E11β is also excluded during evolution. With the small residue Ala in place of the bulkier Val, decreased steric hindrance augments oxygen affinity and favors autooxidation, thus making this mutant evolutionarily unfit.

B. Protein Structure–Function Relationships

Application of these strategies to the understanding of protein structure–function relationships have yielded fruitful results, particularly in the analysis of enzymes. Proteases, for example, are enzymes important in many areas: (a) In medicine, they play normal and pathological functions: (b) In industry, they find widespread use in detergents: (c) In agriculture, they are used as animal feed supplement: (d) In biochemistry, they are indispensable reagents. Trypsin, subtilisin, and carboxypeptidase are among the best characterized enzymes both structurally and kinetically; not surprisingly, they have been the objects of much tinkering.

1. Trypsin

Craik et al. (32) probed the role of Asp102 in the catalytic triad Asp102-His57-Ser195. They have substituted Asp102 with Asn with the result of reducing k_{cat} by 5000 fold. Crystallographic data showed that Asn102 instead of Asp102 in the charge relay complex Asp102:His57:Ser195 at the active site greatly decreases the nucleophilic character of Ser195, thus justifying the catalytic inefficiency. An attempt by Graf et al. (33) was also made to change the substrate specificity of trypsin. They changed Asp189 at the bottom of the substrate-binding pocket to a Lys residue with the expectation that trypsin specificity for cleavage at Lys and Arg residues would be altered to a preference for cleavage after acidic residues. While the enzyme failed, as expected, to cleave after Arg and Lys residues, it failed also to show the hoped for new property, i.e., cleavage after the acidic residues. In this case the 3D structure offered an "after-the-fact" explanation of the phenomenon. Lys in place of Asp189 assumes an extended configuration that produces a very favorable hydrogen-bonding situation with the carbonyl groups of residues 219, 224, and 225 (Fig. 5). The ε-ammonium group of Lys would not be available to the carboxyl group of an acidic substrate and therefore would not contribute to desolvation of charged side groups.

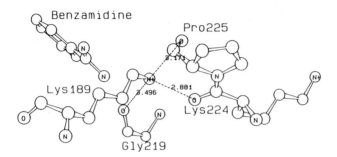

Figure 5 View of the bottom of the substrate binding pocket of trypsin. The Asp189 residue has been replaced by a Lys residue showing favorable H bonding to residues 219, 224, and 225. Atomic coordinates were modified from the crystallographic structure of Bode and Schwager (34). (Drawn using the Mosiac system, which was developed in part at the Upjohn Company, and which is based on the MacroModel system developed by Clark Still at Columbia University.)

2. Subtilisin

Substilisin finds widespread use in detergents and it would be desirable to improve the enzyme's thermal stability. Bryan et al. (35) have prepared several thermally stable variants by site-directed mutagenesis. They found that replacement of Asn218 with Ser resulted in a fourfold increase in thermal stability. The X-ray model of subtilisin at 1.8-Å resolution shows that Asn218 is not at the active site. It is in a β-sheet region that is stabilized when the chains forming the β sheet come 0.2 Å closer together as a consequence of the N218S mutation.

In experiments concerning the catalytic mechanism of subtilisin, Wells et al. (36) and Bryan et al. (37) investigated the role of Asn155. In serine proteases, the transition state of the carbonyl group of the peptide bond to be cleaved is thought to adopt a tetrahedral configuration as it forms a covalent bond with the -OH group of Ser221 (38). Crystallographic studies suggest that the amide of Asn155 stabilizes the transition state by the donation of a hydrogen bond to the carbonyl oxygen of the peptide substrate. Bryan et al. (37) tested this hypothesis by changing Asn155 to Leu (N155L), an isosteric mutation. The K_m of the mutated enzyme was unchanged, but k_{cat} decreased by a factor of 200–300 in a peptide cleavage assay. These investigators concluded that the results are consistent with the hypothesis that Asn stabilizes the activated complex.

3. Carboxypeptidase A

This zinc-dependent enzyme hydrolyzes aromatic and branched aliphatic amino acids from the carboxy terminus of peptides and proteins. The 3D structure of the enzyme complexed to various ligands showed that Tyr248 is within hydrogen-bonding distance of the cleaved peptide bond (39). On the basis of this observation and results obtained from chemical modification of Tyr248, the phenolic hydroxyl was suggested to act as the general acid catalyst in the hydrolysis of the peptide bond (40,41). Hilvert et al. (42) and Gardell et al. (43) tested this hypothesis by changing Tyr248 to Phe (Y248F). The results can be summerized as follows: k_{cat} values for the mutated form and naturally occurring enzyme were comparable. In the mutant Y248F, K_m of peptide substrates and K_i of the potato carboxypeptidase inhibitor increased six- and 70-fold, respectively. These investigators concluded that the phenolic hydroxyl group of Tyr248 is not indispensable as a general acid catalyst but is involved in substrate binding. This example underlines the superiority of site-directed mutagenesis over chemical modification to investigate the role of specific amino acids in catalysis.

4. Tyrosyl-tRNA synthetase

This enzyme catalyzes the amino acylation of $tRNA^{Tyr}$. The reaction takes place in two steps (44): The first step consists of the activation of tyrosine to produce enzyme-bound tyrosyl adenylate, and the second step is the transferring of Tyr to $tRNA^{Tyr}$.

$$E + ATP + Tyr \rightleftharpoons E \cdot Tyr - AMP + PP_i \qquad (1)$$

$$E \cdot Tyr - AMP + tRNA \rightarrow Tyr - tRNA + E + AMP \qquad (2)$$

Protein engineering combined with crystallographic data has already provided a wealth of information concerning the roles of individual amino acids in the formation of tyrosyl adenylate (45). As an example of this fruitful alliance we describe the study of Lowe et al. (46) concerning the nature of the interactions between His48 and ATP. The 3D structure of the enzyme-bound tyrosyl adenylate shows that His48 is in close proximity to the ribose ring oxygen of ATP (47). However, the 3D structure could not reveal which of the two nitrogens of the imidazole ring forms a hydrogen bond with the ribose ring of oxygen. Lowe et al. (46) approached the problem by site-directed mutagenesis. They changed His48 to either asparagine (H48N) or glutamine (H48Q). They reasoned that an asparagine side chain may be superimposed on that of a histidine ring, so that the amide group of asparagine occupies the same position as the N_δ of histidine. Similarly, the equivalent amide group of glutamine may be superimposed on the N_ϵ of the histidine. Asn and Gln therefore have the ability to form hydrogen bonding in a way similar to the δ and ε nitrogens of His, respectively. They found that H48N has the same catalytic activity as the wild-type enzyme whereas H48Q yields a much less active product. Thus the authors concluded that only the δ nitrogen is involved in hydrogen bonding.

Interestingly, an Asn residue has been found as a replacement for His49 in a tyrosine tRNA synthetase from *B. caldotenex* (46) as well as in yeast methionine tRNA synthetase (48).

5. DNA Polymerase I

This enzyme is a single-chain protein (M_r 103,000) with three structural domains and three enzymatic activities (Fig. 6): (a) a DNA polymerase that allows DNA chain elongation; (b) a 3',5'-exonuclease that is proposed to edit out missmatched terminal nucleotides; and (c) a 5',3'-exonuclease that removes DNA ahead of the growing point of a DNA chain. A high-resolution, 3D structure of the large proteolytic fragment (Klenow fragment) has been determined (50). A combination of results obtained through crystallographic and molecular biology strongly supports the view that each of the three domains has a separate enzymatic activity (49). As one aspect of these structure-to-function relationships we mention the work of Steitz and coworkers (51) on the 3',5'-exonucleolytic activity of DNA polymerase I.

Specifically, these investigators wanted to answer the following questions:

1. Is the proposed (49) exonuclease active site located on the small domain of the Klenow fragment, spatially separated from the polymerase active site?

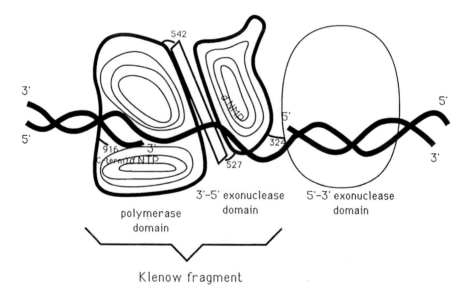

Figure 6 Schematic representation of the domain structure of *E. coli* DNA polymerase. The Klenow fragment three-dimensional structure has been determined; the 5',3'-exonuclease domain has yet to be determined. (Adapted from Ref. 49.)

2. Since the exonuclease active site has binding sites (called A and B) for two divalent metal ions (Fig. 7) (probably Mg^{2+} in vivo), what is the role of the two metal ions?

The metal ion at site A is coordinated to the carboxylate groups of Asp355, Glu357, and Asp501, and to the 5'-phosphate of dTNP (a substrate of the enzyme). The metal ion at site B lies between the carboxylate group of Asp424 and the dNMP 5'-phosphate.

Accordingly, these investigators prepared, by site-directed mutagenesis, two mutants: (a) Asp424 to Ala (D424A) and (b) Asp355 and Glu357 both to Ala (D355A, E357A) in the same molecule; all the mutations are located in the small domain of the Klenow fragment. Question 1 above was answered by the observation that both mutants were devoid of exonuclease activity but retained normal polymerase activity. No alteration in the 3D structure occurred, except of course for the mutated amino acids.

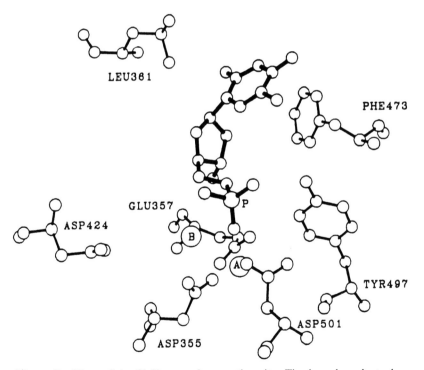

Figure 7 View of the 3',5'-exonuclease active site. The bound product, deoxycytidine monophosphate (dCMP), is shown with thickened bonds, and the two binding sites for divalent metal ions are labelled A and B. Also shown are the side chains of residues interacting with the metal ions and dCMP. (Taken from Ref. 51 with permission.)

By Fourier difference mapping of the 3D structures of the D355A, E357A protein and the wild-type Klenow fragment, they observed that the mutated protein didn't bind dTMP or either of the metal ions in sites A and B, while the native form bound the nucleotide and the metal ions. On the other hand, the mutant D424A protein bound dTMP and metal ion at site A, but not metal at site B. They concluded that loss of activity of the mutant D424A missing the metal ion at site B points to a role of the metal ion at site B in catalysis. Failure of the D355A, E357A mutant protein to bind either dTMP or metal ions suggests a role for the metal ions at site A in substrate binding. Taken together, these findings provide an answer to the second question.

C. Proteins as Targets for Drugs

With the continued unraveling of complexities surrounding biological systems, specific proteins have been identified as playing crucial roles as catalysts, hormones, receptors, etc. Such proteins then may serve as targets for small molecules rationally designed to serve in the therapeutic treatment of disease states.

1. Human Ras p21

The catalytic domain of human *ras* p21 (c-Ha-ras p21), the most common type of oncogene found in human tumors, belongs to the *ras* family. All the products of the *ras* genes are proteins with 188 or 189 amino acids and $M_r \sim 21,000$; they are called p21 proteins. The 165 amino acid residues at the N termini are more than 85% identical. In contrast, the 24 amino acid C-terminal sequences have little homology, with the exception of the conserved Cys186 residue. In eukaryotes, *ras* genes exist ubiquitously either as inactive protooncogenes or as active oncogenes capable of transforming certain cell lines into tumors. Knowledge of the 3D structure of *ras* gene products is fundamental to understanding the functions of normal p21 and the transforming mechanism of oncogenic p21. Furthermore, such information is indispensible for the design of drugs to abolish the transforming properties. Recently DeVos et al. (52,53) reported the crystal structure at 2.7-Å resolution of the normal human c-Ha-ras oncogene protein (amino acids 1–171) complexed to guanosine diphosphate (GDP). Furthermore, Pai et al. (54) reported the crystal structure at 2.6-Å resolution of amino acids 1–166 complexed to the guanosine triphosphate analog guanosine-5'-(β,γ-imido) triphosphate (Gpp Np) (Fig. 8). The structure is composed of six strands of β sheet, five α helices, and nine connecting loops. Normal p21 hydrolyzes GTP to GDP at a much higher rate than many oncogenic forms. The highly reduced rates of GTP hydrolysis in the oncogenic p21 suggests that the GTP form is the one with transforming power, while the GDP form is inactive. The in vitro and in vivo transforming properties can be conferred on p21 by single-point mutations at one of the following positions: 12, 13, 15, 16, 59, 61, 63, 116, and 119. These mutations have direct effects on GDP or GTP binding and GTP hydrolysis because they are

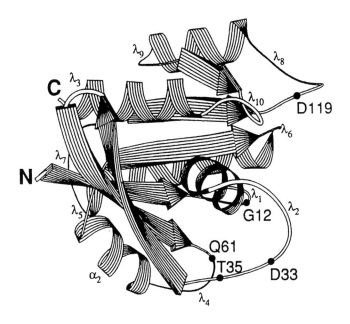

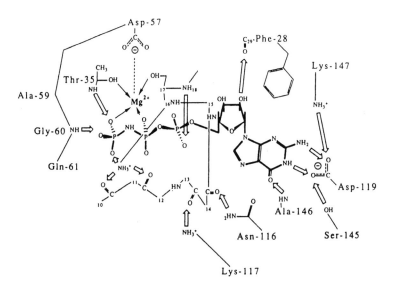

Figure 8 Upper panel: Schematic representation of the three-dimensional structure of the N-terminal 166 residues of c-Ha-ras p21. Lower panel: schematic drawing of protein–nucleotide interactions. Hydrogen bonds are indicated by open arrows (→y), bonds between the magnesium ion and its ligands are shown as solid arrows (→). Some important interactions between side-chain atoms are also included. (Taken from Ref. 54 with permission.)

localized in regions involved in substrate binding and/or in interactions with GTPase-activating protein (GAP), a protein capable of stimulating GTP hydrolysis in normal p21 but not in p21 mutants such as p21 Val12 and p21 Leu61 (55,56). The mutations at amino acids 12 and 61 are the ones most frequently found in human tumors. If the Gly of normal p21 is substituted by a Pro there is no effect; any other amino acid confers transforming capacity. Gly12 is located in a Gly-rich loop, a sequence highly conserved in nucleoside-binding enzymes (57):

$$\text{Gly-Ala-Gly-Gly-Val-Gly-Lys-Ser}$$
$$10 \quad 11 \quad 12 \quad 13 \quad 14 \quad 15 \quad 16 \quad 17$$

This loop is in close spatial interaction with the phosphate residues, the loop λ_2 interacting with GAP, and the loop containing Gln61. These sterical features are perturbed if Gly12 is mutated. Gln61 is in close proximity to the γ-phosphate, and not surprisingly its mutation with residues containing hydrophobic side chains such as Gln61→Val and Gln61→Leu have high transforming capacity (58,59); one can infer that destabilization of interactions with the γ-phosphate and possibly with other residues takes place when these mutations occur.

Interestingly, the mutations at residues 12 and 61 have opposite effects both on p21 affinity for nucleotide substrates and GAP. The mutants Gly12→Val, →Asp, and →Arg have higher affinities, while Gly61→Leu has lower affinity for the nucleotide than the wild type. One the other hand, Gly12→Val mutant has a lower affinity for GAP then the wild type, while Gln→Leu has a higher affinity; however, neither Val12 nor Leu61 mutants are activated by GAP. Pai et al. (54) pointed out that because Gly12 and Gln61 are in close proximity in the 3D structure of p21c'–GppNp complex, the oncogenic activation would therefore affect the same region of the protein.

The forthcoming 3D structure of p21 at higher resolution will provide a more detailed picture of the interaction of substrate and enzyme. Furthermore, it will help to locate, in greater detail, regions on the surface of p21 that are important for recognition by particular target proteins (e.g., GAP) and signal receptor molecules. In conclusion, what we hope to discover from X-ray crystallography in this case is a delineation of unique and subtle structural features which accompany mutagenesis to transforming capability in order to have a basis for designing an ideal drug for certain types of cancer.

2. Interleukin-1β

Another example of a target for drug therapy is interleukin (IL-1). Interleukin 1 (IL-1) is a protein produced by stimulated macrophages and monocytes. It has numerous biological functions including the control of differentiation and activation of lymphocytes, the stimulation of lymphokine production, the stimulation of hepatocytes to elaborate acute phase proteins, and the increase of prostaglandin activity in macrophages and several other cell lines (60). Two types of IL-1 molecules have been isolated and characterized: IL-1α and IL-1β. IL-1β, the native form of which has 153 amino acids, doesn't contain carbohydrates or other cofactors, so it is not surprising that the molecule cloned and expressed in *E. coli* is indistinguishable from the natural one (61). Priestle et al. (62) and Finzel et al. (63) crystallized human IL-1β and reported the X-ray three-dimensional structure at 3.0 and 2.2 Å, respectively. A schematic drawing of IL-1β is given in Fig. 9. The striking structural features of IL-1β are the six β strands which form the core of the protein, with three additional pairs of β strands and no α-helix regions. The overall folding of IL-1β presents a three-fold internal structural pseudosymmetry which has also been noted in the soybean trypsin inhibitor (64). The interior of the molecule

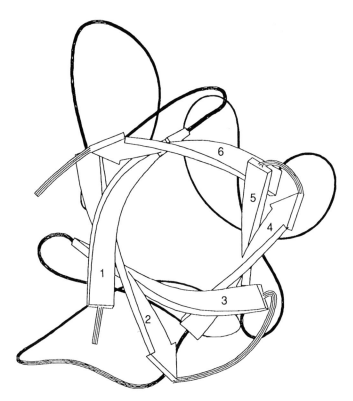

Figure 9 Schematic representation of the three-dimensional structure of Il-1β. The six β strands that form the core of the protei are illustrated as arrows with numbered starting at the N-terminal end. (Drawn from the atomic coordinates used in Ref. 63.)

is predominantly filled with hydrophobic leucine and phenylalanine residues and bears no charged residues. The charged side chains are on the surface of the protein exposed to the solvent. The 3D model indicates that the molecule is made up of about 65% β sheet and 35% random coil/reverse turn, in agreement with circular dichroism studies (65). IL-1β is a potential target for drugs that block its interaction with receptors. Polypeptides constructed from the surface domains can be synthesized and tested as antagonists. In view of a recent report on the cloning of the IL-1 receptor (66), the 3D structure of IL-1β may also allow location of the site(s) of binding and pave the way for improved antagonist molecules. Knowledge of the 3D structure is therefore basic to the development of inhibitors of IL-1β activity as therapeutic agents in treatment of inflammatory diseases such as arthritis.

3. HIV Protease

Because of their enormous chemical versatility in the binding of substrates and their function in the catalytic conversion of them to product molecules, these enzymes are a most attractive target for drug design. Indeed, it is hard to imagine a disease state that is not correlated in some way with enzyme catalysis. There are, of course, many examples here. Two familiar cases include the enzymes ascribed to the renin-angiotensin system, renin itself, and the angiotensin-converting enzyme. Inhibitors of the latter enzyme are on the market and have shown great efficacy in the treatment of hypertension. A relatively recent addition to this class of targets is the protease encoded in the viral genome of human immunodeficiency virus (HIV), the etiological factor in acquired immunodeficiency disease (AIDS). HIV protease is a small enzyme that is related to members of the mechanistic set of proteolytic enzymes including pepsin and renin, i.e., the aspartyl proteases (67). The viral protease is indispensable for processing of the fusion polypeptide products of the *gag* and *pol* genes (68); without this maturation process infective virus cannot be produced (69). Inhibition of the HIV protease is an obvious strategy for therapeutic intervention in AIDS and at the time of this writing there has been a virtual explosion of information in the literature regarding the structure and function of this enzyme. These studies have already produced compounds able to kill the AIDS virus in cell cultures via inhibition of the protease (70,71). HIV protease has been cloned and expressed in bacteria (72,73) and its small size of 99 amino acids has allowed its chemical synthesis (74). The crystallographic structural analysis of HIV-1 protease has been reported by several groups (75–77). Wlodawer's group has also published the 3D structure of HIV-1 protease complexed to a competitive inhibitor (78). All structures reveal a dimeric enzyme with an active site that resembles closely those seen in x-ray analysis of other aspartyl proteases (Fig. 10). Armed, therefore, with a

RECOMBINANT DNA TECHNOLOGY AND CRYSTALLOGRAPHY

sophisticated understanding of the protease at a molecular level, bacterial expression systems that are easily manipulated for site-directed mutagenesis, convenient methods for evaluating the catalytic activity of the enzyme, and inhibitors capable of killing the AIDS virus in cell culture, workers around the world are in hot pursuit of inhibitors that will lead to an effective treatment of AIDS.

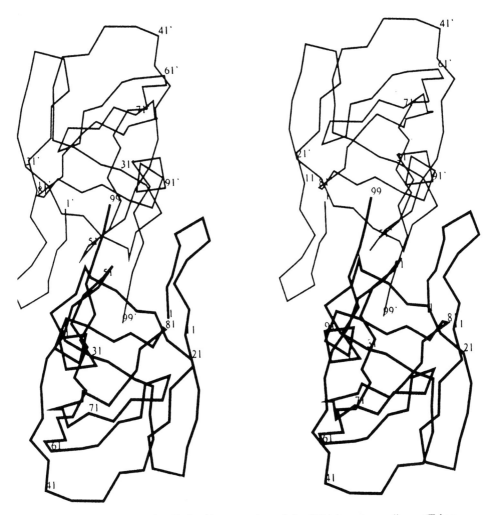

Figure 10 A view of the C_α backbone tracing of the HIV-1 protease dimer. (Taken from Ref. 75 with permission.)

IV. CONCLUDING REMARKS

Many examples may be given of proteins which are in themselves of therapeutic interest. Growth hormones, Plasminogen activators, erythropoietin, interferons, insulin, etc., are either currently in use in humans or are being used in clinical trails. Of the members of the list above, only the structures of insulin and growth hormone (79) have been solved crystallographically. Indeed, it must be said that most of the therapeutically interesting molecules are glycoproteins, or proteins that have not yielded to X-ray analysis. Our ability to determine the tertiary structures for such proteins provides another challenge for future development and improvement of proteins as therapeutics.

We are at the threshold of a new age in the biological sciences that will continue to witness the facilitation of protein three-dimensional structure analysis and its coordination with genetic engineering as a means of producing newly designed proteins. The revolution in molecular biology is changing our approaches toward medical, industrial, and agricultural problems, and toward basic biochemical research itself. It is not hard to predict that in the next decade, the majority of new developments in the biological sciences will be made via the approaches available through biotechnology.

ACKNOWLEDGMENT

We would like to thank Mrs. Paula Lupina and Susan K. Lanting for their diligent preparation of the manuscript, and Heidi Zürcher-Neely for her careful reading of the manuscript.

REFERENCES

1. Pals DT, Thaisrivongs S, Lawson JA, Kati WM, Turner SR, DeGraaf GL, Harris DW, Johnson GA. An orally active inhibitor of renin. Hypertension 1986;8:1105–1112.
2. Duane WC, Hunninghake DB, Freeman ML, Pooler PA, Schlasner LA, Gebhard RL. Simvastatin, a competitive inhibitor of HMG-CoA reductase, lowers cholesterol saturation index of gall bladder bile. Hepatology 1988;8:1147–1150.
3. Jelen P. Industrial whey processing technology. J Agr Food Chem 1979; 27:658–661.
4. Sreekrishna K, Dickson RC, Construction of strains of *Saccharomyces cerevisiae* that grow on lactose. Proc Natl Acad Sci USA 1985; 82:7909–7913.
5. Russell I. Will a recombinant DNA yeast be able to solve whey disposal problems? Tib Tech 1986;4:107–108.
6. Jaynes JM, Yeng MS, Espinosa N, Dodds JH. Plant protein improvement by genetic engineeering: Use of synthetic genes. Tib Tech 1986;4:314–320.
7. Rao AS, Singh R. Improving grain protein quality by genetic engineering: Some biochemical considerations. Tib Tech 1986;4:108–109.

8. Barton KA, Brill WJ. Prospects in plant genetic engineering. Science 1983;219:671–676.
9. Terzaghi BE, Laing WA, Christeller JT, Petersen GB, Hill DF. Ribulose 1,5-bisphosphate carboxylase: Effect on the catalytic properties of changing methionine-330 to leucine in the *Rhodospirillum rubrum* enzyme. Biochem J 1986;235: 839–846.
10. Egner U, Tomasselli AG, Schulz GE. Structure of the complex of yeast adenylate kinase with the inhibitor P_1,P_5-di(adenosine-5')Pentaphosphate at 2.6Å resolution. J Mol Biol 1987;195: 649–658.
11. Deisenhofer J, Epp O, Miki L, Huber R, Michel H. X-ray structure analysis of a membrane protein complex. Electron density map at 3Å resolution and a model of the chromophores of the photosynthetic reaction center from *Rhodopseudomonas viridis*. J Mol Biol 1984;180: 385–398.
12. Garavito RM, Jenkins JA, Jansonius JN, Karlsson R, Rosenbusch JP. X-ray diffraction analysis of matrix porin, an integral membrane protein from *Escherichia coli* outer membranes. J Mol Biol 1983;164: 313–327.
13. McPherson A, Shlichta P. Heterogeneous and epitaxial nucleation of protein crystals on mineral surfaces. Science 1988;239: 385–387.
14. Littke W, John C. Protein single crystal growth under microgravity. Science 1984;225: 203–204.
15. Glusker JP, Trueblood KN. Crystal structure analysis. 2nd ed. London: Oxford Univ. Press, 1985.
16. Ladd MFC, Palmer RA. Structure determination by X-ray crystallography. 2nd ed. New York: Plenum Press, 1985.
17. Blundell TL, Johnson LN. Protein crystallography. London: Academic Press, 1976.
18. Brunger AT, Kuriyan J, Karplus M. Crystallographic R factor refinement by molecular dynamics. Science 1987;235:458–460.
19. Watenpaugh KD. Conformational energy as a restraint in refinement. In: Hermans J. ed. Molecular dynamics and protein structure. Western Springs, IL: Polycrystal Book Service, 1985, pp 77–80.
20. Fink AL, Petsko GA. X-ray cryoenzymology. Adv Enzymol 1981; 52: 177–246.
21. Hope H. Cryocrystallography of biological macromolecules: A general applicable method. Acta Crystallogr 1988;B44:22–26.
22. Moffat K, Szebenyi D, Bilderback D. X-ray Laue diffraction from protein crystals. Science 1984;223:1423–1425.
23. Muchmore SW, Watenpaugh KD. Low-temperature crystal mounting and data collection techniques for macromolecular crystals, Abstr., Am. Cryst. Assoc., 1988;16:43.
24. Hajdu J, Acharya KR, Stuart DI, Barford D, Johnson LN. Catalysis in enzyme crystals. TIBS 1988;13: 104–109.
25. Hajdu J, Acharya KR, Stuart DI, McLaughlin PJ, Bradford D, Klein H, Johnson LN. Time-resolved structural studies on catalysis in the crystal with glycogen phosphorylase b. Biochem Soc. Trans 1986;14: 538–541.
26. Tomasselli AG, Schulz GE. Structural and functional studies in the adenylate kinase family. Chimica Oggi 1987;Gen-Feb: 11–18.

27. Knowles JR. Tinkering with enzymes: What are we learning? Science 1987; 236: 1252–1258.
28. Nagai K, Luisi B, Shih D, Evolution of haemoglobin studied by protein engineering. BioEssays 1988;8: 79–82.
29. Luisi BF, Nagai K. Crystallographic analysis of mutant human haemoglobins made in *Escherichia coli*. Nature 1986;320: 555–556.
30. Shaanan B. Structure of human oxyhaemoglobin at 2.1Å resolution. J. Mol Biol 1983;171: 31–59.
31. Kendrew JC. The three-dimensional structure of a protein molecule. Sci Am 1961;205(6): 96–110.
32. Craik CS, Roczniak S, Sprang S, Fletterick R, Rutter W. Redesigning trypsin via genetic engineering. J. Cell Biochem 1987;33: 199–211.
33. Graf L, Craik CS, Patthy A, Roczniak S, Fletterick RJ, Rutter W. Selective alteration of substrate specificity by replacement of aspartic acid-189 with lysine in the binding pocket of trypsin. Biochemistry 1987;26: 2616–2623.
34. Bode W, Schwager P. The refined crystal structure of bovine β-trypsin at 1.8Å resolution. II. Crystallographic refinement, calcium binding sites, benzamidine binding site and active site and pH 7.0. J Mol Biol 1975;98:693–717.
35. Bryan PN, Pantolinao MW, Wood J, Finzel BC, Galliland GL, Howard AJ, Poulos TL. Proteases of enhanced stability:characterization of a thermostable variant of subtilisin. Proteins: Struct Funct Genet 1986; 1:326–334.
36. Wells JA, Cunningham BC, Graycar TP, Estell DA. Importance of hydrogen-bond formation in stabilizing the transition state of subtilisin. Phil Trans R Soc London 1986;A317: 415–423.
37. Bryan PN, Pantoliano NW, Quill SG, Hsiao H-Y, Poulos T. Site-directed mutagenesis and the role of the oxyanion hole in subtilisin. Proc Natl Acad Sci USA 1986;83: 3743–3745.
38. Kraut J. Serine proteases: Structure and mechanism of catalysis. Ann Rev Biochem 1977;46: 331–358.
39. Rees DC, Lipscomb WN. Binding of ligands to the active site of carboxypeptidase A. Proc Natl Acad Sci USA 1981;78:5455–5459.
40. Lipscomb WN. Structure and catalysis of enzymes. Ann Rev Biochem 1983;52:17–34.
41. Vallee BL, Galdes A. The metallobiochemistry of zinc enzymes. Adv. Enzymol 1984;56:283–430.
42. Hilvert D, Gardell SJ, Rutter WJ, Kaiser ET. Evidence against a crucial role for the phenolic hydroxyl of Tyr-248 in peptide and ester hydrolyses catalyzed by carboxypeptidase A: Comparative studies of the pH dependencies of the native and Phe-248-mutant forms. J Am Chem Soc 1986;108: 5298–5304.
43. Gardell SJ, Craik CS, Hilvert D, Urdea NS, Rutter WJ. Site-directed mutagenesis shows that tyrosine 248 of carboxypeptidase A does not play a crucial role in catalysis. Nature 1985;317: 551–555.
44. Fersht AR, Jakes R. Demonstration of two reaction pathways for the amino acylation of tRNA. Application of the pulsed quenched flow technique. Biochemistry 1975;14:3350–3356.

45. Fersht AR, Leatherbarrow RJ, Wells TNC. Binding energy and catalysis: A lesson from protein engineering of tyrosyl-tRNA synthetase. TIBS 1986;11:321–325.
46. Lowe DM, Fersht AR, Wilkinson AJ, Carter P, Winter G. Probing histidine-substrate interactions in tyrosyl-tRNA synthetase using asparagine and glutamine replacements. Biochemistry 1985;24: 5106–5109.
47. Rubin J, Blow DM. Amino acid activation in crystalline tyrosyl-tRNA synthetase from *Bacillus stearothermophilus*. J Mol Biol 1981;145:489–500.
48. Barker DG, Winter G. Conserved cysteine and histidine residues in the structures of the tyrosyl and methionyl-tRNA synthetases. FEBS Lett. 1982; 145:191–193.
49. Joyce CM, Steitz TA. DNA polymerase I: From crystal structure to function via genetics. TIBS 1987;12: 288–292.
50. Ollis DL, Brick P, Hamlin R, Xuong NG, Steitz TA. Structure of large fragment of *Escherichia coli* DNA polymerase I complexed with dTMP. Nature 1985; 313:762–766.
51. Derbyshire V, Freemont PS, Sanderson MR, Beese L, Friedman JM, Joyce CM, Steitz TA. Genetic and crystallographic studies of the 3',5' exonucleolytic site of DNA polymerase I. Science 1988;240: 199–201.
52. DeVos AM, Tong L, Milburn MV, Matias PM, Jancarik J, Noguchi S, Nishimura S, Miura K, Ohtsuka E, Kim S-H. Three-dimensional structure of an oncogene protein: Catalytic domain of human c-H-ras p21. Science 1988; 239: 888–893.
53. Tong L, Milburn MV, DeVos AM, Kim S-H. Structure of ras protein. Science 1989;245: 244.
54. Pai EF, Kabsch W, Krengel U, Holmes KC, John J, Wittinghofer A. Structure of the guanine-nucleotide-binding domain of the Ha-ras oncogene product p21 in the triphosphate conformation. Nature 1989;341: 209–214.
55. Trahey M, McCormick F. A cytoplasmic protein stimulates normal N-ras p21 GTPase, but does not effect oncogenic mutants. Science 1987;238: 542–545.
56. Vogel US, Dixon RAF, Schaber MD, Diehl RE, Marshall MS, Scolnick EM, Sigal IS, Gibson JB. Cloning of bovine GAP and its interaction with oncogenic ras p21. Nature 1988;335: 90–93.
57. Möller W, Amons R. Phosphate-binding sequences in nucleotide-binding proteins. FEBS Lett 1985;186: 1–7.
58. Der CJ, Finkel T, Cooper GM. Biological and biochemical properties of Human ras genes mutated at codon 61. Cell 1986;44: 167–176.
59. Feig LA, Cooper GM. Relationship among guanine nucleotide exchange, GTP hydrolysis, and transforming potential of mutated ras proteins. Mol Cell Biol 1988;8: 2472–2478.
60. Oppenheim JJ, Kovacs EJ, Matsushima K, Durum SK. There is more than an interleukin 1. Immuno Today 1986;7: 45–56.
61. Wingfield P, Payton M, Tavernier J, Barnes M, Shaw A, Rose K, Simona MG, Demczuk S, Williamson K, Dayer J-M. Purification and characterization of human interleukin-1β expressed in recombinant *Escherichia coli*. Eur J Biochem 1986;160: 491–497.
62. Priestle JP, Schär HP, Grütter MG. Crystal structure of the cytokine interleukin-1β. EMBO J. 1988;7: 339–343.

63. Finzel BC, Clancy LL, Holland DR, Muchmore SW, Watenpaugh KD, Einspahr HM. The crystal structure of recombinant human interleukin-1β at 2.2Å resolution. J Mol Biol 1989;209: 779–791.
64. McLachlan AD. Three-fold structural pattern in the soybean trypsin inhibitor (kunitz). J Mol Biol 1979;133: 557–563.
65. Craig S, Schmeissner U, Wingfield P, Pain RH. Conformation, stability, and folding of interleukin-1β. Biochemistry 1987;26: 3570–3576.
66. Sims JE, March CJ, Cosman D, Widmer MB, MacDonald HR, McMahan CJ, Grubin CE, Wignall JM, Jackson JL, Call SM, Friend D, Alpert AR, Gillis S, Urdal DL, Dower SK. cDNA expression cloning of the IL-1 receptor, a member of the immunoglobulin superfamily. Science 1988;241: 585–589.
67. Pearl LH, Taylor WR. A structural model for the retroviral proteases. Nature 1987;329: 351–354.
68. Kramer RA, Schaber MD, Skalka AM, Ganguly K, Wong-Staal F, Reddy EP. HTLV-III *gag* protein is processed in yeast cells by the virus *pol*-proteases. Science 1986;231: 1580–1584.
69. Kohl NE, Emini EA, Schleuf WA, Davis LJ, Heimbach, JC, Dixon RAF, Scolnick EM, Sigal IS. Active human immunodeficiency virus protease is required for viral infectivity. Proc Natl Acad Sci USA 1988;85: 4686–4690.
70. McQuade TJ, Tomasselli AG, Liu L, Karacostas V, Moss B, Sawyer TK, Heinrikson RL, Tarpley WG. A synthetic HIV-1 protease inhibitor with antiviral activity arrests HIV-like particle maturation. Science 1990;247: 454–456.
71. Meek TD, Lambert DM, Dreyer GB, Carr TJ, Tomaszek TA Jr, Moor ML, Strickler JE, Debouch C, Hyland LJ, Matthews TJ, Metcalf BW, Petteway SR. Inhibition of HIV-1 protease in infected T-lymphocytes by synthetic peptide analogues. Nature 1990;343: 90–92.
72. Darke PL, Leu C-T, Davis LJ, Heimbach JC, Diehl RE, Hill WS, Dixon RAF, Sigal IS. Human immunodeficiency virus protease. Bacterial expression and characterization of the purified aspartic protease. J Biol Chem 1989;264: 2307–2312.
73. Tomasselli AG, Olsen MK, Hui JO, Staples DJ, Sawyer TK, Heinrikson RL, Tomich C-S. Substrate analogue inhibition and active site titration of purified recombinant HIV-1 protease. Biochemistry 1990;29: 264–269.
74. Schneider J, Kent SBH. Enzymatic activity of a synthetic 99 residue protein corresponding to the putative HIV-1 protease. Cell 1988;54: 363–368.
75. Wlodawer A, Miller M, Jaskolski M, Sathyanarayana BK, Baldwin E, Weber IT, Selk LM, Clawson L, Schneider J, Kent SBH. Conserved folding in retroviral proteases: Crystal structure of a snythetic HIV-1 protease. Science 1989;245: 616–621.
76. Lapatto R, Blundell T, Hemmings A, Overington J, Wilderspin A, Wood S, Merson JR, Whittle PJ, Danley DE, Geoghegan KF, Hawrylik SJ, Lee SE, Scheld KG, Hobart PM. X-ray analysis of HIV-1 proteinase at 2.7Å resolution confirms structural homology among retroviral enzymes. Nature 1989;342: 299–302.
77. Navia MA, Fitzgerald PMD, McKeever BM, Leu C-T, Heimbach JC, Herber WK, Sigal IS, Darke PL, Springer JP. Three-dimensional structure of aspartyl protease from human immunodeficiency virus HIV-I. Nature 1989; 337: 615–620.

78. Miller M, Schneider J, Sathyanarayana BK, Toth MV, Marshall GR, Clawson L, Selk L, Kent SBH, Wlodawer A. Structure of complex of synthetic HIV-1 protease with a substrate-based inhibitor at 2.3 Å resolution. Science 1989;246: 1149–1152.
79. Abdel-Meguid SS, Shieh H-S, Smith WW, Dayringer HE, Violand BN, Bentle LA. Three-dimensional structure of a genetically engineered variant of procine growth hormone. Proc. Natl. Acad. Sci. USA 1987;84: 6434–6437.

Index

A

Affinity chromatography, 13–14
  basis for separation of proteins by, 13
  biospecific, 35
  group specific, 35
  immuno-, 14, 37, 46, 267–268
    engineering proteins for, 247–250
    removal of antibody from the final product after, 46
  leaching of ligands during, 46
  ligands used in, 36–37
  metal chelate, 21, 35, 38
    designing proteins with chelation sites for, 256–258
    engineering proteins for, 253–262
    engineering somatotropins with chelation sites for, 258–262
    principle of, 254–56
    purification of recombinant human renin using, 14
  receptor (*see* Receptor affinity chromatography)
  removal of ligands from final products after, 46

α1-Proteinase inhibitor:
  production in and purification from yeast of, 185–187
  use of affinity chromatography in purification of, 186–187
Aminopeptidase:
  from *Aeromonas proteolytica*, 154
    cleavage of N-terminal methionine with, 154
  from *Escherichia coli*, 148–154
    cleavage of N-terminal methionine with, 154
    hyperproduction of, 152–153
    specificity of, 148–150
Antibodies (*see also* Monoclonal antibodies):
  physical characterization of, 214

B

Bacillus subtilis:
  protease-deficient strains of, 98

317

[*Bacillus subtilis*]:
  proteases of, 90–91
β-Endorphin, secretion from *Escherichia coli* of, 174–175
β-Galactosidase, intracellular production in yeast (*Pichia pastoris*) of, 199

C

Carboxypeptidase A, structure-function relationships in, 300–301
Cell disintegrators (*see also* Homogenizers):
  ultrasonic, 69–70
Cell disruption (*see also* Cells, Cell disintegrators, and Homogenizers):
  by chemical agents, 78–79
  devices for, 57–73
  by freeze-fracturing, 72
  by freezing and thawing, 72–73
  by grinding, 70–71
  by meat mincer, 73
  by paddle blender, 72
  techniques for, 57–82
Cells:
  autolysis of, 73–74
    heat, pH, and osmotic shock induced, 75–76
    solvent induced, 74–75
  culture of, 7
  dehydration of, 78
  disruption of, 7 (*see also* Cell disruption)
  extraction of, 7 (*See also* Cell Disruption)
  permeabilization of, 79–80
  programmed self-destruction of, 80
Chromatography:
  adsorption/elution, 10
  affinity (*see* Affinity chromatography)
  cartridges for, 39–40
  choice of stationery phase in, 39–40
  gel filtration/molecular sieve (*see* Gel filtration chromatography)
  hydrophobic interaction (HIC) (*see* Hydrophobic interaction chromatography)
  ion-exchange (*see* Ion-exchange chromatography)
  reverse-phase (*see* Reverse phase HPLC)
Colony Stimulating factor, purification using immuno-affinity chromatography of, 250

D

Disulfide bonds (bridges), reduction of, 8
DNA (*see also* nucleic acids):
  contamination of proteins with, 5, 40
  determination of, 40
  removal of, 40–42
DNA polymerase I, structure-function relationships in, 302–304

E

Endotoxins (*see* Pyrogens)
Epidermal growth factor (EGF), secretion from *Escherichia coli* of, 173–174
Escherichia coli:
  amino peptidase of (*see* Aminopeptidase)
  deformylase of, 147–148
  half-lives of cloned proteins in, 130–131
  inclusion bodies in, 7–9, 121–145
  periplasmic-leaky mutants of, 177
  periplasmic space of, 164

protease-deficient strains of, 97–98, 131–132
proteases of, 88–89
secretion of recombinant proteins from, 163–177
stability of cloned proteins in, 127–129

G

Gel filtration/molecular sieve chromatography, 8, 11
 purification of recombinant proteins using, 8, 11, 21, 169, 274
 use as final step in the purification of recombinant proteins, 276–277
Genetic engineering (*see also* Recombinant DNA technology):
 new alliance between protein crystallography and, 296–310
 use in protein evolution studies of, 297–299
Granulocyte-macrophage colony-stimulating factor (GM-CSF):
 secretion from *Escherichia coli* of, 170–171
 secretion and purification from yeast (*Saccharomyces cerevisiae*) of, 188
Growth hormone (*see* Recombinant human growth hormone)
Growth hormone releasing factor (GRF), secretion and purification from yeast *(Saccharomyces cerevisiae)* of, 187–188

H

Hepatitis B surface antigen, intracellular production in yeast (*Pichia pastoris*) of, 199–201

Hirudin, secretion from *Escherichia coli* of, 172–173
Homogenizers (*see also* Cell disintegrators):
 bead mill, 58
 laboratory scale, 59–60
 pilot plant and large-scale, 60–62
 blade, 64–65
 freeze pressing, 67–68
 high-pressure, 65–66
 laboratory scale, 66–67
 pilot plant and large-scale, 67
 nitrogen decompression, 68
 pestle and tube, 71
 rotor stator, 62–63
 laboratory size, 63–64
 pilot plant and large-scale, 64
Human immunodeficiency virus (HIV):
 production in and purification from yeast (*Saccharomyces cerevisiae*) of the envelope gene product of, 184–185
 protease of, 308
 use as target for drugs of, 308–309
 reverse transcriptase of, 22
 production in and purification from *Escherichia coli* of, 22
 production in and purification from yeast of, 185
Human ras p21, use as target for drugs of, 304–306
Human superoxide dismutase (h-SOD), production in yeast of, 21–22
Hydrophobic interaction chromatography (HIC), 12–13, 34–35
 basis for separation of proteins by, 12
 purification of monoclonal antibodies by, 227–228

I

Insulin, production in and purification from *Escherichia coli* of, 18–20

Insulin-like growth factor-1 (IGF-1):
  secretion from *Escherichia coli* of, 171–172
  secretion into *Escherichia coli* culture medium of, 171–72
Interferon(s):
  recombinant α-2, 142
    temperature effects in inclusion body formation of, 142
  recombinant human consensus, 15
    production in *Escherichia coli* of, 16–17
    purification from yeast (*Saccharomyces cerevisiae*) of, 186–187
Interleukin-1β, use as target for drugs of, 306–308
Interleukin-2 (murine), secretion from *Escherichia coli* of, 165, 167–170
Invertase, secretion from yeast (*Pichia pastoris*) of, 205–207
Ion-exchange chromatography, 12, 34–35
  basis for separation of proteins by, 12
  DNA removal using, 41
  engineering proteins for, 250–253
  pyrogen removal using, 43–44
  use in purification of monoclonal antibodies, 222–223
  use in purification of recombinant proteins, 20, 22, 169, 173, 185–187

M

Methionine amino peptidase from *Escherichia coli*, 148
  hyperproduction of, 152–153
  N-terminal methionine removal by, 147–150, 176
  specificity of, 148–150
Monoclonal antibodies:
  in vivo and in vitro methods for production of, 17, 214–216
  methods for clarifying solutions of, 216–220
  precipitation of, 220–221
  purification of, 17–18, 31–32, 213–231
    use of dye ligand affinity chromatography in, 225
    use of gel filtration chromatography in, 230–231
    use of hydrophobic interaction chromatography (HIC) in, 227–228
    use of hydroxylapatite chromatography in, 18, 225
    use of ion-exchange chromatography in, 18, 222–223
    use of protein A affinity chromatography in, 17, 223–224
    removal of nonprotein impurities during the purification of, 231

N

N-terminal formyl methionine:
  deformylase and, 147–148
  retention in recombinant proteins of, 151
  problems due to, 151
N-terminal methionine:
  in recombinant proteins, 147–148
  problems due to, 150–151
  removal from recombinant proteins of, 147–156, 176
    chemical method for, 155–156
    in vitro methods for, 153–156
    in vivo methods for, 151–153
    use of aminopeptidase in, 154–155
    use of dipeptidyl aminopeptidase I in, 155
    use of ubiquitin protease for, 156
Nucleic acids (*see also* DNA):
  contamination of protein inclusion bodies by, 136–138

INDEX

P

Phospholipase A2 (PLA2), secretion and purification from yeast (*Saccharomyces cerevisiae*) of, 187
Proinsulin, 18
  production in and purification from *Escherichia coli* of, 18–20
Protease(s):
  of *Bacillus subtilis*, 90–91
  biological function of, 85
  classification of, 86
  detection of, 95–96
  inhibitors
    oligopeptide, 102–103
    protein, 103–106
    synthetic low-molecular-weight, 102
  of *Escherichia coli*, 88–89
  of mammalian cells, 93–95
  removal of, 97, 100–102
  specificity of, 87
  strategies for the inhibition of, 102–106
  of yeast (*Saacharomyces cerevisiae*), 91–93
Proteinases (*see* Proteases)
Protein(s) (*see also* Recombinant proteins):
  as targets for drugs, 304–310
  crystallization of (*see also* Protein crystallography), 288
    methods for, 288–290
  DNA and nucleic acid contamination of, 5, 40
  dynamics of, 295
    use of crystallographic methods in the study of, 295–296

  filtration of, 35
  posttranslational modification of, 147–148, 183, 193
  precipitation of, 10
  primary structure of, 25
  purity assessment of, 15
  renaturation of, 133

  structure-function relationships in, 24, 26, 299–309
  recombinant DNA technology and crystallography in the study of, 23–26, 299–309
  therapeutic grade, 29
    contaminants in, 30
    production and purification of, 29–50
    removal of nonproteinaceous contaminants from, 40–45
    SDS-PAGE analyses of, 31
    selection of chromatographic for the production of, 34–35
  therapeutic value of, 286, 310
    examples of, 286, 310
  three-dimensional structure of, 4–5
  solution and refinement by X-ray crystallography of, 291–295
  ultrafiltration of, 10, 37
  use as targets for drugs of, 304–309
  X-ray crystallography of, 25–26, 285–310
    data collection in, 290–291
    recombinant DNA technology and (*see* Protein crystallography)
    three-dimensional structure solution and refinement in, 291–294
Protein A:
  use in protein secretion of, 165, 172
  use in purifying monoclonal antibodies of, 223–224
Protein crystallography:
  alliance between genetic engineering and, 296–309
  methodology of, 287–296
Protein engineering, 25, 240–241
  improved analysis using, 263
  improving chromatographic properties by, 246–247
  isolation of active proteins by, 242–245
  low-cost purification using, 262
  simplification of purification process by, 245–263

Proteolysis:
  avoidance of, 97–100
  prevention of, 96–97
Pyrogens, removal of, 42–44
Purification protocols:
  development of, 30–40
  objectives in the development of, 5–6
    cost of final product and, 5–6
    purity of final product and, 5
  scaling-up of, 14–15
  strategies and techniques in development of, 6–15
  validation of, 47–50

R

Receptor affinity chromatography:
  advantages over immunoaffinity chromatography of, 267–268
  choice of polymer supports for, 270
  choice of receptors for, 269–270
  concept of, 268–269
  cost effectiveness of, 280
  definition of, 267–268
  design parameters in, 269–279
  detection and prevention of protein leaching during, 273–274
  immobilization methods in, 270–271
  longevity and stability of receptor adsorbents during, 278–279
  optimum receptor-coupling conditions for, 272–273
  problem areas of, 279
  procedures for, 274–277
  purification of IL-2 using, 269–277
  scale-up and automation of, 279
Recombinant bovine somatotropin (bST) (*see also* Recombinant human growth hormone):
  engineering for metal affinity chromatography of, 258–262
  half-life in *Escherichia coli* of, 129–130
  production in *Escherichia coli* of, 123–127
  use of reverse-phase HPLC in estimating the production in *Escherichia coli* of, 125–127
Recombinant DNA technology:
  engineering of proteins and (*see* Protein engineering)
  importance in protein research of, 3–4, 285–287
  protein crystallography and, 25–26, 285–310
  structure-function relationships in proteins and, 23–26, 299–310
Recombinant human growth hormone (*see also* Recombinant bovine somatotropin):
  N-terminal methionine processing of, 149–151
  secretion from *Escherichia coli* of, 166–169
Recombinant proteins (*see also* Proteins):
  development of purification protocols for (see Purification protocols)
  FDA's points to consider bulletin and, 277
  inclusion bodies and, 7–9, 121–122
    components of, 133–139
    endotoxin contamination of, 138
    growth conditions and, 141–142
    host cell conditions and, 141–142
    nucleic acid contamination of, 136–138
    plasmid-encoded proteins and, 138
  pharmaceutical grade of, 277–278
    procedures for determining purity of, 277
  production in and purification from methylotropic yeast (*Pichia pastoris*) of, 193–209
  purification from *Escherichia coli* of, 3, 18–23, 163–177
  purification from yeast (*Saccharomyces cerevisiae*) of, 183–190

scaling-up purification of, 14–15
secretion of, 163–177
  advantages of, 163–165
  signal sequences in, 164
  use of alkaline phosphatase (phoA) signal sequence in, 165, 166, 172–174
  use of lamB signal sequence in, 165, 171
  use of ompA signal sequence in, 165–171
  use of ompF signal sequence in, 174–175
strategies for purification of, 6–15
Reverse-phase high-performance liquid chromatography (rp-HPLC) (*see also* Chromatography):
  basis for separation of proteins by, 12–13
  use in purification of recombinant proteins, 6, 18, 20, 34, 171, 173–175, 186–188

S

Somatotropin(s) (*see* Recombinant bovine somatotropin)
Streptokinase, intracellular production in yeast (*Pichia pastoris*) of, 201–203
Subtilisin, structure-function relationships in, 300

T

Tissue plasminogen activator (tPA), production and purification of, 19–21
Trypsin, structure-function relationships in, 299–300

Tumor necrosis factor (TNF), intracellular production in yeast (*Pichia pastoris*) of, 204–205
Tyrosyl-tRNA synthetase, structure-function relationships in, 301–302

V

Viruses:
  inactivation of, 45
  removal of, 44–45

Y

Yeast
  advantages of producing proteins in, 183, 193
  cell disruption of (*see also* Cell disruption), 184
  Pichia pastoris strain of, 193–195
  development of expression system in, 195–199
  host strains of, 195
  intracellular production of heterologous proteins in, 199–205
  methanol-inducible regulatory sequences in, 195
  N-linked oligosaccharides in, 207
  secretion of heterologous proteins from, 205–207
  protease-deficient strains of, 98
  proteases of (*see* Proteases)
  purification of recombinant proteins from 183–190, 193–209 (*see also* Proteins)
  Saccharomyces cerevisiae strain of, 183
  purification of proteins expressed intracellularly in, 184–187

[Yeast]
    purification of proteins secreted from, 187–189
    secretion using mammalian pre-pro signal sequences in, 207
    selectable markers of, 196
    vectors for expression of heterologous proteins in, 196–199